Automotive Paints and Coatings

Edited by
Hans-Joachim Streitberger
and Karl-Friedrich Dössel

Further Reading

H. Lipowsky, E. Arpaci
Copper in the Automotive Industry

2007
ISBN: 978-3-527-31769-1

G. Pfaff (Ed.)
Encyclopedia of Applied Color

2009
ISBN: 978-3-527-31551-2

G. Buxbaum, G. Pfaff (Eds.)
Industrial Inorganic Pigments
Third, Completely Revised and Extended Edition

2005
ISBN: 978-3-527-30363-2

E. B. Faulkner, R. J. Schwartz (Eds.)
High Performance Pigments
Second, Completely Revised and Extended Edition

2008
ISBN: 978-3-527-31405-8

S. K. Ghosh (Ed.)
Functional Coatings
by Polymer Microencapsulation

2006
ISBN: 978-3-527-31296-2

Automotive Paints and Coatings

Edited by
Hans-Joachim Streitberger
and Karl-Friedrich Dössel

Second, Completely Revised and Extended Edition

WILEY-VCH

WILEY-VCH Verlag GmbH & Co. KGaA

The Editors

Dr. Hans-Joachim Streitberger
Market & Management
Sustainability in Business
Patronatsstr. 13
48165 Münster
Germany

Dr. Karl-Friedrich Dössel
Dupont Performance Coatings
Christbusch 25
42285 Wuppertal
Germany

1. Edition 1995
2. Completely Revised and Extended Edition 2008

■ All books published by Wiley-VCH are carefully produced. Nevertheless, authors, editors, and publisher do not warrant the information contained in these books, including this book, to be free of errors. Readers are advised to keep in mind that statements, data, illustrations, procedural details or other items may inadvertently be inaccurate.

Library of Congress Card No.:
applied for

British Library Cataloguing-in-Publication Data
A catalogue record for this book is available from the British Library.

Bibliographic information published by the Deutsche Nationalbibliothek
Die Deutsche Nationalbibliothek lists this publication in the Deutsche Nationalbibliografie; detailed bibliographic data are available in the Internet at <http://dnb.d-nb.de>.

© 2008 WILEY-VCH Verlag GmbH & Co. KGaA, Weinheim

All rights reserved (including those of translation into other languages). No part of this book may be reproduced in any form – by photoprinting, microfilm, or any other means – nor transmitted or translated into a machine language without written permission from the publishers. Registered names, trademarks, etc. used in this book, even when not specifically marked as such, are not to be considered unprotected by law.

Typesetting Laserwords Private Limited, Chennai, India
Printing Strauss GmbH, Mörlenbach
Binding Litges & Dopf GmbH, Heppenheim
Cover Design Grafik-Design Schulz, Fußgönheim

Printed in the Federal Republic of Germany
Printed on acid-free paper

ISBN: 978-3-527-30971-9

Contents

Preface *XVII*
Abbreviations *XIX*
List of Contributors *XXIII*

1 **Introduction** *1*
Hans-Joachim Streitberger
1.1 Historic Development *1*
1.2 Legislation *7*
1.3 Automotive and Automotive Paint Market *9*

2 **Materials and Concepts in Body Construction** *13*
Klaus Werner Thomer
2.1 Introduction *13*
2.2 Methods of Body Construction *15*
2.2.1 Monocoque Design *15*
2.2.2 Space Frame *18*
2.2.3 Hybrid Type of Construction *19*
2.2.4 Modular Way of Construction *19*
2.3 Principles of Design *20*
2.3.1 Conventional Design *20*
2.3.2 Design Under Consideration of Light-Weight Construction *22*
2.3.3 Bionics *23*
2.4 Materials *26*
2.4.1 Steel *26*
2.4.1.1 General Remarks *26*
2.4.1.2 Low-Carbon Deep Drawing Steels *29*
2.4.1.3 Higher-Strength Steels *29*
2.4.1.4 High-Strength Steels *31*
2.4.1.5 High-Grade (Stainless) Steels *32*
2.4.1.6 Manganese–Boron Steels *32*
2.4.1.7 Light-Construction Steel with Induced Plasticity (TWIP Steel) *33*
2.4.2 Aluminum *34*

2.4.2.1	General Remark	*34*
2.4.2.2	Further Treatment	*35*
2.4.2.3	Aluminum Alloys	*38*
2.4.2.4	Aluminum as Light-Weight Construction Material	*39*
2.4.3	Magnesium	*40*
2.4.4	Titanium	*42*
2.4.5	Nonmetallic Parts – Fiber Composites	*43*
2.5	Manufacturing Methods	*47*
2.5.1	Tailored Products	*47*
2.5.1.1	General Remarks	*47*
2.5.1.2	Tailored Blanks	*47*
2.5.1.3	Tailored Tubes	*48*
2.5.1.4	Tailored Strips	*48*
2.5.1.5	Patchwork Blanks	*48*
2.5.1.6	Future	*49*
2.5.2	Hydroforming	*49*
2.5.3	Press Hardening	*50*
2.5.4	Metal Foam	*50*
2.5.5	Sandwich Structures	*51*
2.5.6	Roll Forming to Shape	*52*
2.6	Joining Methods	*52*
2.6.1	Bonding	*53*
2.6.2	Laser Welding	*54*
2.6.3	Others	*55*
2.6.3.1	Clinching	*55*
2.6.3.2	Riveting	*56*
2.6.3.3	Roller Hemming	*56*
2.7	Outlook	*57*
2.8	Surface Protection	*59*
2.8.1	Precoating of Sheets	*59*
2.8.2	Corrosion Prevention in the Design Phase	*59*
3	**Pretreatment of Multimetal Car Bodies**	*61*
	Horst Gehmecker	
3.1	Introduction	*61*
3.2	Car Body Construction Materials	*61*
3.2.1	Sheet Materials	*61*
3.2.2	Surface Conditions/Contaminations	*63*
3.3	Pretreatment Process	*65*
3.3.1	Sequence of Treatment	*65*
3.3.2	Degreasing	*65*
3.3.3	Activation	*69*
3.3.4	Zinc Phosphating	*70*
3.3.5	Passivation	*75*
3.3.6	Pretreatment of Aluminum – Steel Structures	*75*

3.3.7	Pretreatment of Magnesium	76
3.3.8	Pretreatment of Plastic Parts	77
3.4	Car Body Pretreatment Lines	77
3.4.1	Spray Lines	80
3.4.2	Continuous Horizontal Spray/Dip Line	80
3.4.3	RoDip3 Line	80
3.4.4	Vario Shuttle Line	81
3.4.5	Other Types of Lines	82
3.4.6	Construction Materials	82
3.4.7	Details on Process Stages	82
3.4.7.1	Precleaning Stage	82
3.4.7.2	Spray degrease	83
3.4.7.3	Dip Degrease	83
3.4.7.4	Rinsing	84
3.4.7.5	Activation	84
3.4.7.6	Phosphating	84
3.4.7.7	Rinsing	85
3.4.7.8	Passivation	85
3.4.7.9	Deionized water rinsing	85
3.4.7.10	Entering the Electrocoat Line	85
3.5	Properties and Specifications of Zinc Phosphate Conversion Layers	85
3.6	Environmental Legislations	86
3.7	Outlook	86
4	**Electrodeposition Coatings**	**89**
	Hans-Joachim Streitberger	
4.1	History and Introduction	89
4.2	Physico-chemical Basics of the Deposition Process	90
4.3	Data for Quality Control	95
4.3.1	Voltage, Current Density, Bath Temperature, and Bath Conductivity	96
4.3.2	Wet Film Conductivity	96
4.3.3	Solid Content, Solvent Content, and pH	97
4.4	Resins and Formulation Principles	98
4.4.1	General Remarks	98
4.4.2	Anodic Electrodeposition Paints	99
4.4.3	Cathodic Electrodeposition Paints	99
4.5	Film Performance of Cathodic Electrocoatings	101
4.5.1	Physical Film Data	101
4.5.2	Corrosion Protection	102
4.5.3	Chip Resistance	104
4.5.4	Surface Smoothness and Appearance	105
4.6	Design of Cathodic Electrocoating Lines	106
4.6.1	Integration into the Coating Process of Cars and Trucks	106
4.6.2	Pretreatment	107

4.6.3	General Functions and Equipment of an Electrocoat Line	*107*
4.6.4	Tanks, Filters, Heat Exchanger, and Power Supply	*108*
4.6.5	Replenishment and Anode Cells	*111*
4.6.6	Ultrafiltration and Rinsing Zones	*114*
4.6.7	Baking Oven	*118*
4.7	Defects During Application and their Prevention	*119*
4.7.1	Dirt	*119*
4.7.2	Craters	*120*
4.7.3	Surface Roughness	*121*
4.7.4	Film Thickness/Throwing Power	*122*
4.7.5	Other Defects	*123*
4.8	Electrocoating and Similar Processes Used in Automotive Supply Industry	*124*
4.9	Outlook	*125*
5	**Primer Surfacer** *129*	
	Heinrich Wonnemann	
5.1	Introduction	*129*
5.2	Requirement Profile	*132*
5.2.1	Legislative Requirement	*132*
5.2.2	Technological Requirements	*133*
5.2.2.1	Film Properties	*133*
5.2.2.2	Product Specifications	*136*
5.2.2.3	Application	*136*
5.3	Raw Materials	*138*
5.3.1	Resin Components	*139*
5.3.2	Pigments and Extenders	*141*
5.3.2.1	Titanium Dioxide	*143*
5.3.2.2	Barium Sulfate	*144*
5.3.2.3	Talc	*144*
5.3.2.4	Silicon Dioxide	*145*
5.3.2.5	Feldspar	*146*
5.3.2.6	Carbon Blacks	*146*
5.3.3	Additives	*146*
5.3.3.1	Pigment Wetting and Dispersion Additives	*148*
5.3.3.2	Defoaming and Deaerating Agents	*148*
5.3.3.3	Surfactants and Additives for Substrate Wetting	*149*
5.3.3.4	Rheology Additives	*149*
5.3.4	Solvents	*150*
5.3.4.1	Aromatic Hydrocarbons: Diluents/Thinners	*151*
5.3.4.2	Alcohols, Cellosolves, and Esters: Solvents	*151*
5.3.4.3	Tetralin or Pine Oil: Very High Boiling Additive Diluents	*152*

5.4	Liquid Primers *152*	
5.4.1	Formula Principles *152*	
5.4.1.1	Application *152*	
5.4.1.2	Rheology *153*	
5.4.2	Manufacturing Process *156*	
5.4.3	Application *158*	
5.5	Powder Primer Surfacers *160*	
5.5.1	Formula Principles *160*	
5.5.2	Manufacturing Process *162*	
5.5.3	Application *166*	
5.6	Process Sequence *170*	
5.7	Summary and Future Outlook *171*	
6	**Top Coats** *175*	
	Karl-Friedrich Dössel	
6.1	Introduction *175*	
6.2	Pigments and Color *175*	
6.3	Single-Stage Top Coats (Monocoats) *180*	
6.4	Base Coats *181*	
6.4.1	Base Coat Rheology *184*	
6.4.2	Low and Medium Solids Base Coat *185*	
6.4.3	High Solids (HS) Base Coats *186*	
6.4.4	Waterborne Base Coats *186*	
6.4.5	Global Conversion to Waterborne Base Coat Technology *187*	
6.4.6	Drying of Base Coats *189*	
6.5	Clear Coat *189*	
6.5.1	Market *189*	
6.5.2	Liquid Clear Coats *190*	
6.5.2.1	One-Component (1K) Acrylic Melamine Clear Coat *190*	
6.5.2.2	Acrylic Melamine Silane *192*	
6.5.2.3	Carbamate-Melamine-Based 1K Clear coat *192*	
6.5.2.4	One-Component Polyurethane (PUR) Clear Coat *192*	
6.5.2.5	One- (and Two-) Component Epoxy Acid Clear Coat *192*	
6.5.2.6	Two-Component (2K) Polyurethane Clear Coat *193*	
6.5.2.7	Waterborne Clear Coat *194*	
6.5.3	Powder Clear Coat *195*	
6.5.4	Top Coat Performance *198*	
6.5.4.1	Enviromental Etch *198*	
6.5.4.2	UV Durability of Clear Coats *199*	
6.5.4.3	Scratch Resistant Clear Coats *201*	
6.5.4.4	Application Properties *204*	
6.5.5	Future Developments: UV Curing *206*	

6.6	Integrated Paint Processes (IPP) for Top Coat Application	208
6.6.1	Wet-On-Wet-On-Wet Application (3 Coat 1 Bake) of Primer Surfacer–Base Coat–Clear Coat	208
6.6.2	Primerless Coating Process	209

7 Polymeric Engineering for Automotive Coating Applications 211
Heinz-Peter Rink

7.1	General Introduction	211
7.2	Polyacrylic Resins for Coating Materials in the Automotive Industry	214
7.2.1	Managing the Property Profile of the Polyacrylic Resins	214
7.2.2	Manufacturing Polyacrylic Resins	218
7.2.2.1	Manufacturing Polyacrylic Resins by Means of Solution Polymerization	218
7.2.2.2	Polymerization in an Aqueous Environment	222
7.2.2.3	Mass Polymerization	224
7.3	Polyester for Coating Materials for the Automotive Industry	224
7.3.1	Managing the Property Profile of Polyesters	224
7.3.2	Manufacturing Polyesters	228
7.4	Polyurethane Dispersions in Coating Materials for the Automotive Industry	231
7.4.1	Managing the Property Profile of Polyurethane Resins and Polyurethane Resin Dispersions	232
7.4.2	Manufacturing Polyurethane Resin Dispersions	234
7.5	Polyurethane Polyacrylic Polymers in Coating Materials for the Automotive Industry	238
7.5.1	Introduction	238
7.5.2	Managing the Property Profile of Polyurethane Polyacrylic Polymers	239
7.5.3	Manufacturing Polyurethane Polyacrylic Polymers	240
7.6	Epoxy Resins	241
7.6.1	Managing the Property Profile	242
7.6.2	Manufacturing Polyepoxy Resins	243
7.7	Cross-Linking Agents and Network-Forming Resins	244
7.7.1	Introduction	244
7.7.2	Cross-Linking Agents for Liquid Coating Materials	245
7.7.2.1	Melamine and Benzoguanamine Resins	245
7.7.2.2	Tris(Alkoxycarbonylamino)-1,3,5-Triazine	248
7.7.2.3	Polyisocyanates and Blocked Polyisocyanates	249
7.7.2.4	Other Cross-Linking Agents for Liquid Coating Materials	252
7.7.3	Cross-Linking Agents for Powder Coatings in the Automotive Industry	252

8	**Paint Shop Design and Quality Concepts** *259*	
	Pavel Svejda	
8.1	Introduction *259*	
8.2	Coating Process Steps *260*	
8.2.1	Pretreatment *261*	
8.2.2	Electrocoating (EC) *261*	
8.2.3	Sealing and Underbody Protection *262*	
8.2.4	Paint Application *263*	
8.2.4.1	Function Layer and Primerless Processes *263*	
8.2.4.2	Powder *264*	
8.2.5	Cavity Preservation *265*	
8.3	General Layout *266*	
8.4	Coating Facilities *268*	
8.4.1	Process Technology *268*	
8.4.2	Automation in the Paint Application *269*	
8.4.2.1	Painting Robot *270*	
8.4.3	Application Technology *271*	
8.4.3.1	Atomizer *272*	
8.4.3.2	Paint Color Changer *275*	
8.4.3.3	Paint Dosing Technology for Liquid Paints *277*	
8.4.3.4	Paint Dosing Technology for Powder Paints *278*	
8.4.4	Paint-Material Supply *279*	
8.4.4.1	Paint Supply Systems for the Industrial Sector *280*	
8.4.4.2	Paint Mix Room *280*	
8.4.4.3	Container Group *280*	
8.4.4.4	Circulation Line System *282*	
8.4.4.5	Basic Principles for the Design of the Pipe Width for Circulation Lines *282*	
8.4.4.6	Paint Supply Systems for Small Consumption Quantities and Frequent Color Change *283*	
8.4.4.7	Small Circulation Systems *283*	
8.4.4.8	Supply Systems for Special Colors *284*	
8.4.4.9	Voltage Block Systems *285*	
8.4.4.10	Voltage Block Systems with Color-Change Possibility *285*	
8.4.4.11	Installations for the High Viscosity Material Supply *286*	
8.4.5	Conveyor Equipment *287*	
8.5	Paint Drying *288*	
8.6	Quality Aspects *290*	
8.6.1	Control Technology *290*	
8.6.1.1	Process Monitoring and Regulation *292*	
8.6.2	Automated Quality Assurance *293*	
8.6.2.1	Process Optimization in Automatic Painting Installations *296*	

8.7	Economic Aspects 298
8.7.1	Overall Layout 298
8.7.2	Full Automation in Vehicle Painting 298
8.7.3	Exterior Application of Metallic Base Coats with 100% ESTA High-Speed Rotation 300
8.7.4	Robot Interior Painting with High-Speed Rotation 301
9	**Coatings for Plastic Parts** 305
9.1	Exterior Plastics 305
	Guido Wilke
9.1.1	Introduction 305
9.1.1.1	Ecological Aspects 305
9.1.1.2	Technical and Design Aspects 306
9.1.1.3	Economical Aspects 307
9.1.2	Process Definitions 307
9.1.2.1	Offline, Inline, and Online Painting 307
9.1.2.2	Process-Related Issues, Advantages, and Disadvantages 307
9.1.3	Exterior Plastic Substrates and Parts 310
9.1.3.1	Overview 310
9.1.3.2	Basic Physical Characteristics 311
9.1.3.3	Part Processing and Influence on Coating Performance 315
9.1.4	Pretreatment 315
9.1.5	Plastic-Coating Materials 318
9.1.5.1	Basic Technical Principles of Raw-Material Selection 318
9.1.5.2	Car-Body Color 320
9.1.5.3	Contrast Color and Clear Coat on Plastic Systems 324
9.1.6	Technical Demands and Testing 324
9.1.6.1	Basic Considerations 324
9.1.6.2	Key Characteristics and Test Methods 325
9.1.7	Trends, Challenges, and Limitations 329
9.1.7.1	Substrates and Parts 329
9.1.7.2	Paint Materials 330
9.1.7.3	Processes 332
9.2	Interior Plastics 334
	Stefan Jacob
9.2.1	Introduction: the 'Interior' Concept 334
9.2.2	Surfaces and Effects 335
9.2.3	Laser Coatings 337
9.2.3.1	Substrate Requirements 339
9.2.3.2	Requirements to Be Fulfilled by the Paint Systems and Coating 339
9.2.3.3	Demands Expected by the Inscription Technique 340
9.2.4	Performances of Interior Coatings 341
9.2.4.1	Mechanical and Technological Demands 341
9.2.4.2	Substrates and Mechanical Adhesion 342
9.2.4.3	Ecological and Economical Requirements 343

9.2.4.4	Equipment for the Application of Interior Paint Systems	344
9.2.5	Raw-Material Basis of Interior Paints	346
9.2.6	Summary/Outlook	347

10 Adhesive Bonding – a Universal Joining Technology 351
Peter W. Merz, Bernd Burchardt and Dobrivoje Jovanovic
10.1 Introduction 351
10.2 Fundamentals 351
10.2.1 Basic Principles of Bonding and Material Performances 351
10.2.1.1 Types of Adhesives 351
10.2.1.2 Adhesives are Process Materials 355
10.2.1.3 Advantages of Bonding 356
10.2.1.4 Application 359
10.2.2 Surface Preparation 359
10.2.2.1 Substrates 361
10.2.2.2 Adhesion 361
10.2.2.3 Durability and Aging of Bonded System 362
10.3 Bonding in Car Production 366
10.3.1 Body Shop Bonding 367
10.3.1.1 Antiflutter Adhesives 367
10.3.1.2 Hem-Flange Bonding 368
10.3.1.3 Spot-Weld Bonding 369
10.3.1.4 Crash-Resistant Adhesives/Bonding 369
10.3.2 Paint Shop 370
10.3.3 Trim Shop 371
10.3.3.1 Special Aspects of Structural Bonding in the Trim Shop 371
10.3.3.2 'Direct Glazing' 371
10.3.3.3 Modular Design 372
10.3.3.4 Other Trim Part Bondings 372
10.4 Summary 374

11 In-plant Repairs 377
Karl-Friedrich Dössel
11.1 Repair After Pretreatment and Electrocoat Application 377
11.2 Repair After the Primer Surfacer Process 378
11.3 Top-Coat Repairs 378
11.4 End-of-Line Repairs 380

12 Specifications and Testing 381
12.1 Color and Appearance 381
Gabi Kiegle-Böckler
12.1.1 Visual Evaluation of Appearance 381
12.1.1.1 Specular Gloss Measurement 382
12.1.1.2 Visual Evaluation of Distinctness-of-Image (DOI) 385
12.1.1.3 Measurement of Distinctness-of-Image 385

12.1.1.4	Visual Evaluation of 'Orange Peel' *386*
12.1.1.5	Instrumental Measurement of Waviness (Orange Peel) *388*
12.1.1.6	The Structure Spectrum and its Visual Impressions *390*
12.1.1.7	Outlook of Appearance Measurement Techniques *392*
12.1.2	Visual Evaluation of Color *393*
12.1.2.1	Solid Colors *393*
12.1.2.2	Metallic and Interference Colors *397*
12.1.2.3	Color Measurement of Solid Colors *398*
12.1.2.4	Color Measurement of Metallic and Interference Coatings *401*
12.1.2.5	Typical Applications of Color Control in the Automotive Industry *402*
12.1.2.6	Color Measurement Outlook *403*
12.2	Weathering Resistance of Automotive Coatings *405*
	Gerhard Pausch and Jörg Schwarz
12.2.1	Introduction *405*
12.2.2	Environmental Impact on Coatings *406*
12.2.2.1	Natural Weathering *408*
12.2.2.2	Artificial Weathering *414*
12.2.2.3	New Developments *421*
12.2.3	Standards for Conducting and Evaluating Weathering Tests *423*
12.2.4	Correlation Between Artificial and Natural Weathering Results *426*
12.3	Corrosion Protection *427*
	Hans-Joachim Streitberger
12.3.1	Introduction *427*
12.3.2	General Tests for Surface Protection *429*
12.3.3	Special Tests for Edge Protection, Contact Corrosion, and Inner Part Protection *432*
12.3.4	Total Body Testing in Proving Grounds *433*
12.4	Mechanical Properties *434*
	Gerhard Wagner
12.4.1	General Remarks *434*
12.4.2	Hardness *435*
12.4.2.1	Pendulum Damping *435*
12.4.2.2	Indentation Hardness *436*
12.4.2.3	Scratch Hardness *439*
12.4.3	Adhesion and Flexibility *441*
12.4.3.1	Pull-Off Testing *441*
12.4.3.2	Cross Cut *441*
12.4.3.3	Steam Jet *444*
12.4.3.4	Bending *445*
12.4.3.5	Cupping *446*
12.4.3.6	Impact Testing by Falling Weight *448*
12.4.4	Stone-Chip Resistance *449*
12.4.4.1	Standardized Multi-Impact Test Methods *450*
12.4.4.2	Single-Impact Test Methods *451*
12.4.5	Abrasion *454*

12.4.5.1	Taber Abraser	*454*
12.4.5.2	Abrasion Test by Falling Abrasive Sand	*455*
12.4.6	Scratch Resistance	*456*
12.4.6.1	Crockmeter Test	*456*
12.4.6.2	Wet-Scrub Abrasion Test	*457*
12.4.6.3	Simulation of Car Wash	*458*
12.4.6.4	Nanoscratch Test	*460*
12.4.7	Bibliography, Standards	*462*

13 Supply Concepts *467*
Hans-Joachim Streitberger and Karl-Friedrich Dössel
13.1 Quality Assurance (QA) *467*
13.2 Supply Chain *468*
13.2.1 Basic Concepts and Realizations *468*
13.2.2 Requirements and Limitations of a System Supply Concept *473*

14 Outlook *475*
Hans-Joachim Streitberger and Karl-Friedrich Dössel
14.1 Status and Public Awareness of the Automotive Coating Process *475*
14.2 Regulatory Trends *476*
14.3 Customer Expectations *478*
14.4 Innovative Equipments and Processes *479*
14.5 New Business Ideas *481*

Index *483*

Preface

It is now over 10 years since the first publication of "Automotive Paints and Coatings" in 1996. The original publication was made possible thanks to the untiring effort of Gordon Fettis to bring out a book dedicated to automotive OEM coatings. In a changing business environment, the automotive industry is always reevaluating its core competencies and has transferred many technical developments and manufacturing tasks to the supplier industry. Painting, so far, has remained one of its core competencies and is a "value added" process in the car manufacturing industry. This fact underlines the necessity for a new and completely revised edition of the book. We thank Mr. Fettis for taking this up, and giving us the opportunity to publish this book. We also thank all the contributors for their articles that demonstrate the expertise of each of them in his respective field.

The book covers the painting process of passenger cars and "light trucks", as they are called by the American automotive industry. These vehicles are mass manufactured and therefore their coating requirements in terms of processing and coating performance are similar in nature.

The key performance drivers in automotive coatings are quality, cost, and environmental compliance. Quality, in this context, relates to corrosion protection and long lasting appearance. Quality and cost are addressed by more efficient coating processes and a higher degree of automatization. This has led to reduced use of paint and lesser waste per body. With the widespread use of high solid, waterborne, and powder coatings, solvent emission from paint shops have been reduced by more than 50% over the last ten years.

We are proud to contribute to this development by publishing the second edition of "Automotive Paints and Coatings", in which we try to describe the state-of-the-art technology of automotive coating processes and the paint materials used in these processes.

Automotive coating processes represent the cutting edge of application technology and paint formulations. They are the most advanced processes in regard to volume handling, sophisticated body geometries and speed in mass production. Colored coatings offer mass customization and product differentiation at an affordable cost.

This second edition may help many readers to understand these high-performance coating processes and the related materials, and also appreciate the scope for new ideas and innovations.

The contents and the main focus of this edition differ from those in the first edition. Processes, technology, and paint formulations have been weighted equally, reflecting that legislation and cost were the driving forces of the past.

We hope that not only experts but also technically interested readers will find this book useful.

Münster and Wuppertal, Germany *Dr. Hans-Joachim Streitberger*
December 2007 *Dr. Karl-Friedrich Dössel*

Abbreviations

1K	1-component
2K	2-component
ABS	Acrylonitrile-Butadiene-Styrene
AED	Anodic Electro Deposition
AFM	Atomic Force Microscopy
APEO	AlkylPhenolEthOxylate
ASTM	American Society for Testing and Materials
ATRP	Atom Transfer Radical Polymerization
BAT	Best Available Technology
BSC	Balanced Score Card
BMC	Blow Molding Compound
BPA	BisPhenol-A
CASS	Copper Accelerated Salt Spray test
CCT	Cyclic Corrosion Test
CED	Cathodic Electro Deposition
CIE	Comite International d'Eclairage
CMC	Critical Micelle Concentration
CPO	Chlorinated PolyOlefines
CPU	Cost Per Unit
CSM	Centre Suisse d'electronique et de Microtechnique
CTE	Coefficient of Thermic Extension
DBP	Di-ButylPhtalat
DDDA	DoDecane-Di-Acid
DDF	Digital Dichtstrom Förderung (digital dense flow transporation)
DIN	Deutsche IndustrieNorm (German industrial normation office)
DMPA	DiMethylol-Propronic Acid
DOI	Distinctness Of Image
DOS	DiOctylSebacate
DPE	DiPhenylEthane
EDT	Electro Discharge Texturing
EDTA	Ethylene Diamino Tetra Acid
EEVC	European Enhanced Vehicle safety Committee

Automotive Paints and Coatings. Edited by H.-J. Streitberger and K.-F. Dössel
Copyright © 2008 WILEY-VCH Verlag GmbH & Co. KGaA, Weinheim
ISBN: 978-3-527-30971-9

EINECS	European Inventory of Existing commercial Chemical Substances
EIS	Electrochemical Impedance Spectroscopy
EN	European Norms
EO	EthOxy
EP	EPoxy
EPA	Environmental Protection Agency
EPDM	Ethylene-Propylene-Diene-Monomer
ERP	Enterprise Resource Planning
ESCA	Electron Spectroscopy for Chemical Analysis
ESTA	ElectroSTatic Application
FEM	Finite Element Method
FMVSS	Federal Motor Vehicle Safety Standards (USA)
GMA	GlycidylMethAcrylat
HALS	Hindered Amine Light Stabilizer
HAPS	Hazardous Air Polluting Substances
HDI	1.6-HexamethyleneDiIsocanate
HDI	HexamethyleneDiIsocanate
HLB	Hydrophilic Lipophilic Balance
HMMM	HexaMethoxyMethylMelamine
HVLP	High Volume Low Pressure
ICEA	IsoCyanatoEthyl(meth)Acrylate
IPDI	IsoPhoronDiIsocyanate
IPN	InterPenetrating Network
IPP	Integrated Paint Process
ISO	International Standard Organisation
KPI	Key Performance Indicator
LEPC	Low Emission Paint Consortium
MCC	Mercedes Compact Car
MDI	Diphenyl-Methan-DiIsocyanate
MDF	Medium Density Fibreboard
NMMO	N-MethylMorpholine-n-Oxid
NAFTA	NorthAmerican Free Trade Agreement
NTA	NitroloTriAceticacid
OEM	Original Equipment Manufacturer
PA	PolyAmide
PA-GF	Glas Fiber reinforced PolyAmide
PBT	PolyButyleneterephtalate
PC	PolyCarbonate
PFO	Paint Facility Ownership
PMMA	PolymethylmethAcrylate
PO	ProprOxy
PP	PolyPropylene
PPE	Poly Phenylene Ether
PUR	PolyURethane

PVC	PolyVinylChloride
QC	Quality Control
QA	Quality Assurance
QUV	Tradename for test chambers (Q-Lab)
RAFT	Reversible Addition Fragmentation chain Transfer
RCA	Rheology Contol Agent
RRIM	Reinforced Rejection InMold
RT	Room Temperature
R-TPU	Reinforced ThermoPlastic Polyurethane
SAE	Society of Automotive Engineers (USA)
SCA	Sag Control Agent
SCM	Supply Chain Management
SDAT	Short safe Drive Away Time
SEA	South East Asia
SMC	Sheet Molding Compound
SNIBS	Sodium NItroBenzeneSulfonate
TACT	Tris(AlkoxyCarbonylamino)-1.3.5-Triazine
TDI	Toluylene DiIsocaynate
TEM	Transmission Electron Microscopy
TEMPO	TEtraMethylPiperidin-n-Oxid
TGIC	TriGlycidylIsoCyanurat
m-TMI	3-isoporpenyl-dimethylbenzyl-isocyanate
TMXDI	Tetra-Methyl-Xylylene-DiIsocyanate
ToF-SIMS	Time of Flight Secondary Ion Mass Spectroscopy
TPO	ThermoPlastic Olefines
UP	Unsaturated Polyester
UV	UltraViolette
UVA	UV-Absorber
VDA	Verband der Deutschen Automobilindustrie (German association of car manufactureres)
VOC	Volatile Organic Compounds

List of Contributors

Bernd Burchardt
Sika Schweiz AG
OEM Adhesives & Sealants
Tüffenwies 16
8048 Zürich
Switzerland

Karl-Friedrich Dössel
Dupont Performance Coatings
Christbusch 25
42285 Wuppertal
Germany

Horst Gehmecker
Chemetall GmbH
Trakehnerstr. 3
60407 Frankfurt
Germany

Stefan Jacob
Mankiewicz Gebr. & Co.
Georg-Wilhelm-Str. 189
21107 Hamburg
Germany

Dobrivoje Jovanovic
Sika Schweiz AG
OEM Adhesives & Sealants
Tüffenwies 16
8048 Zürich
Switzerland

Gabi Kiegle-Böckler
BYK-Gardner GmbH
Lausitzerstr. 8
82538 Geretsried
Germany

Peter W. Merz
Sika Schweiz AG
OEM Adhesives & Sealants
Tüffenwies 16
8048 Zürich
Switzerland

Gerhard Pausch
Pausch Messtechnik GmbH
Nordstr. 53
42781 Haan
Germany

Heinz-Peter Rink
BASF Coatings AG
CTS Trailers, Trucks & ACE
Automotive Refinish / Commercial
Transport Coatings Solutions
Glasuritstr. 1
48165 Münster-Hiltrup
Germany

Jörg Schwarz
Daimler AG
Werk Sindelfingen – B 430 – PWT/VBT
71059 Sindelfingen
Germany

Automotive Paints and Coatings. Edited by H.-J. Streitberger and K.-F. Dössel
Copyright © 2008 WILEY-VCH Verlag GmbH & Co. KGaA, Weinheim
ISBN: 978-3-527-30971-9

Hans-Joachim Streitberger
Markt & Management
Patronatsstr. 13
48165 Münster
Germany

Pavel Svejda
Dürr Systems GmbH
Application Technology
Vertrieb/Sales
Rosenstr. 39
74321 Bietigheim-Bissingen
Germany

Klaus Werner Thomer
Adam Opel AG
ITDC/ME
Friedrich-Lutzmann-Ring
65423 Rüsselsheim
Germany

Gerhard Wagner
Dupont Performance Coatings
Qualitätsprüfung
Christbusch 25
42285 Wuppertal
Germany

Guido Wilke
Hochschule Esslingen
University of Applied Sciences
Labor Polymerwerkstoffe
Kanalstr. 33
73728 Esslingen
Germany

Heinrich Wonnemann
BASF Coatings AG
CO/XEH – B325
48165 Münster
Germany

1
Introduction

Hans-Joachim Streitberger

1.1
Historic Development

The car painting industry has undergone incredible changes by way of materials and processes development following the general progress of manufacturing technology from the start of the twentieth century until today. Early coating processes, that is, during the first half of the twentieth century, involved the use of air drying paints, sanding of each layer and polishing, all of which needed weeks for completion. All the coating steps were executed manually (Figure 1.1).

Different driving forces behind the development of better and more efficient processes have brought in dramatic changes over the last 100 years. Introduction of mass production requiring faster curing paints, better film performance in terms of corrosion and durability of colors, improved environmental compatibility, and fully automated processes for better reliability characterize the most important milestones in this field (Table 1.1).

The status of mass production of cars during the 1940s required new coatings providing faster drying and curing: the result – the birth of enamels! At the same time, owing to their limited availability, the natural raw materials used in the manufacture of the paints had to give way to synthetic chemicals. Crosslinking of paints became state of the art. The coating process could be reduced to a day including all necessary preparation time for the car body like cleaning, sanding, repairing, and so on.

The number of applied coatings had been reduced to four or five layers, all hand sprayed this time (Figure 1.2). The function of these layers were corrosion protection for the primers, smoothness and chip resistance for the primer surfacers (which are often applied at the front ends and exposed areas in two layers), and color and weather resistance for the final top-coat layer. In the 1950s the process of applying the primer changed to dip coating, a more automated process, but a hazardous one owing to the solvent emission of the solvent-borne paints. Explosions and fire hazards then forced automotive manufacturers to introduce either waterborne paints or electrodeposition paints. The latter, which were introduced during the

1 Introduction

Fig. 1.1 Painting of cars in the 1920s and 1930s.

Table 1.1 Milestones and driving forces in the car coating process

Year	Topics/Driving forces	Aspects
1920	Manual painting	Time-consuming process : weeks
1940	Mass production	Enamels/oven/time : day
1970	Improved film performance	CED/2-layer top coat/new materials
1980	Environmental compliance	Waterborne coatings/powder/transfer efficiency
2000	Automated processes	First time capability/time : hours

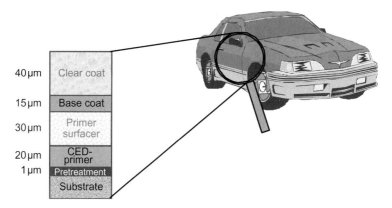

Fig. 1.2 Scheme of the multilayer coating of cars.

late 1960s, are more efficient in terms of material transfer as well as throwing power that is necessary for improved corrosion protection of the inner parts of the car body.

In the 1970s the anodic deposition coatings, mostly based on maleinized polybutadiene resins, quickly gave way to cathodic ones owing to better corrosion protection by their modified epoxy resin backbones and reactive polyurethane-based crosslinkers, increased throwing power, and higher process reliability. At the same time, the single layer top coats were gradually replaced by two-layer top coats consisting of a thin base coat and a thicker clear coat applied wet-on-wet. The base coats are responsible for color and special effects (for example, metallic finish), whereas the clear coats provide improved durability using specially designed resins and formula ingredients like UV-absorber and radical scavengers. Today, most clear coats in Europe are based on two-component (2K-) formulation consisting of an acrylic resin with OH-functions and a reactive polyurethane crosslinker. The rest of the world still prefers the one-component technology based on acrylic resins and melamine crosslinkers. An interesting one-component technology based on carbamate functionality has been recently introduced in the United States [1].

All these developments contributed to an improved film performance resulting in better corrosion protection and longer top-coat durability – for example, gloss retention for up to 5–7 years was observed in Florida.

Furthermore, raw material development in the pigment section, with improved flake pigments based on aluminum and new interference pigments that change color depending on the angle in which they are viewed, has resulted in enhanced brilliance and color effects of automotive coatings [2].

Along with this development of coating and paint technology, spray application techniques also underwent significant improvements. Starting with simple pneumatic guns and pressure pots for paint supply, today, craftsmanship in painting is no longer needed. Several factors have contributed to the development of coating machines and robots and the state of automation that is present today. The first factor was the health risk to the painters who were exposed to solvent emission from paints in the spray booth and the investment in safety equipment, which was often unsuitable for them. The second factor was the hazards of the electrostatic application technique. Yet another factor was the lack of uniform quality in a manual painting job.

Because of the latest developments in wet-on-wet coating technology, coating machines, automated cleaning processes, and modern paints, the time taken today for the coating process, including pretreatment, can be as short as 8 hours for a car body leaving the body shop and entering the assembly line (Figure 1.3) [3].

Together with the continuous improvements of the application technology, new water-based materials were developed to contribute toward the legally enforced environmental compliance of the processes. The first water-based base coats were introduced at Opel in Germany in the 1980s, followed by water-based primer surfacers in the 1990s.

Investments in modern paint shops vary from €200 up to €600 million for coating 1000 units a day. Today, painting technology for the car industry has been more

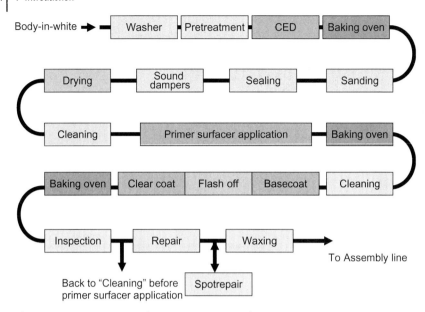

Fig. 1.3 Process steps in modern automotive paint shops.

or less standardized all over the world. Inorganic pretreatment, cathodic electrodeposition, liquid or powder primer surfacer, liquid base coats, and one-component or two-component solvent-borne clear coats are mostly used today. This is a result of the consolidation of the engineering and coating line manufacturers and paint producers into just a handful of major players. In 2002, 70% of the car coating market was in the hands of Dupont, PPG, and BASF.

As of today, the technology of powder coatings has reached a point where many car manufacturers have decided to introduce environmentally compliant technology more aggressively. Today, powder is established as the primer surfacer in North America: at Chrylser in all actual running plants, at GM for their truck plants, and in all new paint shops. In Europe, in a number of plants at BMW, powder is also used as a clear coat. [4].

Over the same period of time, body construction materials have also changed significantly. Starting with pure steel bodies, today the share of aluminum and magnesium, as well as plastic, as raw materials for specific car components or hang-on parts can account for about 30% of the weight of the car. On an average, half of this is plastic. The main focus is on weight reduction, design variability, and cost (Figure 1.4).

Starting with bumpers in the 1970s, and moving on to fenders and hoods, many of the exterior parts are now made of specially designed plastics. The coating process is predominantly outsourced for these parts, even though some car manufacturers assemble the parts in the coating line and apply the original base coat–clear coat technology on these parts to overcome color match problems.

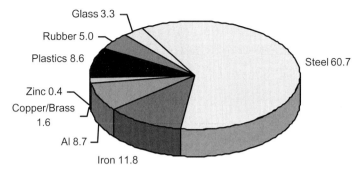

Fig. 1.4 Share (%) of different material classes for car manufacturing (source : Ward's Communications, Facts & Figures 2002).

Interior parts made of plastic materials were introduced in the 1960s and are today increasingly being coated by specially formulated paints for providing the 'soft feel' touch that gives the plastic an improved acceptance by the customers. These parts were mostly painted by the part manufacturers and supplied to the automotive industry as complete modules to the assembly line. Special laser coatings have been developed to inscribe symbols of functions on coated dash board or other units in the interior of a car.

Another increasing application of coatings is connected with the modern design of head lamps made of blow molding compounds(BMC) for the head lamp reflectors covered by a polycarbonate lens. Both parts are coated dominantly by UV coatings.

Many other exterior parts like hoods and trunk lids, as well as other body segments, are increasingly made of aluminum or sheet molding compounds (SMC). In addition to the aspect of careful construction in respect of building galvanic elements, pretreatment chemicals and the process of multimetal bodies had to be developed. This was a demanding task that needed avoidance of any harmful heavy metal ions like chromic-VI. The complete aluminum body still remains a niche product.

Connected with an increased share of multicomponent parts of different substrates, the welding process for manufacturing cars has been partially replaced by other assembling techniques like clinching and riveting. Glewing, a technology known from the air transportation industry, has become very important. This technique is based on surface treatment and product application similar to coatings. On the basis of the use of interface with coating layers like gluing back-windows on electrocoat layers and its increasing application, this will be described in Chapter 10.

The performance requirements for an OEM (Original Equipment Manufacturer) coating of passenger cars are many and diverse. They can be attributed to corrosion protection, durability, including stone chip resistance, and appearance. One should bear in mind the extent of extreme stress that cars are exposed to throughout their life span. High temperatures up to 70 °C on dark colors in Florida and similar regions, as well as low temperatures of −50 °C in polar regions, permanent temperature fluctuation of 10–20 °C daily, stone chip attacks on unpaved roads, high loads of salt in coastal regions, as well as in wintertime, high ultraviolet radiation in combination with dry and humid periods, the action of acid or alkaline air pollutants from many sources, and the physical and mechanical stress in vehicle washing installations are the most important among the many factors. Recently, the gloss, appearance, and effect coatings became important factors for selling cars by underlining the image and personality of the car and its driver.

Because of the fact that many resources have been directed to the environmental improvement of paint application as well as toxic aspects of the paint formulas, the performance of automotive coatings has increased significantly. Even the film thickness of a car coating today is only 100–140 µm, which needs, in most coating processes, about 9–16 Kg deposited paint per car; the corrosion protection and the long time durability of color and gloss is about two times higher than what it was 25 years ago. Three main factors have contributed to this: new substrates, introduction of cathodic electrodeposition paints, and the two-layer top-coat system with a special designed clear coat for long term durability. The life time of a car is no longer related to the corrosion or durability of the coating and color.

The color of cars has become a very important design tool, significantly supporting the purchasing habits of customers. For this reason, color trends are being observed by the paint and automotive industry together to develop trendy colors for the right cars.

The color variability has been increased at the same time. Today's customers, especially in the premium car level, can demand whichever colors they want at a cost. New color effects like 'color flop', which is a coating providing different color impressions to the customer depending on the angle of vision, have entered the scene [5].

The latest significant milestone in the history of the development of car painting is the combination of highly efficient application techniques like high rotational mini- or micro-bells and the painting robots (Figure 1.5). This has lead to the highest degree of transfer efficiency and reliability, resulting in an efficiency of 90% and more of defect free coatings in modern paint shops [6].

The application of sealants, sound deadeners and underbody protection is very often a part of the coating process. These materials need to be dried. In an efficiency move, the primer baking ovens are mostly used for this process.

Shorter cycle time of car models requires faster planning and realization periods for designing new paint shops. Additionally, globally competitive business has brought into the focus of car manufacturers not only the respective investment costs, but also the running cost of a coating line [7].

At the end of the paint line, the application of transport coatings, wrap up of coated cars for company design on a commercial fleet, as well as safety measures, all become part of the coating processes connected with the manufacture of cars [8].

Quality assurance has reached a new dimension for automotive OEM paints and coatings. While time-consuming batch-to-batch approvals of the customer's specifications has been the order of the day, supporting-system approaches, for example, audit-management systems according to DIN-ISO 9001 and 14001, ISO/TS 16949, QS 9000 or VDA 6.1, have become mandatory for the paint industry

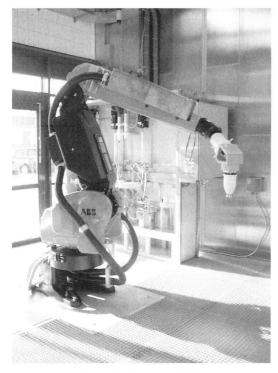

Fig. 1.5 Paint robot with electrostatic bell (source: BASF Coatings).

to be approved as a supplier to the automotive industry. So, right from the beginning, the development of new products has to be quality oriented to 'zero defect' levels through all the steps leading up to the delivery status.

At the same time, the testing and physical methods of describing the performance of paint and coatings have become much more precise and value based not only for the performance of the films, but also the performance during the application and film forming processes. In top-coat color specifications and in general physical color matching, especially, colorimetric data will replace visual inspections, which have reliability problems. The so-called 'finger print' methods have been developed to improve the performance of paints during the application processes [9]. Film performances like durability and corrosion protection, and mechanical performance like scratch, mar, and gravel resistance are very predictable in short term testing procedures today.

1.2
Legislation

Another driving force for finding new coating formulas is the passing of legislative acts all over the world calling for a ban on toxic components in all the formulae to the maximum possible extent [10]. So lead- and chrome-free paint formulas are today 'state of the art'. Also, the emission of volatile organic compounds (VOC) has been restricted, especially in Europe and the NAFTA (North American Free Trade Agreement) region for the last 20 years, to numbers that are defined in various ways and controlled in NAFTA and Europe, and are about five to ten times lower than what they were 30 years ago. This was reached by simultaneous material and process development as described herein. The industry has to deal with moving targets set by the authorities according to the best available technology. In general, the emission of any type of 'greenhouse gases' is in worldwide focus based on the 'Kyoto' protocol. Life cycle assessments are gaining awareness [11].

Owing to the fact that the most important application processes of automotive paints release solvents and that the composition of coatings can have more than 15 ingredients, together with the global agreements on environmental targets, the political scenario in North America and Europe was to focus first on paint consumers and manufacturers for improved environmentally compatible products and application processes.

The response in the 1980s in North America was to increase the solid contents and decrease the solvents of the actual solvent-borne paint formulations. New resins had to be developed and formulation as well as application conditions had to be optimized in the direction of sagging resistance and surface and film properties. The relative increase of solids by about 20% generated a completely new technology. The legislation controlled the progress by measuring the solvent input in the factories.

In Europe the resources for research and development were directed to waterborne coatings first, resulting in the introduction of waterborne base coats. Solvent-borne base coats were the paints providing the highest amount of solvent emission owing to the very low solid content of 12–18% at that time. Later in the 1990s waterborne primer surfacer and some waterborne clear coats were introduced. The legislation in Germany supported abatement technology to meet its requirements. The new European approach of the VOC directive of 1999 now combines after-treatment with the pragmatic approach of North America, mostly the United States.

Both paint development directions resulted in rather different paint formulations, creating problems for the South East Asia (SEA) region, which still has to decide which way to go. Signals and recent decisions of Toyota, Honda, Nissan, and Mitsubishi favor the waterborne products.

Safety standards for the handling of paints in the paint kitchen as well as application booths are quite uniform around the world. The personal safety equipment in spray booths now mostly consists of complete overall, mask, and respiratory systems protecting the worker from any contamination. A greater use of robots keeps workers out of the paint booth. They become engineers who program and run the robots and look into other booth parameters.

Harmful chemicals and additives have been tested comprehensively during the last 30 years. Many products have been abandoned either by the car manufacturers themselves or by legislation. Among these are lead, in electrocoatings and pigments, chromium, in primers and electrocoatings, cadmium, in pigments, many solvents qualified as HAPS (Hazardous Air Polluting Substances), and monomers like acrylonitrile, acrylamide, and so on, which belong to the cmr (carcinogenic, mutagenic, reproduction toxic) products. Special awareness has been created with respect to the biozides in Europe [12]. Other VOCs may contribute to the generation of ozone in the lower atmosphere and so they have been limited by legislation, step by step, leading up to the best state-of-the-art technology of the paint application process [13] as well as paint product development. Significant steps have been made in this respect by the introduction of waterborne base coats, waterborne primer, and the slurry-clear coat. Further reduction in VOC can be achieved by powder primer surface and powder clear coat [14].

In recent years, legislation in Europe focuses on the harmful and environmentally significant impact of chemicals. This will also lead to further replacement of ingredients and components in paint formulation [15].

1.3
Automotive and Automotive Paint Market

In 2004, car manufacturers, worldwide, produced about 58 million cars and light trucks (vans, mini-vans, pickups). Europe, North America, and Japan are traditionally the largest producing regions, accounting for about 78% of all cars (Figure 1.6).

SEA including Korea, China, and India, is gaining importance as a consumer market as well as a producing region. This is a change from the past and will change in the future owing to two main reasons:
1. Today's quality concepts in manufacturing as well as painting cars allow car manufacturers to produce wherever workforce is available. This drives most of them to regions with low labor costs.
2. Most car manufacturers produce world class cars, which can be exported and brought into all market places around the world.

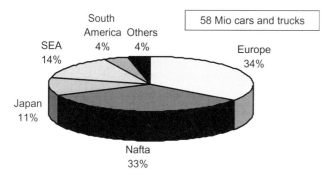

Fig. 1.6 Contribution of the regions to world light vehicle car manufacturing (Automobilprod.Juni/2004).

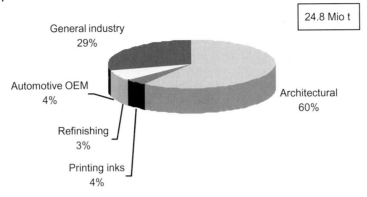

Fig. 1.7 Volume of the worldwide paint market as of 2002.

Among the limiting factors are the supply bases of components and hang-on parts, especially when it comes to just-in-time deliveries. This leads to certain regions where the car manufacturers would prefer to settle in the future like North Carolina, South Korea, Eastern Europe, and Shanghai, for example.

Since the 1980s, the outsourcing of manufacturing components and integrated modules like headlamps, fenders, bumpers, doors, and roof elements for direct delivery to the assembly line represents the global trend in worldwide car manufacturing so that the 'value added' is transferred to the automotive supplier industry. Since 2000, the supplier industry, which has a 40% share of the worldwide production value in the car industry and which includes more of design work, is increasingly focusing on innovation efforts and is consolidating to become a global industry.

The worldwide market volume of car paints of about 1.0 Mio t consists of electrodeposition coatings, primer surfacers, base coats, and clear coats, as well as speciality coatings consumed in the automotive coating lines and in the supplier industry, only counts for about 4% of the total paint market (see Figure 1.7). The biggest share contributes to the architectural market, followed by general industry (OEM market), which, in other statistics, like in North America, includes the contribution of car paints. From a technological standpoint, car paints as well as their coating processes are valued as the most advanced technologies both by coating performance, and the efficiency and reliability of the coating process. These high standards and requirements forced many players out of business, resulting in the fact that today just three main paint suppliers dominate the worldwide automotive OEM paint market. Regional paint manufacturers exist mostly in Japan and SEA.

In recent years, the method of conducting business activities has begun to change significantly. With the 'single sourcing' concept in which one paint supplier delivers all products and takes responsibility for all paint-related problems of the coating process, the business arrangements between paint manufacturers and the automotive industry has reached a point where the responsibility of running paint lines as well as cost targets have been taken over by the paint supplier [16]. The so-called *Tier 1'* suppliers manage the various degrees and levels of cooperation with

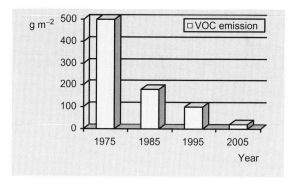

Fig. 1.8 Reduction potential of volatile organic compound (VOC) emission of the automotive coating process in gram per square meter body surface based on paint and process development (best available technology).

the customer. *Tier 2* suppliers manage the total paint shops including engineering, logistics, auxiliaries, quality control, and staffing [17].

Targeting the cost per unit and not cost per paint, together with new processes [18], modern application techniques, and paint formulations has brought the efficiency and environmental compliance of the coating processes in terms of VOC to a level that could not have been imagined 30 years ago (Figure 1.8). This has happened with an increased performance level of coatings in terms of corrosion protection, the top coat's long-term durability in respect to chip resistance, and color and gloss retention.

References

1 Green, M.L. (**2001**) *Journal of Coatings Technology*, **73** (981), 55.
2 Rochard, S. (**2001**) *Modern Paint and Coatings*, **91** (9), 28.
3 Anonymous, *Automobilproduktion* (12/2000) 38.
4 Könneke, E. (**2002**) *Journal für Oberflächentechnik*, **42** (3), 64.
5 Cramer, W.R., Gabel, P.W. (7–8/2001) *European Coatings Journal*, 34.
6 Hoffmann, A. (**2002**) *Journal für Oberflächentechnik*, **42** (3), 54.
7 Gruschwitz, B. (**2004**) *Metalloberfläche* **58** (6), 22.
8 Vaughn-Lee, D. (**2004**) *Polymers Paint Colour Journal*, **194** (4478), 24.
9 Voye, C. (**2000**) *Farbe Lack*, **106** (10), 34.
10 Shaw, A. (**2001**) *Modern Paint and Coatings*, **91** (2), 13.
11 Papasauva, S., Kia, S., Claya, J., Günther, R., (**2002**), *Journal of Coatings Technology*, **74** (925), 65.
12 Hagerty, B. (**2002**) *Polymers Paint Colour Journal*, **192** (4451), 13.
13 Drexler, H.J., Snell, J. (04/2002) *European Coatings Journal*, 24.
14 Klemm, S., Svejda, P. (**2002**) *Journal für Oberflächentechnik*, **42** (9), 18.
15 Scott, A. (**2007**) *Chemical Work*, **169** (11), 17.
16 Esposito, C.C. (**2004**) *Coatings World*, **9** (3), 21.
17 (a) Cramer, W.R. (**2005**) *Fahrz. + Kaross.*, **58** (5), 34. (b) Bloser, F. (**2004**) *Coatings Yearbook*, Vincentz, Hannover.
18 Wegener, E. (**2004**) *Coatings World*, **9** (10), 44.

2
Materials and Concepts in Body Construction

Klaus Werner Thomer

2.1
Introduction

From the days of the carriage to the efficient car bodies of today, in the course of over 120 years, the automotive industry has undergone enormous changes. Technological progress, the use of new and better materials, the basic conditions of transportation and the politics surrounding it, as well as general, social, and cultural changes have all, more or less, contributed to the evolution of the automobile to what it is today (Figure 2.1).

Whereas, in the beginning, bodies were built on frames, today the body frame integral (BFI) or self-supporting body is used in the manufacture of passenger cars. The body on frame (BOF) mode of construction is used only for trucks and off-road vehicles. The BFI is a complex way of construction, with a large number of requirements to be met. All parts of a car are attached to the body, which constitutes the core carrier and is called the *body-in-white*. Design, space conditions, occupant protection, and crash performance are important criteria for the customer to make the purchase decision. To summarize, the body structure fulfills the following functions:

- It meets all the structural requirements for all static and dynamic forces.
- It provides a crash resistant passenger compartment.
- It provides peripheral energy conversion areas.
- It provides the platform to mount all drive units and the axle modules.

In addition, there are other requirements to be met during the body manufacture, for example, part forming, assembly sequence, use of existing manufacturing equipment, low part proliferation, optimized material utilization, design of parts, body-in-white coating, modular method of construction, dimensional accuracy, and low production cost (Table 2.1, Figure 2.2) [1, 2].

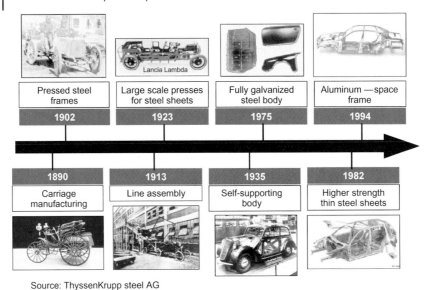

Source: ThyssenKrupp steel AG

Fig. 2.1 Milestones in car body manufacturing.

Table 2.1 Materials and Manufacturing Methods for Advanced Car Body Design

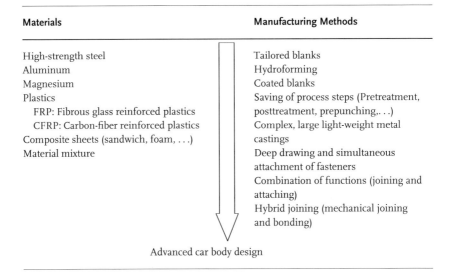

Materials	Manufacturing Methods
High-strength steel	Tailored blanks
Aluminum	Hydroforming
Magnesium	Coated blanks
Plastics	Saving of process steps (Pretreatment, posttreatment, prepunching,...)
FRP: Fibrous glass reinforced plastics	
CFRP: Carbon-fiber reinforced plastics	Complex, large light-weight metal castings
Composite sheets (sandwich, foam,...)	
Material mixture	Deep drawing and simultaneous attachment of fasteners
	Combination of functions (joining and attaching)
	Hybrid joining (mechanical joining and bonding)

Advanced car body design

The specifications for a certain body type cover hundreds of details and parameters that can increasingly be handled only by computerization. The demand for shorter development periods, starting with the original idea, covering design concepts, and up to development of prototypes and saleable cars, can safely be realized only by using analytical methods.

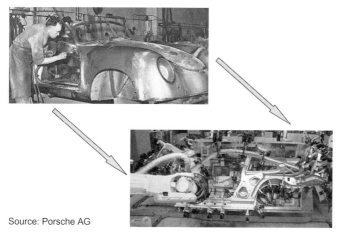

Source: Porsche AG

Fig. 2.2 Evaluation in car body manufacturing.

2.2
Methods of Body Construction

In principle, a distinction can be made among various types of body construction. The most common form is the monocoque design. In the last few years, new developments have emerged in this field, among which the space-frame concept is of great importance. There are some hybrid forms also, which combine both methods of construction.

2.2.1
Monocoque Design

In the construction of self-supporting bodies, steel bodies have been used above all others, in the monocoque design (Figure 2.3). Here, formed sheet-metal panels are joined, preferably by weld spots. The body-in-white consists of the underbody assembly made up of the front compartment, the center floor and the rear compartment. In addition, there are the side panels, which together with the underbody assembly and the roof, form the box-type body-in-white. The add-on parts like doors, trunk lid, and fenders complement the body-in-white [3].

The lower part of the underbody is formed by longitudinal and cross members. At the front and the rear ends, these are terminated by floor panels. The two longitudinal front members, which are made of the longitudinal front and its extension, are provided with a closing plate at the front end, on which the bumper is fastened. For the remaining longitudinal members, locators are welded at the lower side to position the chassis member. In addition, the engine mounts are accommodated on these front longitudinal members. The front frame, which takes the engine and the front axle, also plays a major role as an energy-absorbing component during a frontal crash, and is reinforced at the points with the highest load impact. In most cases, the extension of the front frame ends at the connection

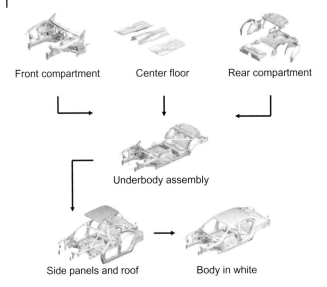

Fig. 2.3 Monocoque design.

of the seat cross member to the floor panel. The extension primarily serves to avoid a deflection of the floor panel during driving, and in case of a crash. The seat cross member ensures a stiff connection of the front seats. The inner panels needed for the rear underbody are formed by the two longitudinal rear frame members and three cross members for the rear floor panels, the rear axle, and the back panel. To obtain a better flow of forces, the rear frame is directly connected to the rocker panel, thus increasing the stiffness of the entire body. During driving, the cross member situated at the rear area ensures that the floor panel in the area of the rear seats does not deflect too much. In case of a rear-end impact, it counteracts deformations of the underbody. In addition, the cross member of the rear axle is decisive for the torsional stiffness. Moreover, it stabilizes the rear frame in case of a rear-end crash, and prevents an uncontrolled buckling. As a girder subject to bending, it counteracts strong torsions of the rear frame that may be produced as a result of a counterrotating deformation of the rear frame sections. The back panel cross member serves to prevent deflection of the floor panel in the trunk area. The wheelhouse is welded inside the vertical flange of the rear frame. The reinforcement of the back panel takes up the extensions of the wheelhouse. These extensions are important for the torsional stiffness of the entire body, and the local stiffness of the rear opening. In this manner, an additional load path is created for a rear-end crash.

Front parts are attached to the underbody. The upper front part takes up the locking for the hood and the radiator mounting, and the lateral front parts take up the headlamps and the front wheelhouse. The locator of the shock-absorbing strut, which takes up all reaction forces of the strut, is positioned on the wheelhouse. To achieve sufficient stiffness, the wheelhouse is connected to the front frame through a preferably straight element acting as a tension rod. Above the wheelhouse, there is a strut, which in case of a frontal crash, induces forces in the A-pillar and the door window channel, before it absorbs energy by a controlled forming process. At the rear end, the wheelhouse terminates with the lateral part of the dashboard panel. Cross to it, is the dashboard panel, which together with the closing plate and the extension of the front floor, separates the engine compartment from the passenger compartment. Furthermore, the closing plate protects the passenger compartment from noise and contamination, and prevents the engine or other components from penetrating the interior in case of a crash. Dashboard panel penetration after a crash is therefore a measure of the structural quality of the body, and it is often used for comparison. Above the dashboard panel there is a reinforced cover, which forms the windshield opening together with the inner A-pillar and the front roof frame. The side panel assembly essentially consists of an inner and an outer part. In the door opening area, these two shells form the pillar cross sections together with the inner A-pillar. For many cars, the A-pillar cross sections include additional reinforcements. In the case of convertibles, some of these reinforcements are made of high-strength pipe cross sections or hydroformed parts. The B-pillar cross section is provided with door weather strips on both sides. In addition, the B-pillar accommodates the hinges for the rear door with further local reinforcements. These reinforcements also serve to acoustically reduce diagonal deformations in the door openings during normal driving. Moreover, the reinforcement of the B-pillar is important during a lateral impact because it prevents a penetration into the passenger compartment. In the rear area, there is an extension at the side panel. An extension piece surrounds the opening for the tail lamps. As in the case of the extension of the wheelhouse, this measure is important to augment the torsional stiffness and the rear-end diagonal stiffness.

The roof consists of the outer panel, which is stabilized by bonded bows, and the rear frame. Whereas the roof bows do not influence the total torsional stiffness of the body, the rear roof frame plays an important part.

Add-on parts are usually all parts which belong to the car body, but which are not welded to the body-in-white, that is, doors, lid, fenders, and bumpers. The development of the add-on parts is characterized by two approaches:

- Modular construction – As a result of outsourcing the development and production, add-on parts are increasingly delivered as complete modules to the production line of the car manufacturer.
- Use of light-weight materials – The use of new body materials with a lower specific weight than steel has begun in large volumes, above all for add-on parts. Today, bumpers and fenders are often made of plastics, and the lid and the hood, of aluminum.

2.2.2
Space Frame

The space-frame concept is based on the skeleton design which has been in use since the beginning of the industrial production of cars. A steel or wood skeleton forms a solid structure onto which the secondary and tertiary body parts are mounted as noncarrying parts by means of different joining methods. In the 1960s, a method of construction known as *Superleggera* was patented by the Carrozzeria Touring Company. Here, a tubular space frame made of steel was covered with aluminum sheets. This method of construction was used for the construction of small-volume sports cars like the Austin Martin (1960) or the Maserati 3500GT (1957).

Today, Morgan, the British company, still uses body skeletons made of wood. In the case of super sports cars (like the Lamborghini), the body skeletons are made of steel.

For the manufacture of different wrought products, the light-weight material, aluminum has evident advantages over steel. Extruded aluminum sections and cast parts provide low-cost wrought products that can be combined to create a design suitable for aluminum. The use of these products for the construction of car bodies leads to a new design concept, the Audi space frame (ASF) [4], illustrated in Figure 2.4.

With the optimum design, the framework structure can be distinguished by its high static and dynamic stiffness, combined with high strength. Compared to today's steel bodies in monocoque design, approximately 40% of the weight can be saved by using this concept. A special feature of the ASF is the mixture of wrought aluminum products made of castings, sections, and sheet-metal parts. These parts form a self-supporting frame structure into which each load-bearing surface has

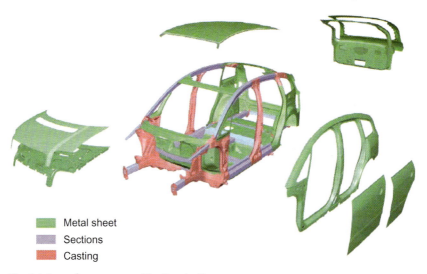

Fig. 2.4 Space-frame concept (Quelle : Audi).

been integrated. A characteristic feature is the very large and complex die-cast parts. By ribbing and load-adjusted wall thicknesses, a construction mode which meets the requirements of strength and weight can be achieved. The large number of possibilities in designing cast parts permits a functional integration and thus, a reduction of the number of parts as well as joining operations. Functions such as connections, location of hinges, or bolted connections can be integrated without any additional parts or joining operations. Another type of wrought products, in addition to castings and sheet-metal parts, are extruded sections that are used for the body structure.

By extrusion, these sections are created with a high degree of self stiffness. Any changes of the wall thickness section or flanges that are required for stiffness or strength reasons can be realized almost at will. The sheet-metal parts of the structure are joined to the body by mechanical methods (e.g. riveting/clinching) or thermal joining processes (e.g. laser or laser metal-inert-gas (MIG) hybrid welding).

Another interesting aspect in the manufacture of the space frame is the modular way of construction and the associated separate design, try out, and manufacture of single vehicle modules.

2.2.3
Hybrid Type of Construction

In the hybrid type of construction, the monocoque and the space frame concept are applied in one body structure by integrating cast, extruded, or sheet-metal parts as assembly groups or single components, into the car body structure.

2.2.4
Modular Way of Construction

Owing to a steady proliferation, the modular way of construction has gained in importance in the automotive industry. On the one hand, it is expected that clear advantages can be achieved by a standardization of assembly groups, and more important, standardizing the interface between modules. This in turn will shorten the time of development by simultaneous engineering. On the other hand, simultaneous fabrication of parts and assembly groups also leads to a decrease in the time needed for developing the vehicles.

If one considers the many types of variants available in large automotive companies, it becomes obvious that there are an almost uncountable number of possibilities. The definition of suitable interfaces in the vehicle allows for the possibility of dividing the assembly groups into various modules. An essential advantage of such modules is that the interfaces can be defined independent of body styles. This permits a higher output of components. Moreover, the storage cost of style-specific components can clearly be reduced, and the retooling costs incurred for model changes can be minimized. The modular way of construction aims at simplifying the body-in-white manufacture by defining modules with certain interfaces. There are various possibilities. For some time now, almost all manufacturers have applied the platform strategy. This means that different car styles can be built on one engineering basis. This basis is usually the floor panel group of the body-in-white, on which the so-called *hat*, in this case, the rest of the body, is placed.

The second possibility refers to the modules themselves. This means that certain parts together form one system. In the automotive industry, this refers to the parts such as fenders or entire units like van modules. Usually, they are added in the final assembly stage.

The modular way of construction offers more flexibility to car makers. By outsourcing certain assembly groups, that is, procuring them from external suppliers, modules can be ordered just in time, depending on the receipt of the specific order. Owing to the modular concept, changes concerning the body structure may need to be carried out, which frequently do not comply with the light-weight idea, for example, a stiff frame structure may need to be created, with defined contact surfaces. On the other hand, the modules offer a large number of possibilities such as the addition of different materials to the basic structure, for example, paneling with plastic components. Moreover, such systems make it easier to try out light-construction variants on derived products before integrating them into high-volume models.

2.3
Principles of Design

2.3.1
Conventional Design

The design starts with sheet-metal panels which are formed into carriers and fields. These are assembled by various joining methods.

The sections of the body-in-white that are decisive for the carrying capacity are designed as hollow sections. Reinforcements are achieved by beads, reinforcing sheets, or spatial curved surfaces. On the one hand, these sheet-metal reinforcements are required for reasons of stability and acoustics, on the other hand, they are needed for manufacturing reasons, to achieve a constant shape of the single parts before assembling them onto car bodies.

Actually, interior passenger space and cargo space play a large role in styling a vehicle. The design is based on the structural draft, where the sheet-metal parts are determined, but at the same time, the assembly of the single elements is to be considered as well. Moreover, the body structure is to be divided into assembly groups. Here, the body shape is determined by the plan of the outer panels. It is made up of the body surface as developed during styling in connection with aerodynamics. But it does not constitute a supporting structure; so the design department first determines a supporting structure, taking light-construction principles into consideration.

To present a simplified view, a passenger cell is a box that consists of frames and sheet-metal fields. The front and rear ends consist of girders subject to bending and sheet-metal fields. Forces are applied through carriers, which may be designed to be open or closed. Closed carriers are characterized by a considerable torsional stiffness, in addition to the bending stiffness. The points of energy application are designed at the structural nodes. Here, especially, loads are investigated taking into consideration their bending, torsion, and crash behavior.

Sheet-metal parts are designed according to manufacturing principles. The assembly should have as few parts as possible. In addition, the parts are to be divided so that deep drawing or any other type of forming is viable. A carrier structure is formed by joining the parts. Otherwise, the design is based on the following general principles:

- as far as possible a direct application and compensation of forces
- realization of the maximum of planar moment of inertia and drag torque
- microstructuring
- use of the natural supporting effect of a curvature.

A special task is to design the nodal points of carrying members. The lateral faces of carriers are to be continued in the nodal points. It is possible to transfer the torques and forces properly only in this way. Particular care is needed while dividing the sheet-metal panels at the nodal points. In addition to the solid connection of the compounds, deep drawing requirements are also to be considered.

Special care is also required for the bead pattern. The position and the form of the beads should not permit any straight line passing through free space because the moment of inertia is applied only through the sheet-metal thickness, and so stiffening is not possible. To avoid a high local degree of forming, the run out dimensions of the beads should also be observed. As an alternative to beads, crowning can be incorporated into the sheet-metal field. This approach is of special interest, especially for the outer panels. Another possible method of reinforcement of the sheet-metal fields is the addition of bonded reinforcing bars.

In the field of light construction, there is a constant search for new methods of reinforcing that do not stiffen the material. An interesting variant is given by holes surrounded by collars.

Considering designs of different sheet-metal thicknesses, which frequently is the case in connection with light constructions, attention must be directed to the concentration of stress due to different degrees of stiffness, for example, a step must be designed at the transition from a small sheet-metal thickness to a larger one. This is especially critical in the area of the door hinges at the pillars. Here, sheet-metal thicknesses of approximately 1.6 mm coincide with thicknesses of approximately 8 mm, making additional reinforcements necessary.

Another aspect to be taken into consideration during the design stage is to take precautions for a compensation for tolerances at the assembly.

2.3.2
Design Under Consideration of Light-Weight Construction

The improvement of passive passenger safety is one of the many focuses in developing car bodies. The static and dynamic properties form the basis of crash optimization. In the course of only a few years, the requirements have increased from 48 km h^{-1} with a 100% overlapping and a rigid barrier (referring to the US standard FMVSS-208) to 64 km h^{-1} with a 40% overlapping and a deformable barrier as a design criterion. Here, the stricter occupant injury criteria are fulfilled with a greater distance, and the remaining survival space is further enlarged. Since the beginning of the 1990s, the energy to be absorbed per structural unit with reference to the side of the car has increased by more than 80% and continues to increase.

Side crashes, according to the European Enhanced Vehicle Safety Committee (EEVC) with 50 km h^{-1} with the greater barrier height of 300 mm, and the US Federal Motor Vehicle Safety Standards (FMVSS-214) with the increased Lateral Impact New Car Assessment Program (LINCAP) velocity of 40 mph are the required standards today. In the important market relevant, but not legal tests such as the Euro-New Car Assessment Programme (NCAP) and the US Insurance Institute for Highway Safety (IIHS), similar configurations and speeds are used, although different criteria are applied to the evaluations. As a consequence of these different requirements, especially for side impact, global car manufacturers are faced with great challenges. In the last 20 years, increasing the passive safety was the eminent, driving factor for the body-in-white. The aim is to destroy as much energy as possible during a crash by changing the structural form.

As aluminum has a lower strength and less elongation at fracture compared to steel, it can transform less energy by changing form, and the compensation is achieved by means of crash boxes. Given the same build conditions, a steel structure is more favorable in relation to a minimum-occupant load.

For designing bodies from the view point of crash relevance, the design of the load paths is very important, especially with reference to front and side crashes, because it determines the course of the carriers and the nodal points. In the case of a front crash, the load is mainly applied over the front longitudinal members and the subframe to the lower vehicle structure situated behind them. The forces are distributed through the front-end cross member as a supporting element of the passenger cell, and through the central longitudinal member, which in turn applies the load onto the floor structure, the lower nodal points of the B-pillar, and the rocker panel. The crash load is introduced through the connection of the front longitudinal member to the lower roof frame, the so-called A-pillar member, into the upper load path. In the case of a side crash, the rocker panel and the B-pillar transfer the energy. Here, the lateral impact beams located in the door are the most important connection for transfer of the load onto the A-pillar and the B-pillar.

Today, in addition to safety, another question that is asked in car building, is the one about light-construction aspects. In relation to the design, this means basic changes in the selection processes and methods, because the problem is not solved solely by substituting materials. There are a large number of choices in the implementation of the idea of light construction, especially in the area of design. The aim of a light-construction design is to increase the stiffness as far as possible, without any additional material.

As regards stiffening, a distinction is made between:
- beads
- bottom boom members made of sections
- shell-type design
- forming of ribs
- integration of beams
- design of marginal reinforcements
- bonding of metal foams.

The most frequent application, in addition to beads, is the use of hollow sections versus solid sections. Apart from their lower weight per meter, they have higher bending and torsional stiffness, a good functional adjustment due to shaping, and a high capacity of integration. However, the forming operation is critical, with reference to the cross section and the longitudinal section, the distortion after the heat transfer during welding, the corrosion, and the compliance with the tolerance ranges. Furthermore, a number of elastomechanic problems are to be taken into consideration, such as the application of forces, the flow of forces in the shear center, and local or entire instability by buckling, tipping, and bulging.

In the case of shell-type forming, the aim is to manufacture parts or sections with curved surfaces.

Recently, sandwich designs have gained more and more importance. In the meantime, the first serial applications have also taken place.

2.3.3 Bionics

The term *bionics* was first used in 1960 by Major Jack E. Steele of the US Air Force at a conference held in Dayton, Ohio. It is a combination of the words 'biology' and 'technics', which means the use of biological principles for engineering purposes. The general idea of bionics is to decipher 'inventions of the animate nature', and their innovative application to engineering. Therefore, bionics is an interdisciplinary subject where biologists, engineers and designers work together. Biological structures and organization are used either directly as a sample referred to as *analogy bionics*, or in an abstract way, detached from the biological archetype, referred to as *abstraction bionics*, and as an idea or an inspiration for solving engineering problems. Although, there are some examples of the use of bionics in history, for example, the analysis and attempted transmission of the flight of birds onto flying machines was performed by Leonardo da Vinci, nevertheless, bionics has established itself as science only in the recent decades, primarily due to new and improved methods such as computer capacity and manufacturing processes.

While developing functional elements from an engineering point of view, we should be aware of the fact that parallel phenomena in nature have not always been known, for example, the half-timbered building was developed without knowing the microstructure of the trabeculae. In these cases, one cannot speak of copying a certain idea that already existed, but rather of a certain correspondence between nature and technology. In contrast to this, bionics, as a science, involves searching for structures in nature that can be significant as engineering models. Frequently, this procedure can be considered as a mere search for analogies. In these cases, however, the innovation cannot be called *a quantum leap* because the application to engineering must already be recognizable. An example of bionic part optimization is the principle of the light construction in the microstructures of bones as an analogy to the weight reduction of structural components. Today, we apply a considerable number of procedures with the help of computers.

The computer aided optimization (CAO) method (CAO optimization in topology), for example, is used to investigate the elimination of peak stress on vehicle parts subject to stress, as a result of a notch effect. In this case of topology optimization, the computer program works based on the existing space. Another parameter, the engineer can ask for is, for example, a weight reduction. The computer calculates which areas of the part may be omitted without affecting the required properties. It aims at demonstrating the ratio of weight and stiffness in the best case, in the given circumstances. The software used for it passes through several iterations, draws conclusions based on mathematical intermediate results, and starts to recalculate with changed factors.

The formulation used in this example is the stress-controlled growth principles, based on the biological principle of tree growth. It is recognizable that in nature, material is added to highly stressed areas, whereas the residual areas remain as they are.

A similar result is obtained when taking the human femur as a model for the weight reduction of parts. The idea is to consider stress-controlled growth based on the biological principle of femur growth. In this way, it is possible to recognize that material is added to highly stressed areas, and removed from points with a low demand of stress. An illustration of this stress-controlled distribution of Young's modulus of elasticity in the bone is the frame type construction made of very fine trabeculae. Their orientation shows considerable accordance with the force flow in the femur. This principle uses the soft kill option (SKO) method which means that material which is not loaded, is successively removed.

At present, metal experts are investigating the natural light-construction structures that are found in fauna that have ideally evolved over the course of several hundred million years. The car body structure, for example, carries many other parts of the vehicle, similar to the skeleton of a mammal. An interesting aspect, in this connection, is the weight of the supporting frame compared to the total weight. In the case of the BMW 3, the body accounts for approximately 20% of the total weight. The weight of the human skeleton is approximately 18% of the total weight. In horses, this percentage is perfect – the bone weight is only 7–10% of the total weight, and the rest primarily serves for movement and energy. This is the reason that, over the centuries, horses have gained recognition as efficient working animals and beasts of burden. The secret of their efficient weight ratio is the bone structure and density.

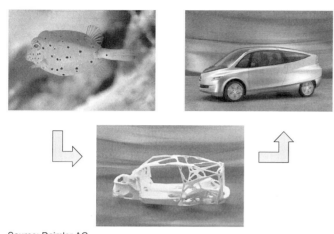

Source: Daimler AG

Fig. 2.5 Concept 'Bionic Car'.

The application of bionics can easily be illustrated by means of the Mercedes concept car called *Bionic Car* (Figure 2.5) [5]. In June 2005, the result of this study was made public and the following characteristics were highlighted:
- The vehicle study was made with the most favorable contours of the boxfish as seen from a fluid mechanics point of view.
- The air drag coefficient of 0.19 was a peak in aerodynamic performance.
- The diesel engine was equipped with a particle filter, and modern selective catalytic reduction (SCR) technology.
- A bionic design approach was used for intelligent light constructions.

The study was based on the boxfish. It lives in coral reefs, lagoons, and seaweed areas of tropical oceans, where, many aspects that prevail are conditions which are valid for cars. To economize on its energy, the fish needs to move using minimum amount of energy. From the point of view of fluid mechanics, this requires strong muscles and the most favorable form. The fish has to resist the high pressure of the water, and also protect its body from collisions, for both of which a stiff skin is necessary. When searching for food, it is forced to move within a very limited space and this presupposes good maneuverability.

In the construction of car bodies, the hexagonal bone plates of the boxfish correspond to the principle of the highest strength and the lowest weight. In transferring this property to the outer panels of a car door, this natural principle requires a honeycomb design with a 40% higher stiffness. If the complete body-in-white structure is considered according to the SKO method, the weight is decreased by around 30%, while at the same time, the stability, crash safety, and driving dynamics are excellently maintained.

2.4
Materials

2.4.1
Steel

2.4.1.1 General Remarks

Steel and light construction are two terms that, at first glance, seem to be contradictory. When we consider the development of light construction, it becomes evident that the majority of the products have been developed, not by replacing steel, but by developing new varieties of steel that could be formed as desired [6]. The conventional types of steel cannot meet the high requirements of the automotive industry, and therefore, the steel producers have been developing more and more materials with very specific properties, which are suitable for light construction (Figures 2.6 and 2.7).

This is the only way to explain why the weight portion of steel and iron in today's passenger cars, that is, 60–70%, has remained more or less unchanged. The current production in Germany is approximately five million passenger cars and commercial vehicles per annum, for which around 5.3 million tons of steel are delivered to the automotive industry. Here, the share of hot-rolled and cold-rolled flat products in the form of steel strips and sheet steel amounts to around 70%.

So far, steel has been the favored material from the economic point of view, because it can easily be modified by alloying, hardening, and work hardening that offers good stiffness and strength. Moreover, there is hardly any other material which has received as much importance in respect of its properties and processing. A considerable disadvantage, however, is its high density and susceptibility to corrosion.

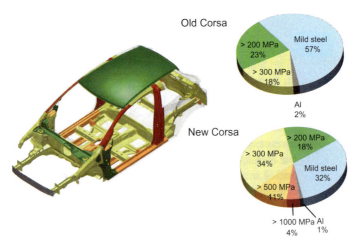

Source: Adam Opel GmbH

Fig. 2.6 Material distribution of body-in-white – new versus old Opel Corsa.

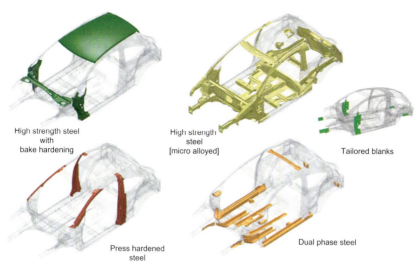

Source: Adam Opel GmbH

Fig. 2.7 Application of steel types in body-in-white of new Opel Corsa.

In the body shop, cold-rolled sheet steel with a thickness ranging from 0.5 to 1.5 mm is mostly used.

In the past, soft unalloyed materials were preferred because they offer a high degree of formability and freedom of design. In addition to deep drawing and stretch forming, the suitability for welding, joining, and painting are significant criteria of processing. These requirements are met by the higher-strength thin sheets (with a minimum yield point of >180 MPa), that have been developed during the past 20 years, and by which, sheet thickness could be reduced, thus reducing the mass (Table 2.2). In addition, higher-strength sheet steels are being developed intensively, some with tensile strengths of over 1000 MPa. At present, there is a tendency toward Dual-phase (DP) steels, partially martensitic, and transformation induced plasticity (TRIP) steels or multiphase (MP) steels. All kinds of sheet metals

Table 2.2 Load- and strain-compatible use of steel grades

Yield strength ($R_{p0,2}$)/Steel grade		Areas of use
<140 MPa	soft	Outer skin parts, parts of complicated geometry
160–240 MPa	high strength	Areas of complicated geometry requiring high strength
260–300 MPa	higher strength	Structural parts requiring high strength
300–420 MPa	very high strength	Crash areas with high level energy absorption
>1000 MPa	hot-formed	Highly deformation resistant load-bearing structural parts

are used with and without metallic or nonmetallic (organic) coatings. The range of quality of cold-rolled thin sheets in body manufacturing consists of:

- soft unalloyed steels suitable for cold working
- higher-strength steels suitable for cold working with augmented yield points
- higher strength, stretch-forming quality categories, and the MP steels.

There are different mechanisms to increase the strength of conventional steels [7]. Seen from an engineering point of view, the following are significant (see Table 2.3).

Steels with tensile strengths of more than 1000 MPa are finding increasing use in design elements and crash-relevant, structural parts of vehicles. The basic problem

Table 2.3 Characteristics and Application of Different Types of Cold-Rolled High-Strength Steels

Type of steel	Special characteristics	Areas of use
High–strength IF steel (interstitial–free ferritic matrix)	Good formability for difficult drawn parts	Door inners, door rings and apertures, fenders
Bake–hardening (BH) steel (ferritic matrix with carbon in solid solution)	Good formability for stretch forming and deep–drawing applications	Flat parts, door outer, hood, trunk lids
High–strength stretch–forming (HS) Steel (ferretic matrix with precipitations)	Good formability in the mid–strength range	Relatively flat parts like doors, hoods, trunk lids
Phosphorus–Alloyed (PH) steel (ferretic matrix with solid strengthening elements like P, Mn)	Good deep drawing formability in the mid–strength range	Wheel arch, doors, hood, trunk lids
Micro–alloyed (MH) Steel (fine–grained microstructure with Ti–and/or Nb–carbonitride precipitations)	Realization of high component strength levels	Structural parts
Dual–phase (DP) steel (mainly ferritic matrix with dispersed mainly martensitic islands	Good isotropic forming properties at the higher strength level, good spring back behavior, BH potential	Flat stretch–formed exposed paneling like doors, structural and crash–relevant parts
Residual austenite (RA) steel (mainly ferritic–bainitic matrix with dispersed residual austenite islands)	Good isotropic forming properties at high–strength levels, BH potential	Pillars, rails, cross members

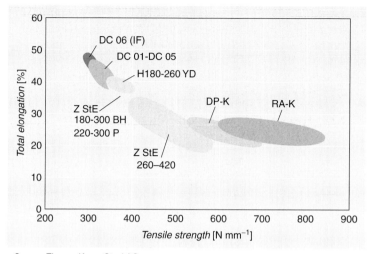

Source: ThyssenKrupp Steel AG

Fig. 2.8 Properties of cold-rolled steels.

with increasing strength is a natural decline in the forming capability. This has led to a new concept, the so-called MP steels. The raising of the strength is based on structural hardening. To a matrix of soft, ferritic portions, a harder portion is added, which consists of one or more other phases, and which should be distributed as evenly as possible. This development started with the DP steels, followed by TRIP steels. Recently, Complex-phase (CP) steels, which are of even higher strength, have been developed (Figure 2.8).

Today, new steel materials are being developed by means of numeric simulation, so that the special requirements of the customers can be taken into consideration, and materials are tailored made.

2.4.1.2 Low-Carbon Deep Drawing Steels

Cold-rolled flat products made of low-carbon steels for cold forming, are killed by aluminum, which means, the oxygen remaining in the steel from the refining process in the converter is fixed by aluminum, and in this way, the steel can continuously be cast. Owing to new developments in the vacuum treatment, it has become possible to keep the content of carbon and nitrogen as low as possible, and fix corresponding alloying elements that are required for the fabrication of DC 05 and 06 (mild unalloyed steels). In the same way, the content of sulfur is set at the lowest values by metallurgical measures, which in turn, permits a reduction of the manganese content. In addition to these, solid solution hardening elements (C, N, Mn), and the grain growth inhibiting elements, such as trace elements, are also to be restricted, which leads to a special selection of scrap.

2.4.1.3 Higher-Strength Steels

Since the development of higher-strength steels in the mid-1970s, more and more high-strength sheet metals are used for the fabrication of vehicles. Through further developments, this grade of steel today amounts to a (weight related) portion of over 40% in bodies-in-white. Owing to their low carbon content and the low amount of alloying additions, these higher-strength cold-rolled steels can still be spot welded. The mechanical properties are barely above those of the low-carbon deep drawing steels.

2.4.1.3.1 Microalloyed Steel

Age hardening of finely distributed carbonitrides results in an increase in the strength and higher-strength drawing properties of the conventional microalloyed steel. Even a small amount of titanium, vanadium, and niobium in the region of about 0.01% in the composition of the alloy, leads to a clear increase of the yield point to 260–540 N mm^{-2} and the tensile strength to 350–620 N mm^{-2}. Owing to the finely distributed precipitates, a slightly lower elongation at fracture can be observed, compared to phosphor-alloyed steels. To reach higher yield point values, solid solution hardening is combined with age hardening.

2.4.1.3.2 Phosphor-Alloyed Steel

Solid solution hardening leads to an increase of the yield point to 220–360 N mm^{-2} and the tensile strength to 300–500 N mm^{-2}. Here, the structure can be compared to that of low-carbon deep drawing steels. For solid solution hardening, elements with deviating atom radii are incorporated, instead of iron atoms. Phosphor is the element with the strongest solid solution hardening effect. Addition of 0.01% affects an increase in the yield point by approximately 8 N mm^{-2}.

2.4.1.3.3 Bake-Hardening (BH) Steel

This refers to phosphor-alloyed cold-rolled steel strips which resist aging at room temperature, and which experience an additional yield point increase of approximately 40 N mm^{-2} by a controlled carbon aging during baking of the automotive paint (at 180 °C). This means that at room temperature, there is a stable carbon supersaturation in the structure, so that the steel is not subject to natural aging as long as it is stored in the usual way. The stabilization of carbon is achieved by microalloying of the steel with diffusion-inhibiting or carbide-forming alloy elements. In the oven, part of the carbon is precipitated and dislocations are blocked, with the effect of increasing the strength. Thus, these steels offer the advantage of a good cold-forming ability in the delivery stage, and of achieving their full strength when the parts have been formed.

2.4.1.3.4 Interstitial Free (IF) Steel

For very complex products like inner door panels, side parts, or inner wheelhouses, microalloyed interstitial free (IF) steels are available. This type of steel is free from interstitially detached atoms like carbon and nitrogen, which results in low yield point values, and are associated with very good deep drawing properties. The content of C and N in IF steels ranges between 20 and 40 ppm or 0.0020 to 0.0040%. Moreover, by an overstoichiometric addition of titanium and/or niobium, a complete fixation of carbon and nitrogen is reached in the form of nitrides (TiN), carbides (TiC), or carbonitrides (TiCN). Because of these extra efforts involved in the fabrication of IF steels, they have experienced an enormous upswing in demand, in the last few years. This is because of their excellent forming ability, low yield points, and at the same time, high forming characteristics r (anisotropy) and n (hardening exponent). As a result of the complete fixation of carbon and nitrogen, the IF steels resist aging, and therefore, they are used largely for the fabrication of special deep drawing products in hot-dip galvanized design, for which resistance to aging would otherwise only be reached by an additional heat treatment.

2.4.1.3.5 **Isotropic Steel** These steels possess unidirectional flow characteristics in the sheet-metal level and therefore, a better deep-drawing property and at the same time an increase in their strength. The minimum yield point in the delivery status of these sheet metals ranges between 210 and 280 MPa. These steels too, show a bake-hardening effect after preforming.

2.4.1.4 High-Strength Steels

2.4.1.4.1 **Dual-Phase (DP) Steel** Dual-phase steels belong to the group of polyphase steels. They are created by adding high-carbon phases to the structure adjacent to low-carbon phases, which leads to an increased strength. In the case of DP steels, the structure is essentially formed by a ferritic matrix with an islandlike embedded martensite portion of up to 20%. Thereby, it possesses good isotropic forming properties with a higher strength level, and a favorable spring-back behavior, and high hardening and energy-absorbing capacity. Moreover, these steels possess a bake-hardening potential, a low yield point ranging between 270 and 380 N mm^{-2}, and a high tensile strength of 500–600 N mm^{-2}. These properties make DP cold-rolled steels useful for application in the flat-outer panels, especially in strength-relevant, structural, and crash-relevant parts (Figure 2.9).

2.4.1.4.2 **Retained Austenite (RA) Steel** These steels essentially consist of a ferritic–bainitic matrix, with retained austenite. During the forming process, the

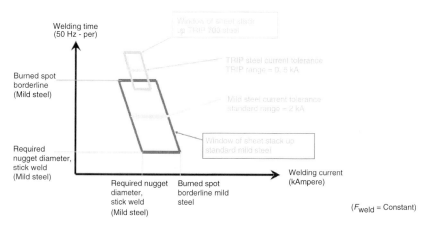

TRIP 700 welding process sensitiveness is basically four times higher compared to mild steel

Source: Adam Opel GmbH

Fig. 2.9 Spot welding resistance of high-strength steel.

remaining austenite parts change to hard martensite, which leads to a considerable hardening, known as TRIP. Retained austenite (RA) steels are characterized by a high elongation without necking, and thus a high hardening capacity, recognizable by the high n-value. Compared to DP steels, RA steels have higher strengths of up to 850 N mm^{-2} at a comparable elongation. As an alternative, higher elongation values of 600 N mm^{-2} are possible with the same strength. These steels possess a bake-hardening potential, and are also used as complex parts, which are relevant to strength, structure, and crash.

2.4.1.4.3 **Complex-Phase (CP) Steels** CP steels are currently fabricated as hot-rolled steel strips, with a minimum thickness of approximately 1.5 mm. Their fabrication as cold-rolled steels is still being developed. The materials known so far in this class of strength have, as a rule, to be hot rolled, and subsequently quenched and tempered. The advantage of the CP steels is that cold forming, without subsequent quenching and tempering, is possible, thus implying a considerable cost-saving potential. Essentially, this kind of steel consists of a fine-grained, ferritic bainitic, and martensitic structure, which in addition to its high strength, has a good cold-forming and welding capacity.

2.4.1.4.4 **Martensite Phase (MS) Steels** These steels are also hot rolled and essentially consist of a martensitic structure. Yield points range between 750 and 900 N mm^{-2} with a tensile strength of 1000–1200 N mm^{-2}. These values, the good cold-rolling and welding ability at a high tensile strength, and resistance to wear, make this material ideal for door-impact beams and crash-relevant parts.

2.4.1.5 High-Grade (Stainless) Steels

Stainless steel is a collective term for rust-resistant steels that are characterized by an increased resistance against chemically aggressive substances. Usually, these materials have a minimum chromium content of 10.5%. A distinction is made between ferritic chrome steels and austenitic chrome-nickel steels, the second group being more applicable to automotive production. These austenitic materials show a good strain-hardening ability, and have elongation values considerably exceeding those of the ferritic kinds. It is possible to reach extremely high forming degrees with these steels. Together with the good strain hardening, a high degrees of part stiffness can be reached. The biggest problem is presented by the cost incurred for these materials. More recent concepts, where part of the alloy elements is replaced by nitrogen, can reduce the costs. This material is especially noteworthy because of its application in crash-relevant areas, which are usually hard to cover using the normal corrosion-prevention measures.

2.4.1.6 Manganese–Boron Steels

For hot forming and hardening, the manganese–boron steels offer the highest strengths of up to 1650 N mm^{-2} in the hardened condition. After having heated the steel to the austenitization temperature, a subsequent controlled cooling leads to a martensitic structure, and thus, to a high strength of the material.

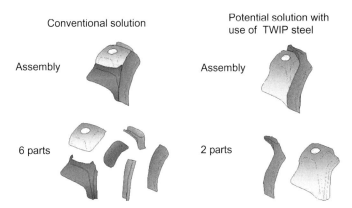

Fig. 2.10 New design possibilities with TWIP steel.

Manganese–boron steels are of special interest for parts with complex geometries, and high demands concerning strength. The mechanical properties, which can be influenced by tempering, correspond to the highest demands, and enable significant weight saving when these steels are used in the production of strength-dimensioned, structural, and safety parts of vehicles, like bumper supports, side impact beams, column, and body reinforcing panels.

2.4.1.7 Light-Construction Steel with Induced Plasticity (TWIP Steel)

With increasing tensile strength of steels, the elongation after fracture decreased correspondingly. The steel manufacturers are in the search for new materials to avoid this undesired effect.

One solution of this problem is offered by the so-called twinning-induced plasticity (TWIP) steels, which have a high alloy content of manganese. The formability is easier than with low-carbon steels (see Figure 2.10). As the material was created in a laboratory, an industrialization concept is still being explored. On the one hand, the range of properties enables very complex parts, on the other hand, it would necessitate asking the vehicle developers for a new design philosophy, with a high weight saving potential. A small series is planned with this material from the year 2006, and large-scale production is anticipated by 2009–2010.

Today, automotive designers can choose from a large number of joining methods. These are, for example, thermal processes such as welding and brazing, and mechanical methods such as clinching, riveting, stamping, screwing, bonding, and hybrid processes, which combine two or more of the joining techniques. However, not every steel material can be joined with any approach. Therefore, it is of great importance to know the characteristic properties, for example, of conventional deep drawing steel or modern MP steel, to be able to ensure high-quality connections and safe processes (Figure 2.11).

In addition to the original material properties, surface coatings may also influence the selection of the most suitable joining method. With an increasing degree of surface refinement, whereas thermal joining methods have reached their limits,

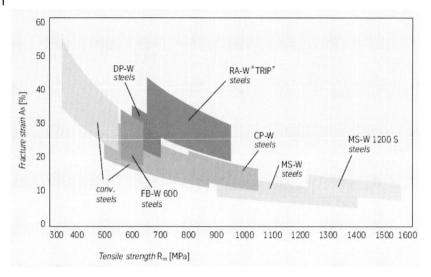

Source: ThyssenKrupp steel AG

Fig. 2.11 Properties of warm-rolled steels.

bonding and mechanical joining methods can be used without almost any material restrictions.

To reduce weight, we also use tailored welded glands, where two sheet steels of different thicknesses are laser welded into a sheet that is then stamped into a single part, with strength only where required. Typical applications are door inner panels and side ring panels.

2.4.2
Aluminum

2.4.2.1 General Remark (Table 2.4)

The most important raw material for the extraction of aluminum is bauxite, whose availability is estimated to be approximately 140 billion tons. It contains the oxygen compound Al_2O_3, with an Al content ranging between 52 and 65%. Bauxite deposits are found in France, Hungary, Romania, former Yugoslavia, Greece, India, Guyana, the United States of America, Brazil, and Russia. Owing to a lower economic efficiency, as a result of the high proportion of SiO_2, and an aluminum content of only 1–8%, Aluminum is not extracted from clay and kaolin. With a share of 8% in the earth's crust or lithosphere, aluminum occupies the second place in the terms of the most prevalent metals, behind silicon, and ahead of iron [7, 8]. The known international stocks amount to 30 billion tons.

Owing to its high affinity to oxygen, aluminum has very good anticorrosive properties. Bare aluminum forms a thin, electrically nonconductive oxide film made of Al_2O_3. In humid air, this blocking layer is covered by another oxide layer. The oxide layer adheres firmly to the surface, and cures itself in case of any

Table 2.4 Properties of aluminum

Properties	Value
Density	2.699 g cm^{-3}
Modules of elasticity	67 GPa
Linear thermal expansion coefficient	23.6×10^{-6} K^{-1}
Thermal conductivity	235 Wm^{-1} K^{-1}
Strength	170–500 MPa
Melting point	660.2 °C

damage. This is known as the *self-healing effect*. In this way, oxidation is retarded and practically stopped, even in aggressive atmospheres.

In regard to the availability of aluminum, a distinction is made between primary and secondary aluminum. Secondary aluminum is obtained by recycling, and primary aluminum, by making use of the Bayer approach with subsequent dry electrolysis by the Hall and Héroult method. In the Bayer approach, bauxite is ground and rendered soluble in the autoclave by means of caustic soda at temperatures up to 270 °C. By inoculation during quenching, Al(OH)$_3$ is precipitated and burnt to Al$_2$O$_3$ (alumina or aluminum oxide) in the calcining furnace, with the production of water. In the dry electrolysis method, aluminum oxide is decomposed with cryolite (Na$_3$AlF$_6$, sodium aluminum fluoride) at 950–980 °C in an electrolytic cell. Oxygen reacts with the carbon of the anode to form the gases CO$_2$ and CO. Owing to its higher density, the aluminum settles under the molten mass at the cathode, from where it is extracted regularly. Each ton of aluminum requires approximately 2 t of aluminum oxide from 4 t bauxite, 500 kg anode carbon, 50 kg cryolite, and 18 MWh of energy. The treatment of secondary aluminum requires only about 5% of the original energy. The primary aluminum obtained in this process, has a purity of 99.7–99.9%. The primary aluminum is transported to the foundry, where it is cleaned, and alloying elements are added to the molten mass. In the so-called degassing, the excess hydrogen is removed either by finely distributed argon bubbles by a process called *spinning nozzle inert flotation* (SNIF), or by chlorine, which has the effect of decreasing the porosity in the cast structure and the content of alkali and earthy base. After a filtration through porous tiles, the molten mass is cast, mostly in the continuous-casting method, into ingots of up to 9 m length, 2.2 m width, and 0.6 m depth.

2.4.2.2 Further Treatment

Further treatment starts with annealing for several hours, which is the first step in the rolling mill. At temperatures varying between 480 and 580 °C, the structure is homogenized. The next manufacturing step is mostly performed in a hot-rolling train, where the castings are hot rolled in several steps under temperature control. In many cases, the hot-rolling train consists of a reversing stand, where the thickness can be reduced to 25 mm. In case of a high output of approximately 700 000 t annually, a tandem manufacturing line would follow, with

Table 2.5 Application potential of aluminum

Subject	Processing technology
Sections	Extrusion
	Bending and hydro forming of extruded sections
Plane load-bearing structures	Sheets
	Tailored blanks/engineered blanks
	Patchwork sheets
	Sandwich panels
	Aluminum foam sheets
Moldings	Pressure die-casting
	Vacuum die-casting
	Chip removal from cast sections
	Squeeze casting
	Thixo-forming
	Forging
Joining methods	MIG, TIG, laser electron-beam welding
	Bonding
	Bonding and spot welding
	Laser brazing/Dip brazing
	Clinching/Punch riveting

several successive reduction stages. For an output of a maximum of 250 000 t per annum, a reversing stand would be used. Here, the strip is spooled between two coils through a mill stand, similar to a cartridge drive. After that, the strip is cold rolled at room temperature to the final thickness, and then solution annealed to achieve a well formable material, which is suitable for automotive body panels.

Aluminum can be treated in many ways. It is very versatile and so there is a whole spectrum of potential applications, especially in light-weight construction (Table 2.5).

A surface treatment of aluminum body panels in a rolling mill serves to improve the forming behavior so that, in the end, even surface structures are achieved.

There are several methods available to texture the surface, such as electro-discharge texturing (EDT) or laser texturing (Lasertex), where the rollers are roughened in a controlled way by electrical discharge machining, or a laser beam (see Figure 2.12). The pits appear as microcraters on the sheet-metal surface, which act as lubricant bags. This is of great significance for the forming behavior (see Figure 2.13).

The tribology of the material plays an important role in preventing adhesion of the steel tools, during the forming of aluminum sheets. Furthermore, the following points should be taken into consideration during the formation of aluminum panels:
- Cleanliness – avoid dust, abrasion, filter, and chips.
- Aluminum should be able to flow to control the blankholder forces.

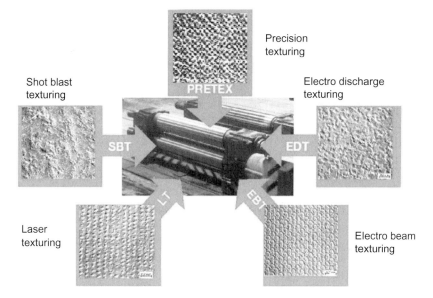

Source: Salzgitter AG

Fig. 2.12 Texturing methods of aluminum.

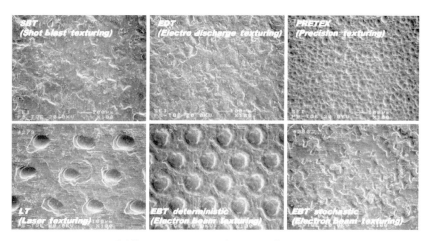

Fig. 2.13 Comparison of different aluminum surfaces by different texturing methods.

- The radius at the punch opening line should be five to seven times larger than the sheet-metal thickness.
- Avoid vertical walls.
- Deep drawing should be done in one step, without relieving holes.
- Polishing or, if necessary, coating of dies should be done.

2.4.2.3 Aluminum Alloys

Compared to pure aluminum, all aluminum alloys demonstrate an increased tensile strength, yield point, and hardness. The alloying elements strongly influence properties such as the coefficient of thermal expansion, the electrical conductivity, stability, and formability. For further treatment, a distinction is made between wrought alloys and cast alloys, and between age-hardenable and nonage-hardenable alloys. By adding Cu and Mn, for example, the strength is increased because solid solutions are formed, which lead to tensions in the crystal lattice and blocking of the sliding planes. The most important alloying elements are Cu, Si, Mg, Zn, and Mn, which are frequently added to pure aluminum (99.5%). Alloys applied onto outer panels are subject to high demands in terms of the surface quality after the deep drawing process. Here, compositions that are needed are those which do not produce any flame-shaped patterns (that is, stretcher strains or Lüders lines of type A), or fine stripes (that is, stretcher strains or Lüders lines of type B), on the surface. These demands are met by the hardenable AlMgSi group of alloys. In Europe, according to the nomenclature of the Aluminum Association and DIN EN 573-1:2005, alloys of the type AA6016 are employed, whereas, in North America, high-strength types such as AA6111 and AA6022 dominate the market (Table 2.6).

Alloys for inner and structural applications are selected because of their appearance, their strength, and their forming behavior. Both AlMg and AlMgSi alloys are used. In addition to a high surface quality, AlMgSi shows hardening properties, which occur during the paint baking process. The strength of the material is further increased, so that thinner sheet thicknesses can be reached. AlMg alloys are very suitable for forming, and are therefore employed for the production of complex parts.

Table 2.6 Examples of aluminum alloys applied in the automotive industry

Alloy	Type (AA)	Typical properties
Outer panels		
Ac-120	6016	Standard for outer panels
Ac 120 PX	6016	Fast curing Ac 120 variant
Ac 140 PX	6016	High-strength variant
Ac 160	6016	Extreme forming operation
Ac 170		Extreme hemming operation
Inner panels		
Ecodal–608	6181 A	Good forming, curable
Pe 440	5182	Very good formability, not curable
Pe 300	5754	Very good formability, medium strength
Crash application		
Ac 300		High energy absorption, good crash behavior

Source: Alcan

2.4.2.4 Aluminum as Light-Weight Construction Material

Aluminum is said to be the classic light-construction material. The density of Al ($2.7\,\text{g cm}^{-3}$) is only one-third of the density of steel. The same applies to the modulus of elasticity, which is only 70 GPa. Further positive properties of aluminum are

- a broad range of strength
- high viscosity
- favorable casting and forming properties
- good weldability (oxidation on the outer Al panels must be controlled through a coating to enable consistent spot welding)
- good corrosion resistance.

Using aluminum, designs which are characterized by a greater mobility and carrying capacity than conventional types of construction, can be manufactured.

Aluminum was used very early in the automotive sector, with the aim of saving weight. As long back as 1899, on the occasion of the international motor show celebrated in Berlin, a small sports car was presented, with a body made of aluminum. From 1912 onward, Pierce Arrows began to build large-surface body parts such as doors, back panels, and roofs made of cast aluminum, a technique which attracted great interest. With increasing engine power and with large-scale production and process safety of steel materials, which were backed by experience, however, aluminum lost its importance in the automotive sector. However, in upper-class vehicles, for example, in the Silver Arrow of Mercedes, aluminum still continued to be used. Owing to the increased environmental-protection requirements in the second half of the twentieth century, aluminum began regaining significance. Engine hoods, hardtops, and tailgates were increasingly made of aluminum. Aluminum was an important constituent in the manufacture of the Land Rover, the Dyna–Panhard, and the Audi. Above all, the development of the ASF for the Audi A8 and A2 models was the trend setter for aluminum bodies.

An enormous amount of experience has been acquired, especially in the sector of the aluminum sheet-metal parts. As a result, steel sheets were consequently placed on the substitution test bench with the lighter aluminum. On this occasion, a distinction is made between partial substitution, where additional advantages can be gained by the use of aluminum, and entire substitution, where designs are completely made for aluminum. In partial substitution, weight saving plays a major role, but there are other advantages as well. This makes itself felt, for example, in the easier handling and the lower center of gravity for aluminum hardtops, an improved weight distribution of aluminum engine hoods, as well as easier opening mechanisms and greater resistance to corrosion for aluminum tailgates. By replacing certain materials of add-on parts, it is possible to achieve local improvements, without changing the design. Here, steel components are often replaced by aluminum, where the integration of aluminum parts in steel bodies hardly presents any problems. Owing to its high specific energy-absorbing capability, aluminum offers itself especially for the design of weight-optimized crash-management systems such as bumpers or entire front modules.

In Europe, the thickness of outer aluminum panels is approximately 1 mm, whereas in the US it is only 0.7–0.8 mm because of the use of a higher-strength alloy like AA6111. The principal characteristic of outer panels is the stiffness against buckling, the resistance against the impact of chips, hail, and so on. The influencing factors for stiffness against buckling are the part geometry, the yield point, and the sheet-metal thickness. The resistance against elastic deformation depends on the stiffness. It is determined by the part geometry, the modulus of elasticity, and the material thickness. In order not to lose the light-weight advantage compared to steel, because of an increased thickness, design solutions are sought out which, for example, allow for more depths in deep drawing areas, and can provide continuous inner parts that can be made thinner. When employing aluminum for body parts, the fatigue behavior and the long-term stability are deemed not critical, in spite of the continuing processes of precipitation. So far, no negative effects have been observed concerning the mechanical properties. The resistance to corrosion, however, does play an important part for the aluminum body. A suitable design can contribute toward avoiding gaps and the formation of galvanic elements. The paint layer has a blocking function, and can be provided with active corrosion-protection pigments. When joining noble metals, the nobler (i.e., has a higher redox potential) contact partner is to be protected by an insulating layer, to avoid corrosion.

The future potential of aluminum as body material is to be seen in connection with new alloys and adjusted heat treatment states, and with manufacturing and processing methods to optimize the material properties like strength, crash behavior, formability, joining, and resistance to corrosion.

2.4.3
Magnesium

Industrial production of magnesium started in Germany, in 1886. In 1900, the entire world production was about 10 t. Magnesium was first obtained from pure magnesium chloride, followed by dry electrolysis at 700–750 °C. Magnesium is also be produced by the reduction of magnesium oxide with carbon, but with higher material and energy efforts (Table 2.7).

Table 2.7 Properties of magnesium

Properties	Value
Density	1.74 g cm^{-3}
Modules of elasticity	45 GPa
Linear thermal expansion coefficient	25×10^{-6} K^{-1}
Thermal conductivity	17 Wm^{-1} K^{-1}
Strength	100 MPa
Melting point	649 °C

Table 2.8 Effect of alloying elements on properties of magnesium [9]

Alloying elements	Effects
Aluminum	Aluminum, being the most important alloying element, augments the tensile strength, the elongation after fracture, and the hardness. However, an aluminum content of more than 6% leads to brittleness, a decreasing strength and elongation
Zinc	A zinc content of up to 3% increases the tensile strength and the fatigue strength under vibratory stresses. In case of higher additions, however, the elongation after fracture is decreased to 1%
Manganese	Manganese increases the resistance to corrosion and the weld ability, a percentage above 1.5% improves also the strength
Silicon	A maximum content of 0.3% augments the casting ability and enables pressure sealed casting. In case of higher additions, the elongation after fracture decreases considerably.
Zirconium	Zircon oxide has the effect of seed crystals which lead to the formation of a fine-grained structure with an increased tensile strength, without a reduction of the elongation.
Cerium	Cerium supports the formation of a fine-grained structure and increases heat resistance.
Thorium	Thorium strongly increases heat resistance.

Source: [4]

As a design material, magnesium is almost exclusively used in the alloy form. Table 2.8 describes the influence of the alloy elements on the behavior of magnesium.

In the 1930 and 1940s, the need for magnesium rose dramatically, to cover military needs and aircraft requirements. During the postwar period, first of all, the most favorable sorts of scrap were used up. After that, owing to the high cost of production, magnesium was displaced by aluminum. Volkswagen was the only manufacturer who relied on the use of magnesium, mainly for transmission casings. In 1971 an average of 20 kg of magnesium was used for each vehicle. But in the following years, the critical properties like corrosion and susceptibility to heat, as well as the high cost, displaced magnesium from the Volkswagen models. In the 1990s, the production volume increased again when high-purity magnesium alloys were developed and the anticorrosive properties improved. Worldwide, a total of 432 000 t of magnesium were used in 2002.

In addition to the pressure die-casting applications, the use of magnesium sheet metal for automotive body building gained special importance. Magnesium sheet metal is characterized by a high availability of material, recycling capability, and high specific-weight related strength and stiffness properties. Compared to steel and aluminum, the weight saving potential is approximately 25 and 60%, respectively. Compared to plastics, magnesium sheets have a higher heat resistance, a lower thermal elongation, and can be recycled more easily.

The cold-forming ability of magnesium alloys is rather low owing to the hexagonal lattice structure. At temperatures above 200 °C, the forming possibility greatly increases. Therefore, a multiple-step rolling process is applied in the manufacture of magnesium sheets, to achieve a fine crystalline sheet with stabilized structure, by optimized forming and temperature. In this way, according to DIN 1729-1:1982, the Magnesium alloy AZ31 (MgAl3Zn1), for example, can be used because its properties are comparable to those of an aluminum body material.

Fusion welding methods, MIG, and laser welding are appropriate joining methods of magnesium components. Using these methods, it is possible to accomplish joints with few faults and of high strength. Mechanical joining methods such as clinching, riveting, flanging, screwing, and bonding are still in the trial stage. The range of applications in the automotive sector focuses on the interior and the nonvisible areas, owing to the low corrosive requirements.

2.4.4
Titanium

Titanium, discovered in the year 1795 by the Berlin chemist Martin Klaproth, derives its name from Greek mythology. The Titans, hated by their father Uranos, were held in the interior of the earth. To produce titanium, it is necessary to expensively reduce the raw material, titanium oxide (TiO_2), to metal. In the Kroll process, TiO_2 is processed to $TiCl_4$. This is then reduced with magnesium to titanium sponge, which is freed by vacuum distillation from the residues of Mg and $MgCl_2$, removed from the reaction tube in small chips, and remelted in the arc under argon or vacuum.

The first alloys of titanium were developed at the end of the 1940s for large-scale application. The best known of the alloys is Ti-6Al-4V, according to DIN 17850:1990-11. Titanium belongs to the group of nonferrous and light metals. Almost twice as heavy as aluminum, with a density of 4.51 g cm^{-3}, titanium is the heaviest of the light metals. Nevertheless, it has only half of the specific weight of iron (Table 2.9). With an estimated 0.6% of all elements deposited in the crust of the earth, it is the fourth metal in order of occurrence. Its price, and the component

Table 2.9 Physical properties of titanium

Properties	Value
Density	4.51 g cm^{-3}
Modules of elasticity	110 GPa
Linear thermal expansion coefficient	9×10^{-6} K^{-1}
Thermal conductivity	17 Wm^{-1} K^{-1}
Strength	170–500 MPa
Melting point	1700 °C

fabrication, are often the main reasons for its low utilization in mass production. The special properties of titanium are its corrosion resistance and the high specific strength of 500–1500 MPa (titanium alloys).

The application possibilities of titanium in the body sector, are primarily in the engine and the chassis, because a good torsional stiffness and a flexural stiffness are of maximum importance here. If, however, a part has to be designed with an optimum stiffness, titanium materials are less suited owing to their comparitively low modulus of elasticity. The decision in favor of titanium is reasonable only if the engineering advantage is convincing, because the cost involved is far higher than that of other materials.

The use of titanium in the automotive industry began in the mid-1950s. For instance, the outer panels of the turbine-driven experimental car, the *Titanium Firebird*, manufactured by General Motors, were made of titanium. In 1998, Toyota was the first producer of engine valves made of titanium.

In the engine area, titanium is preferably applied to reduce rotating or oscillating masses. An example is the connecting rod. In the past, connecting rods were used in sports cars of Honda, Ferrari, and Porsche.

Compared to steel rods, weight reductions up to 20% can be achieved, by using titanium. The production, however, involves a considerable costs because for instance, the machining process is complex, owing to high notch sensitivity. Another example is light-weight valves made of titanium. By achieving mass reduction, the performance of the engine is increased and the fuel consumption reduced. In addition, more weight can be saved, for example, lighter valve springs can be employed, whereby the percentage of weight saved varies between 40 and 50%.

Titanium was first used as an axle spring in the mass production of cars. Today, due to its good anticorrosive properties, it is used for sealing rings, flanges, and brake guiding pins. As a result of weight saving of so-called unbraked masses such as titanium wheels, brakes, wheel carriers, wheel bearings, and axle springs, it is easier to follow the road irregularities, and this contributes to the riding comfort. In car bodies, crash elements made of titanium grade 4 or the alloy TiAl6V4 are of interest, because the material has an even and consequently, an energy-absorbing forming behavior.

2.4.5
Nonmetallic Parts – Fiber Composites

The percentage weight of plastics in motor vehicles is 11.7%, which corresponds to 100–150 kg. Experts are of the opinion that this percentage will grow to 20% by the year 2010. Even today, the Audi A2 has a plastics share of 25%. Here, the use of plastics is focused with 46% on the interior of cars, but in the relatively recent application of outer panels, the percentage is already 22%. Plastics contribute considerably toward weight saving in vehicles. Of the 100 kg plastics in a middle-class car, approximately 40 kg are weight relevant, which means, these plastics have replaced other, heavier materials [10, 11].

With the Smart, a vehicle was launched onto the European market, whose outer panels are made completely of thermoplastics. The nonarmored and reinforced thermoplastics and duroplastics, and the fiber composites like unidirectional fiber-reinforced fabrics, wovens, and knits offer new possibilities of material application. Especially in connection with the paneling of the body structure, these materials can be used as they show properties which cannot be accomplished with the conventional manufacturing methods and materials. Basically, however, the new chemical substances have to adjust themselves to the structures existing in the automotive industry. This is applicable, above all, to body building and painting, because here the producers are sometimes faced with conflicts in regard to the painting and the assembly (see also Chapter 9). To guide plastics through the cathodic electrodeposition painting, special plastics are needed, with a temperature resistance of up to 190 °C. As an alternative, the parts can be painted and assembled separately, and then integrated into the existing system of the manufacturer, or solid-color parts (which implies a limitation of the color flexibility) could be used. Plastics, above all, are characterized by the relative ease with which they can be designed for integration with parts and functions. Primarily, this can be traced back to the great variety of synthetic polymers whose molecular chains can be adjusted to suit the requirements of the finished products. In the next stage too, that is, processing, polymers can be tailored to the engineering requirements. Similar to metals, it is possible to determine certain orders, and make the polymers subject to them, for example, all molecular chains have the same orientation – an effect which increases the strength in the direction of the molecular chain. Further improvement of properties can be achieved by polymer alloys or mixtures, and additional reinforcing methods. These reinforcements may be:

- milled fibers, mostly based on glass fibers
- glass powder
- glass flakes
- marbles
- talc.

Processing can be done using a whole range of methods for example, injection molding, extrusion, thermal deep drawing, thermal pressing, which make multiple forming possible from temperatures of 200 °C onward.

The strength values reach the level of light metals in rare cases only (the tensile strength of fiber-reinforced plastics (FRP)). However, owing to their comparably low weight because of the low specific weight of $0.9-1.8\,\mathrm{g\,cm^{-3}}$, this can often be balanced. An advantage regarding body parts is the stability toward chemical substances and corrosive surrounding conditions. When evaluating the mechanical properties of the thermoplastics and duroplastics, it is to be considered that they depend highly on the temperature, so that in addition to the strength properties, the forming behavior always has to be considered as a function of the temperature and the relative humidity. The thermal forming stability is characterized by the modulus in shear, depending on both the temperature and the material damping.

Considering the scope of application, it is the thermoplastics sprayed in car color that offer the greatest variety of possibilities. Parts such as bumpers, fender, and sill moldings are standard today.

Thermoplastics compete with certain duroplastics, which are equipped with similar or even better properties, when seen from the viewpoint of thermal performance such as temperature resistance, the thermal longitudinal expansion, and the modulus of elasticity as a function of the temperature. Thermoplastics can be processed from the molten mass or from a softened state. After their forming, they can be remelted. There is a great variety of grades in this category of materials, such as polystyrene (PS), polypropylene (PP), polyamide (PA), polyethylene (PE), polycarbonate (PC), polyvinylchloride PVC, or thermosets which can all be processed in a soft state; however, after forming, they cannot be reformed by heating them again. The base material of fiber composites is given by the unsaturated polyester (UP) resins, epoxy (EP) resins, and the continuous filaments. UP resins can be bought at a good price, and are easy to process. But they have low mechanical properties, a low thermal resistance, high volume shrinkage, and high internal stress. In contrast, EP resins are relatively expensive, and difficult to process. On the other hand, they have good mechanical properties especially regarding dynamics, and a low volume shrinkage.

In the manufacture of automotive bodies, unsaturated fiberglass reinforces plastics (FERP) based on polyester resins, (sheet molding compound (SMC), or bulk molding compounds (BMC)), at low volumes (<250 parts per day) have, above all, proven to be an economically favorable alternative to metallic materials. This refers to applications as body structural parts, such as spare wheel well with integrated rear floor panel, and body parts such as hood, tailgate, fender, and bumper support. When using SMC or BMC with outer panel quality (reference is made to the class A surface), the possibility of using in-mold coat (IMC) or powder mold coat (PMC) for visible areas offers good prospects for a quality improvement [10, 12, 13].

In addition to the described standard SMC, with approximately 25% weight related fibers, special SMC types with higher glass fiber content of 50–70%, mass share, or partially directed glass fiber strands are used. In this way it is possible, with regard to the pressing technology, to accomplish a higher degree of stiffness and strength by selecting the right approach to insertion.

Moreover, there are high-performance prepregs with thermoplastics. These glass mat-reinforced thermoplastics (GMT) are processed with PP to semifinished products. After heating, they can further be processed by pressing, among other things, as underbody protection or bumper supports.

The forming behavior is of great importance for the properties and appearance. The vacuum injection technique or the resin transfer molding (RTM) for high-performance fiber composites may open new application possibilities for outer body panels, which have a high value from an optical point of view. The RTM process begins with the insertion of reinforcing fibers, which can be used in the form of continuous filament mats, fabric, or specially developed complexes. After the insertion of the armoring materials, the mold is closed. Once the mold is closed,

Table 2.10 Molding processes of reinforced plastic parts

Manufacturing method	Description
IM	Injection molding, processing of milled fiber reinforced thermoplastics; granulated material with 1–5 mm long fibers; economic process; various geometric forms are possible
RIM	Reaction injection molding method for foamed large parts; largely processing of PUR structural foam; density 1.0–1.1 kg dm^{-3}, tensile strength R_m 15–30 MPa, modulus of elasticity 55–500 MPa
RRIM	Reinforced reaction injection molding method to produce reinforced structural foam; density 1.25 kg dm^{-3}, tensile strength R_m 25 MPa, modulus of elasticity 3300 MPa
RTM	Resin Transfer Molding, resin injection method preferably used to produce prototypes; small volumes of large parts; continuous or glass mats or fabric impregnated with resin, formed in the die and evacuated
Filament winding	Rovings are impregnated with resin and processed according to the lathe principle; preferably for prototypes or small volumes

the mixture of resin and hardener is injected through one or more spots into the mold. Prior to this, a vacuum may be created in the mold, to improve the flow of resin. Usually, curing of the resin is accelerated by heating the mold. In addition to the various armoring materials, resin systems such as polyester, vinyl-ester, epoxy, or phenol resins can be used in the RTM process (Table 2.10).

Plastic foams with a high light-weight potential offer a good solution for the improvement in the behavior of instability of hollow sections, or in the local forming behavior of body parts.

Continuous unidirectional fiber-reinforced high-performance composites are known from aeronautics, aerospace engineering, and from racing. Experts hope that they can be used increasingly to solve the problems of primary body structures. This applies to those body parts which stiffen the thrust field such as the dash panel, front floor, rear floor, or back panel. They are built from high-stiffness carbon fibers reinforced plastic (CFRP) and impact resistant aramid fibers, mostly in a matrix of epoxy resin. Carbon offers a whole range of possibilities, but the manufacturing methods and the cost are still not suitable for large-scale application.

Carbon components are largely created in the autoclave approach. The autoclave is an air tight and vapor tight closable pressure tank. During the process of the fiber-composite production, usually pressures of up to 10 bar and temperatures of up to 400 °C are created. The high pressure in the interior is employed to press the single layers of laminate. In most cases, the parts are evacuated simultaneously, to remove superfluous air from the composite. The synthetic resin in the fiber-composite part is then cured for up to 20 hours, at a high temperature.

Another possible application of plastics as light-weight materials in the automotive industry, is back-foaming. Here, plastics are applied as foam onto a carrier material (for example, Polyurethane (PUR)), or as a plastic backing film (for example, Acrylonitrile-Butadiene-Styrene (ABS)). Thermoplastic composites with a stiffness of up to 10 000 MPa can be used. Furthermore, this process offers the possibility of functional integration (for example, built-in antenna).

2.5 Manufacturing Methods

2.5.1 Tailored Products

2.5.1.1 General Remarks

These products are based on the idea that single parts which are joined by laser welding, are employed to make sheet metal with optimized properties, that is, tailored blanks, tailored tubes, or tailored strips. These semifinished products possess locally defined properties such as thickness, mechanical properties, and surface coating which depend on the space.

The requirements to be fulfilled by the subassembly groups of vehicles vary widely, and often, are contradictory. A low weight may be contrasted by a high stiffness. Usually, a compromise is adopted by adding single parts. The tailored technology can be used to save weight, by replacing spot welds at overlapping sheet-metal edges by continuous laser welds. The continuous joints of the single sheets increase the stiffness of the design [2].

2.5.1.2 Tailored Blanks

The process was first used on a large scale in 1985. Two galvanized sheets of the same thickness and the same mechanical properties were laser welded to form a large board, which subsequently was cold worked to floor panels. In the mid-1990s, the tailored blanks were provided with an additional degree of freedom, so that the weld could be positioned more or less at the sheet-metal level. The possibility of nonlinear structures decreases the gap between the design requirements in relation to the parts, and the available process technology (Figure 2.14).

The manufacturing chain includes processes such as cutting, welding, application of beads, marking, and if necessary, turning, as well as storage, and transport.

For the floor panels of passenger cars, which consist of curved center parts that accommodate parts of the transmission, and the cardan shaft in case of vehicles with rear-wheel drive, and the two flat-outer parts, tailored blanks can contribute to reducing the weight of the vehicle. Three blanks are welded by laser in the form of butt joints. In this process, neither flanges that have been required so far for spot welding, nor sealants are necessary. Various sheet-metal thicknesses also contribute to further weight reduction.

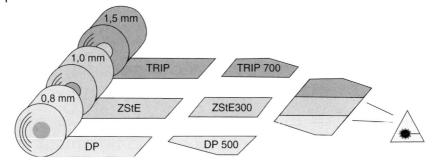

Fig. 2.14 Manufacturing principles of tailored blanks.

2.5.1.3 Tailored Tubes

Tailored blanks for primarily hydroformed body parts, can be converted to closed sections, the so-called *tailored tubes*. They offer advantages by way of an increased torsional stiffness at a low sheet-metal thickness, number of parts, and number of flange faces. The steps for the manufacture of tailored tubes are as follows:

- manufacturing tailored blanks
- forming to slit tubes
- longitudinal tube welding
- tube end processing
- cold drawing and bending
- preforming
- hydro forming
- finish cutting.

Welding single tubes at their faces is another possible application in production.

The application of the tube technology does offer the possibility of reduction in weight, on the other hand, this requires new joining and manufacturing techniques. In the future, it is likely that more hydroformed tailored tubes will be used for automotive body building purposes.

2.5.1.4 Tailored Strips

Here, strips of steel are provided with a continuous weld. The production of small parts is especially economical when tailored strips are employed, for example, rims.

2.5.1.5 Patchwork Blanks

Patchwork blanks are blanks of sheet metal that are partially reinforced by one or more patches. In contrast to the conventional way of reinforcing, the patches are applied before the forming process. In this manner, the technology of the tailored blanks can be enlarged, because even the smallest of areas can be reinforced easily. To connect the patches with the base blanks, methods such as spot welding, laser welding, and bonding are employed. These methods can also be combined.

2.5.1.6 Future

Due to the increasing use of materials with a low density, such as aluminum and magnesium, new joining approaches need to be developed. Another possible method to produce sheets with locally different properties, is flexible rolling (tailored rolled blanks). Here the sheet is provided with variable thicknesses by a certain change of the roll gap. In this case, combination of different materials is not possible, and sheet metal-thicknesses can be performed only cross to the rolling direction. The method, however, is quite economical.

2.5.2 Hydroforming

Fluids have a very low compressibility, therefore their volume changes to a small extent only, even under high pressure. Due to this property, water is employed in metal forming as a medium to transfer pressures. Since the pressures and forces that are necessary to form steel and aluminum can be controlled by hydraulics, valve technology, and press making, the hydroforming process has been adopted in the automotive industry. Hydroforming of metal semifinished products offers an interesting spectrum of solutions for the production of automotive system components for the body, chassis, powertrain, and other groups of automotive assembly. The mechanical properties of the parts are improved by a plastic forming of the material. The omission of weld flanges enables weight saving. By integrating add-on parts of an adjusted design into the work piece, it is possible to diminish the number of components to be joined.

Hydroforming belongs to the method of active means based sheet-metal forming. The most common are hydroforming, hydrostatic stretch forming, and hydromechanical deep drawing. Hydroforming can further be classified as forming of tubes and sections, and of welded and nonwelded twin sheets. By using active means, new light-construction designs can be applied, such as manufacturing of function integrated structural parts, which would not be possible using conventional manufacturing methods. However, in spite of further developments, these methods are subject to certain limitations, some of which are listed below:

- restriction of the ratio between the diameter and the wall thickness, in case of previous bending processes
- limited radii
- limited design of the periphery
- connection and joining of hydroformed parts
- cost – tooling (die) cost significantly lower than stamped parts, but longer cycle time for forming.

For the last 20 years or so, hydroforming has been applied in the automotive industry to an extent worth mentioning. Parts which were manufactured based on this approach, have basically two functions – first, as force and torque-transmitting structural parts in the powertrain, body, and chassis; and second, as media carrying elements in the powertrain, heating, ventilation, and air-conditioning systems.

There are various applications of active hydromechanical forming. In the production of shell-type parts, such as fuel tanks, structural parts, or outer panels, especially this approach offers advantages in regard to the implementation of complex geometries, quality, and properties. It is expected that in the near future, the method will be increasingly used for mass production purposes.

New semifinished products, such as rolled sections with complex cross-section geometries, as well as conical tubes and tailored tubes, will presumably contribute to considerably expand the spectrum of hydroforming application. In connection with the demand for flexible and economic body structures, especially for small volumes, hydroforming is deemed very important, in view of manufacturing frame structures. A further development is hydroforming of wrought aluminum products with warm, active means, which permits a considerable broadening of the part spectrum. Holes can also be pierced in preclosed sections as part of the hydroforming process. For complex parts, the tubes are bent with a computerized numerical control (CNC) bending machine, prior to placement in the hydroformed dies [2].

2.5.3
Press Hardening

Press hardened parts of car bodies are usually those parts made of sheet steel, which are heat treated during the hot-forming process, by a defined quenching at the cold tool. Here, two manufacturing processes are used for hardened structural parts – the direct process, which means that parts are heat treated in one hot-forming process only, and the indirect process, which means that complex part geometries are produced with a cold-forming operation, prior to hot forming.

The advantages of press hardening are
- achieving a high material strength with a good ductility
- very high weight saving potential
- very good dimensional accuracy
- even complex geometries can be manufactured
- excellent crash behavior.

The biggest problem with this type of hardening is scaling of parts, and so they need to be coated in many cases.

2.5.4
Metal Foam

Metal foams have been known since the 1950s, but so far it has not been possible to produce foamed metals in a sufficient quantity and of consistent quality.

In general, a distinction is made among powder metallurgical, smelting metallurgical, and special processes, whereby the powder metallurgical processes are deemed the most suitable ones. In this case, an expanding agent is mixed with the metal powder and consolidated (extrusion molding, hot isostatic pressing (HIP) etc). After this, these semifinished products can be foamed reproducible by heating to temperatures close to their melting point. In addition to the known aluminum foams, other metals and alloys can be also foamed; iron and titanium foams are still being developed. At present, almost all foams are manufactured with closed pores. Metal foams permit the production of a multitude of different forms, such as foamed cavities, or composites, and plates. The plates can be deep drawn before foaming, which increases the dimensional variance. Metal foams can be joined not only by means of laser processes, but also by bonding, screwing, or riveting.

With increasing foam density, the mechanical properties have been strongly augmented, but the thermal expansion coefficient in relation to massive bodies, remains almost unchanged. During the plastic deformation of metal foams, the level of stress is maintained almost constant over a long path of deformation, and it is very similar to an ideal absorber.

2.5.5
Sandwich Structures

A new development in applications in the body area, is the use of formed metal–plastic–metal composites. Here, the positive properties of the known materials, that is, stiffness, weight, and acoustics have been combined to form the optimum material composites. The diversity of variants covers a large number of combinations such as

- metals/PU, PP or PS foam
- FRP or FRP/PU, PP or PS foam
- metals/metal foam.

Sandwich structures are applied to meet the requirement of accommodating high stiffness, in a compact way. The face sheet, called *skin*, serves to accept force and the core sheet to hold the distance. By using different kinds of face sheet and core sheet materials, together with adjusted wall thicknesses, the properties can be adjusted to the requirements. Compared to sheet steels with the same bending strength or aluminum sheets (1 mm), these sandwich sheets (0.2 mm Al/0.8 mm PP/0.2 mm Al) are approximately 60 or 35% lighter.

However, problems occur with further processing. The thermal joining methods with the associated temperature load are not suitable owing to the thermoplastic core sheet, so that the joining process is to be performed mechanically or by bonding. The temperatures of the painting process are also not suitable for the PP core sheet with a melting temperature of 165 °C. In case of deep drawing, the process parameters need to be adjusted to the modified material properties compared to conventional sheets [2, 12].

2.5.6
Roll Forming to Shape

Roll forming to shape, is a continuous forming under bending condition with a turning tool movement, where an even sheet strip is formed to the desired shape, step by step, by successively arranged pairs of bending rollers. According to DIN 8586, this refers to the group of manufacturing processes called *forming under bending conditions with a turning tool movement*.

This process is suitable for manufacturing complex profiles with both open and closed cross sections. By using the design and the material possibilities to the optimum degree, these profiles can contribute essentially to cost saving and weight saving by space-frame solutions.

A roll forming system essentially consists of a decoiler, a strip leveler, a cropping shear with welding fixture to connect the strip ends, and the roll forming element itself with a straightener, and a cutter at the end (Figure 2.15) [1, 2].

The form of the profile, the material, and the specific quality requirements, determine the number of bending steps or forming stages, which are in some cases, as high as 50. The higher and the more complex the profile, the more bending steps are necessary. Hot rolled or cold rolled, coated or uncoated flat stock, with a very high strength and low elongation values, may be used as initial material. Frequently, further processing steps like welding, stamping, and foaming are combined while forming the profile. In case of closed profiles, closing is performed in connection with high frequency (HF), tungsten inert gas (TIG), or laser welding.

2.6
Joining Methods

Many joining methods are involved in actual body-in-white manufacturing, owing to the complex material mix. The appropriate method has to be selected depending on the specific situation, (see Table 2.11).

The most important critera are the adhesive forces which strengthen the specific part and the total construction.

Limits due to pretreatment and coatings are demonstrated in Figure 2.16.

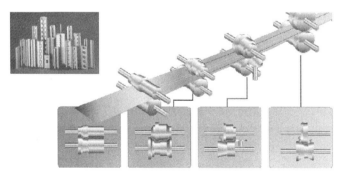

Fig. 2.15 Schematic layout of a roll forming process.

Table 2.11 Evaluation of joining methods

Joining method	Pro and cons
Spot welding	Standard joining methods, proven, favorable price
Gas shielded arc welding	Low extent, mechanization difficult, high heat input
Stud welding	Large application possibilities, proven, favorable price
Laser welding	Increasing application, high investment
Clinching	Increasing application, favorable price, not crash safe
Punch riveting	Increasing application, crash safe, involves more efforts than spot welding
Bonding	Structural adhesives are established, increasing application, crash safe, durable
Combined methods	Increasing application, advantages of single processes are combined, more complex, but quicker part fastening

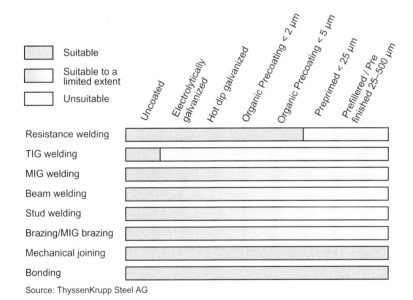

Source: ThyssenKrupp Steel AG

Fig. 2.16 Application limits of joining methods.

2.6.1
Bonding

According to the definition given in DIN 16920, bonding is a kind of joining, which uses an adhesive; this means that parts are joined by adhesion and cohesion, using a nonmetallic material. Bonding, being an independent joining method, is gaining more and more importance in welding and mechanical methods. As a result, today's body assembly process can hardly be imagined without it.

The tasks of bonding, however, cover a broad range and are not only limited to the pure joining process:
- connection of different materials
- sound and vibration damping
- gas and fluid sealing
- increasing the structural stiffness
- corrosion protection (crevice corrosion, contact corrosion)
- low-torsion joining by low temperature load
- application of force onto a large surface.

Considering the large number of potential tasks, a corresponding number of adhesive application systems is available.

For structural bonding, hot melt one-component epoxy adhesives are widely used in the automotive industry. Their advantage essentially, arises from their high mechanical characteristics, good automatic-bonding process, and forgiving processing behavior. PUR and PVC adhesives have proven to be successful as far as sealing, sound, or vibration-damping measures are concerned. The surface quality of the joining materials is of vital importance. This is applicable for all types of bonding. In the case of metals, contaminants, which occur mainly during forming and handling, and corrosion products at the surface, influence the joining quality in a negative way.

For polymeric materials, the molecular structure and the associated surface polarity play an important role during bonding. In many cases, prior to bonding, cleaning (in this case, degreasing) or in the area of plastics, a pretreatment like flaming or plasma treatment is necessary (see Chapter 9).

Depending on the requirements to be fulfilled by the joint, the mechanical properties can be improved by additional chemical processes like using adhesive primers, or mechanical processes like roughening or blasting. Designs for adhesives can also positively influence the bonded connections. Primarily, what should be seen to is that the joint is exposed to tensile load and shearing stress only, and not to peel-off stress.

As all adhesive systems imply polymeric materials, temperature, and aging effects are to be taken into consideration, in addition to design and surface related influences. For these reasons (that is, application of force, temperature, and aging), most of the joints are bonded by using a combination of thermal and mechanical joining methods. In this manner, spot weld bonding can clearly increase the strength of a connection and the stiffness under dynamic loads, with a simultaneous reduction of the number of spot welds. The same applies to a combination of clinching and bonding.

2.6.2
Laser Welding

In laser welding, the welding process is performed by means of a focused laser beam. With the use of high-strength steels, this technique can improve the stiffness of the body. When selecting the type of laser, the following factors play a decisive role – high, adjustable performance, a constant performance, and a good beam quality.

The CO_2 and the Nd : YAG (Nd = neodymium, YAG = yttrium aluminum garnet) lasers are the ones most frequently applied. The CO_2 laser is created from a mixture of CO_2 and other gases. It has a wavelength of 10.6 μm, thus being in the infrared range. The Nd : YAG laser is a solid-state laser, that is made from an artificial monocrystal. The wavelength of 1.06 μm is situated in the infrared range. The light of the laser is focused on the weld head, over a concave mirror and emitted from the nozzle.

For laser welding, a distinction is made between thermal-conduction welding, and deep-penetration welding. In thermal-conduction welding, the laser beam is focused on the joint. The materials in the heat-affected zone are fused by the absorbed energy. The welding depth amounts to some tenth millimeters, which makes this procedure apt, especially for thin sheets.

In penetration welding, the laser evaporates the material at the welding spot.

As a result of this evaporation, and the pressure of the metal vapor, a steam passage is created, also called *keyhole*. The evaporated material absorbs laser energy, thus forming a laser induced plasma, which leads to more energy being supplied to the material, so that thicker sheets can be welded.

The depth of the weld corresponds to 10 times its width.

The advantages of laser welding are
- no wear of the tool
- undercuts
- minimum distortion due to a low heat supply
- weight reduction by a low part overlapping
- single sided joining enabled.

The disadvantages are
- high cost
- gab of joining materials must be controlled within small tolerances.

2.6.3
Others

2.6.3.1 Clinching

Clinching is a technical term, which is used as an umbrella term for a kind of joining. Work-pieces made of sheet metal, tubes, or profiles are joined by exerting pressure, if necessary, during cutting and subsequent upsetting. Compared to other types of forming, clinching does not require any additional ancillary elements (e.g. rivets) or auxiliary substances such as adhesives and fluxing agents.

A distinction is made between single-stage clinching and multiple-stage clinching. In single-stage clinching, the parts (i.e. sheet metal, tubes, and profiles) are joined in a continuous manufacturing process by means of clinching with cutting or sinking and subsequent upsetting, so that a frictional and a positive connection are achieved by expansion or extrusion.

In multiple-stage clinching, different dies are used successively. A distinction is made between clinching with and without cutting share. Clinching without cutting means joining of work-pieces made of sheet metal, tubes, or profiles in a one-step or multiple-step manufacturing process, by clinching from the common sheet-metal level with sinking and subsequent upsetting, so that a frictional and positive connection is achieved by expansion or extrusion.

The process of clinching with cutting means that the material is shifted from the sheet-metal level and upset, so that a positive connection is formed, which is replaced by a frictional connection at the end of the process.

2.6.3.2 Riveting

Whereas, resistance spot welding is mainly applied to steel structures, riveting plays an important role in connection with aluminum body parts. Like clinching, riveting is a mechanical joining process without thermal influence on the parts to be connected. Without previous boring, the parts are joined together in a riveting and a cutting process. Access is to be ensured from both sides.

The use of riveting was promoted to overcome the problems that emerged in connection with the aluminum space frame. Full tubular rivets are most frequently employed. The connection is created by a stamping and a forming process. Here, the parts to be joined are laid on a die, and the punch with the rivet presses the rivet into the upper part and forms the lower part plastically to a closing head. The form of the head is determined by the contour of the die. During this process, the rivet expands, thus creating a frictional and positive connection. When steel rivets are used in aluminum parts, a corresponding coating of the rivets is required for protection from corrosion.

Assuming a proper rivet connection, the strength properties under both shear and tensile loads are clearly better than those of spot welding connections. All kinds of semifinished products can be connected with a high degree of process reliability, provided that access is possible from both sides. Further advantages are the low thermal load, no heat distortions, and the fact that the process can additionally be designed to be gas and water tight.

2.6.3.3 Roller Hemming

Flanging, in sheet-metal forming, is a positive locking, irreversible type of joining where the edges of sheets are placed into each other and pressed so that a firm connection is achieved. Moreover, sealants provide the possibility of obtaining absolutely tight flanged connections. The process has been known for a long time, but in the automotive industry it gained significance in connection with light construction. It is used together with bonding, to produce large-scale parts such as doors, hoods, tailgates, and sunroofs.

In course of time, conventional hemming has developed into roller hemming with industrial robots. Roller hemming is a very flexible flanging method, with a tool guided by a robot. In most cases, the flange of the sheet-metal part is closed by a roller, which is attached to the robot wrist. The position of the roller and the pressure, decisively determine the flanging process. A flanging step is not made simultaneously on the entire part, but locally at a spot where the robot wrist or the roller is situated. This means that to make one step, the contour of the part concerned has to be followed. In front of the roller, the upward pointing flange would still be there, whereas, behind it, the first flanging step would have already been made.

The tool pushes the sheet metal like a wave. This operation is repeated several times. This is the reason that the manufacturing time for single parts is higher than the time needed for steel flanging.

2.7
Outlook

In the past few years, a considerable number of studies and concepts focusing on light construction have been presented. When considering them, attention should, however, be paid to the fact that the background of these concepts was largely the intention to show the core competences of the companies involved. The steel industry, especially, is trying to substantiate the market share of steel with innovative ideas [14].

The ultralight steel auto body (ULSAB) studies include three closely connected development projects covering the period from 1994 to 2001, with the aim of investigating the economics of light-construction bodies with steel materials.

The ULSAB was the first project set up by a consortium of 35 steel companies, in 1994. Porsche Engineering Service (PES) was entrusted with the development and the subsequent studies. Proceeding from a four-door, five-seat middle-class sedan, the task focused on searching for state-of-the-art steel materials for bodies-in-white to save as much weight as possible. To define the functional aims and to determine comparative values, 32 vehicles were benchmarked for properties of weight, stiffness, package, and concept.

In 1997 the ultralight steel auto closures (ULSAC) project was set up, with the same task for vehicle doors and lids. The decision for doors was made in favor of a frameless door with a modular mounting concept.

Two years later (in 1999), the third study was started, that is, the ultralight steel auto body–advanced vehicle concepts (ULSAB–AVC) with a budget of 10 million US dollars, which expanded the area of investigation of the previous projects, by considering the remaining components of a total vehicle development, including the drive and the chassis. Here, the limiting conditions were governed by more recent requirements, taking into consideration the current emission laws and safety requirements. In addition to development, planning, and economic efficiency, the ULSAB and ULSAC projects include the production of prototypes for test and demonstration purposes, and validation of simulation results.

Owing to the employment of state-of-the-art steels, corresponding forming and joining methods, weight orientation and, last but not the least, a design suited to the materials under consideration, the ULSAB study aims to demonstrate that a weight saving by 25% is feasible, compared to the average of the benchmark bodies. The project comprises the design and math-based simulation of the stiffness, strength, and crash behavior. Here, the legal limit values valid for the known crash tests are to be taken into consideration. The proof of the desired static torsion and flexural stiffness is furnished by means of additional tests with simulation.

Regarding the doors, front, and deck lids of the ULSAC project, 10% weight saving are to be proven, compared to the lightest comparative part. During the quasistatic door-impact test, with real dented parts, the same force level should be reached as with the average of the comparative doors.

In the ULSAB–AVC project the interior, exterior, and the drive parts are added to the box-type body-in-white with doors and lids. On the basis of identical parts and a platform concept, the market requirements for economic model diversity are taken into consideration. With slight modifications, one platform can serve to produce two body styles, that is, a hatchback variant of the European compact class (C-class) and an American middle-class notchback sedan of the partnership for a new generation of vehicles (PNGV) class.

The studies of the ULSAB family showed that if today's light-construction methods, state-of-the-art steels, and manufacturing methods are applied consistently, clear weight reductions are possible compared to today's large-scale bodies, without affecting the function and safety requirements and the economic efficiency. In this way, the objective of the consortium was also achieved, that is, to demonstrate that light construction is possible with steel bodies, and that a weight level can be achieved, which is comparable to that of an aluminum body.

In the automotive industry, a uniform tendency toward light construction is not recognized. Each producer is strongly engaged in this subject, and each new generation of vehicles means another step toward a lighter body-in-white.

Steel and plastics producers, especially, steadily offer innovative products on the market, so that light-construction materials such as aluminum, magnesium, or plastics, are always faced with new challenges. Above all, this is due to the tendency toward higher and high-strength materials that are not yet available, for example, for aluminum. Steel is the material that can maintain its market position precisely because of the many innovations, and many experts still do not see the potential being exhausted.

For hybrid bodies, too, the share of aluminum elements seems to be below 25%, above all, because of the extra cost incurred for complex joining methods and the problems involved in contact corrosion. In spite of this, aluminum occupies an important place in body fabrication, mainly for the add-on parts.

Another important aspect is given by the developments of the manufacturing and joining methods, because they make the new, largely sophisticated structures and parts of large-scale series possible. In the next few years, further innovative solutions are expected, in the adhesive application technology, which offers promising possibilities.

New developments are also being offered by the sector of sandwich and hybrid structures. This technology, which has been applied for quite some time in the aircraft industry, could open up new possibilities for light construction in the automotive industry.

2.8 Surface Protection

2.8.1 Precoating of Sheets

In today's automotive industry, bodies are dominantly made of galvanized coils or steel sheets, which are coated with zinc alloy. The zinc layer acts as a corrosion inhibitor, especially on trimmed edges and in case of defects of the painted surface. Owing to the lower electrochemical potential, zinc functions as a galvanic element in the presence of an electrolyte. Zinc acts as the anode and protects the steel from corrosion.

Usually, the zinc coating is applied in the steel mill in a coil coating processes. In addition to the mostly used electrogalvanization and the hot-dip galvanization, so-called *galvannealed sheets* are used, whose zinc layer is converted by a thermal treatment into a zinc–iron alloy, after hot-dip galvanization. These sheets are primarily used in Asian and Latin-American countries. The advantage compared to simple galvanized sheets is the better weldability owing to the inclusion of iron in the zinc layer. A disadvantage is the higher brittleness of the layer. This is crucial for the use of structural adhesives (see Chapter 10) for bonding. In a severe load case, the break occurs in the zinc layer.

Moreover, so-called duplex coatings are increasingly being used. In this case, galvanized sheets or coils are coated additionally by organic materials [15, 17]. As a rule, this is performed in automatic coil coating installations. The zinc pigmented organic coating is applied and cured to approximately 5 µm film thickness. Sheets being treated in this way, can be processed in the body shop, without any significant conversions in the existing press shops. Advantages are expected by an improved corrosion protection in flange areas, the potential reduction of waxing operations in cavities, and at least partial reduction in the hem sealing.

2.8.2 Corrosion Prevention in the Design Phase

Usually three types of corrosion can occur in connection with a car body [6]:
- surface corrosion
- crevice corrosion
- contact corrosion.

If the corrosion appears on a coated and exposed metal surface, it is called *surface corrosion*. Crevice corrosion can occur in flanges, or hems, and under seals [12]. It can largely be traced back to the fact that humidity cannot sufficiently escape from gaps. Residual salt has a hygroscopic effect, and keeps the gap humid for a longer time. If two different metals have an electrical contact, this may produce contact corrosion if an electrolyte is present. The higher the difference of the electrochemical potential of the materials in contact, and the smaller the anode area in relation to the cathode, the higher is the corrosion speed and the damage.

Surface and corrosion protection can essentially be taken care of in the design phase of a car. The selection of suitable materials, and material mating or the avoidance of contacts, can contribute to prevent contact corrosion in this phase. Cavities, for example, in doors, beams, rocker panels, and structural carriers that are exposed to humidity, should be designed so that they can be ventilated well. These vents, at the same time, favor a good paint penetration during the electrodeposition process (see Chapter 3). Moreover, all components of the body should be designed so that water can run off easily. When connecting two sheet-metal parts, sharp edges or sharp-edged transitions are to be avoided. Furthermore, the design engineer has to examine whether cavities are to be protected by wax or whether connections are to be sealed by suitable sealers and adhesives (see Chapter 10).

References

1 Haas, C., Thomer, J. (2005) *Leichtbau von Automobilkarossen*, Master Work, University Bingen.
2 Adam, H. (2002) *Proceedings 12th International Body-in-White (BIW) Expert Conference*, Fellbach.
3 Dullinger, M. (2003) *Proceedings 17th International Body-in-White (BIW) Expert Conference*, Fellbach.
4 Klingler, M. (2002) *Proceedings 13th International Body-in-White (BIW) Expert Conference*, Fellbach.
5 N.N., *Hightech Report Daimler Chrysler*, 2005 (2), 58.
6 Breass, H.-H., Seiffert, U. (2005) *Vieweg Handbuch Kraftfahrzeugtechnik*, Vieweg, Wiesbaden, p. 333.
7 Fischer, U. (2005) *Tabellenbuch Metall (43)*, Europa Lehrmittel, Haan, p. 115.
8 Furrer, P., Bloeck, M. (2001) *Aluminium-Karosseriebleche*, Moderne Industrie, Landsberg.
9 Domke, W. (2001) *Werkstoffkunde und Werkstoffprüfung*, Cornelsens, Düsseldorf.
10 Stauber, R. (2000) *Ingenieur Werkstoffe*, **9**, 3.
11 Wagner, A. (2003) *Proceedings 17th International Body-in-White (BIW) Expert Conference*, Fellbach.
12 Klein, A. (2002) *Proceedings 12th International Body-in-White (BIW) Expert Conference*, Fellbach.
13 Jakobi, R. (2002) *Proceedings 12th International Body-in-White (BIW) Expert Conference*, Fellbach.
14 Schneider, C. (2003) *Proceedings 17th International Body-in-White Expert Conference*, Fellbach.
15 Hickl, M. (2006) *Metalloberflache*, **60**(4), 44.
16 *Journal für Oberflächentechnik*, (2006), **46**(1), 46.
17 Lewandowski, J. (2003) *Journal für Oberflächentechnik*, **43**(9), 52.

3
Pretreatment of Multimetal Car Bodies

Horst Gehmecker

3.1
Introduction

The pretreatment of car bodies manufactured from different metals is mandatory for state-of-the-art corrosion protection and provides best adhesion for electrodeposition coatings. It is an established process and used by almost all car manufacturers in the world.

Car bodies manufactured today frequently contain many different substrates, for example, cold-rolled steel, bake hardening steel, electrogalvanized steel, hot-dip galvanized steel, galvanneal, aluminum sheets made from different alloys, and plastic hang-on parts. Trication zinc phosphating is the standard process worldwide for pretreating multimetal car bodies. The process comprises several stages, namely, degreasing, rinsing, activation, phosphating, rinsing, passivation (optionally), and a final demineralized water rinsing.

Over recent years, pretreatment process has been adapted to a large variety of substrates. Further modifications are constantly initiated by demands for process improvements, environmental constraints, or economic factors.

3.2
Car Body Construction Materials

3.2.1
Sheet Materials

It is generally accepted that the use of zinc-coated steel is an important part of the corrosion protection concept of car bodies. The percentage of bare steel sheets and coils for car body construction has been decreasing continuously over the last few decades. Currently, most cars worldwide are produced from virtually 100% zinc-coated steel and aluminum sheets or coils. Except for low-cost vehicles and/or

Automotive Paints and Coatings. Edited by H.-J. Streitberger and K.-F. Dössel
Copyright © 2008 WILEY-VCH Verlag GmbH & Co. KGaA, Weinheim
ISBN: 978-3-527-30971-9

Table 3.1 Substrates used for car body construction

Number	Material	Thickness of deposited layer	Remarks/Examples
1	Cold-rolled steel	–	Mild steel, high tension steel a.o.
2	Electrogalvanized, one-sided	5–7.5 µm	Zn
3	Electrogalvanized, two-sided	5–7.5 µm	Zn
4	Hot-dip galvanized	10–15 µm	Zn
5	Galvanneal	10–15 µm	Zn/Al layer
6	Zinc–Nickel	7.5 µm	–
7	Electrogalvanized + FeZn layer	7.5 + 3 µm	–
8	Hot-dip galvanized + FeZn layer	10 + 3 µm	–
9	Weldable primer	1.5–6 µm	Pigmented with Zn, FeP a.o.
10	Aluminum	–	AA 6016, AA 6022, AA 5182
11	Magnesium	–	AM60, AZ31, AZ91
12	Plastic	–	PP-EPDM, R-TPU a.o.
13	Stainless steel	–	–
14	Zinc–Magnesium	3 + 0.3 µm	–

emerging markets, cold-rolled steel sheet is only used in areas of the car body that are little exposed, that is, for interior parts or for the roof panel.

Electrogalvanized and hot-dip galvanized steel sheets are widely used in Europe and North America, whereas Japanese manufacturers commonly use galvanneal (Table 3.1).

Aluminum sheet, although used for many years in the luxury and upper-midsize car segments, has been introduced in a large number of compact, lower-midsize and midsize vehicles in Europe over the recent years [1]. AA 6016 (AlMgSi) panels are commonly used for exterior panels and 5000-series (AlMg) for interior parts.

Magnesium parts for exposed areas of the vehicle require an appropriate off-line pretreatment and painting before being assembled to the car body. As this is a costly and not lean manufacturing process, closure parts made from magnesium have not been introduced on a large scale in the automotive industry so far.

Plastic parts are treated and painted before assembly either on-line or off-line, depending on the type of plastic and the process requirements (see Section 9.1). It is widely accepted that plastic parts are advantageous for specific parts on bodies like bumper, trunk lids, and fenders, and for building niche models in small series, requiring high demands of design.

Coil coatings with weldable primer (also known as *corrosion protection primer*) are electrically conductive, thin organic coatings that are used by several vehicle manufacturers to achieve additional corrosion protection in hem flanges and other exposed areas of the car [2]. There may be a cost offset by reducing the amounts of seam sealer and cavity wax used per vehicle. Although applicable also on other substrates, weldable primers are currently applied on electrogalvanized steel sheets only. The film builds applied in the steel mill are in the range of 1.5–6 µm.

3.2.2
Surface Conditions/Contaminations

The *body-in-white*, which is the term for a completed body from the body assembly line entering the pretreatment process, may have the following surface conditions and contaminations:
- corrosion protective oils
- prelubes
- washing oil
- stamping lubricants
- dry film lubricants
- hot melts
- body shop sealants and adhesives
- prephosphate layer applied to the coil as spray prephosphate, no-rinse prephosphate by spray/squeeze or roll-on application without rinsing, or micro-prephosphate
- amorphous sodium phosphate layer
- thin organic coatings (i.e. weldable primer)
- Ti-/Zr-oxide/hydroxide for aluminum sheets applied to a part or a coil
- dioctylsebacate (DOS) for aluminum sheets
- internal or external mold releases for plastic parts
- welding pearls and particles
- metal grains
- other contaminants like chalk marks, dust, fingerprints, marks from pneumatic handling devices, and so on.

Corrosion protective oils are applied in the steel mill in amounts typically less than 1.5 g m^{-2}. They are based on mineral oils and corrosion inhibitors. The products are normally approved by the automotive manufacturer for compatibility with the electrocoating materials used, to avoid cratering (see Chapter 4). The ability to remove the protective oil is tested by standardized procedures, for example, VDA 230–201.

Prelubes are applied in the steel mill and act as corrosion protective oil and stamping lubricant at the same time. They are based on mineral oil and contain corrosion inhibitors and other additives.

Washing oils are used by several automotive manufacturers to remove the protective oil applied in the steel mill and to reoil the coil or blanks for temporary corrosion protection, as well as for supporting the stamping processes.

Dry film lubricants are mostly used to improve transport, storage, and stamping properties of aluminum sheets. They may be more difficult to remove, especially when exposed to high temperatures to bake-harden the alloy or to precure sealers and adhesives.

Hot melts are somewhat similar to dry film lubricants, but can be also used on steel substrates. They contain wax-related raw material, which have melting points of the order of 45 °C.

Prephosphate layers are predominantly deposited in the steel mill on electrogalvanized steel and galvanneal coils. The coatings typically have a layer weight of less than 1.8 g m^{-2} for weldability reasons and have a composition similar to the conversion layers produced in car body lines, that is, Ni-, Mn- modified Hopeite $Zn_3(PO_4)_2 \bullet 4\ H_2O$. Spray prephosphate layers are typically very fine crystalline, whereas no-rinse prephosphate are either crystalline or amorphous, depending on the application process. Prephosphate layers are attacked and dissolved at pH > 11 and pH < 3.

Micro-prephosphate layers are applied by the no-rinse process with coating weights of <0.3 g m^{-2}. All prephosphate layers are oiled in the steel mill. Normally, no additional stamping lubricant is applied in the body shop except for heavy stamping operations.

Amorphous prephosphate layers are based on sodium phosphate with coating weights of about 10 mg m^{-2} phosphor.

Thin organic coatings (i.e. weldable primer) on galvanized steel coils are used by various automotive manufacturers to improve corrosion protection in the box section, hem flanges, and so on . This shifts pretreatment and value-addition upstream to the coil manufacturer (see Chapter 14). Typically, weldable primers contain zinc powder, iron phosphite (FeP) powder, or precious metal powder as corrosion inhibitors. In addition, they make the film electrically conductive. In general, these thin organic coatings are not changed when passing through the car body pretreatment process, except for coatings containing zinc particles close to the surface. In this case, the zinc particles may receive a thin zinc phosphate coating.

Ti-/Zr-oxide/hydroxide layers on aluminum improve the transport, storage, welding and adhesive bonding properties of aluminum sheets. The layers must be thin (<5 mg m^{-2} Ti or Zr) in order to avoid any interference with car body phosphating. Ti-/Zr-oxide/hydroxide layers are applied either on to the coil or on to the stamped parts prior to assembly of the vehicle.

Welding pearls and particles and metal grains may cover a wide range of particle sizes (~1 to ~1000 μm). In modern vehicle production, the body-in-white is contaminated with less than 0.2 kg per body (<2 g m^{-2}) of these particles.

Mold release agents for plastic parts, which help easy removal of the finished part from the manufacturing forms, may be classified into internal and external mold releases. They contain products with very low surface tensions, which cause problems in the coating processes as contaminants.

3.3
Pretreatment Process

3.3.1
Sequence of Treatment

The standard pretreatment process for car bodies comprises the following five operations for optimum preparation of the metal surface [3]:
- precleaning (optional)
- degreasing
- activation
- phosphating
- passivation.

Depending on the degree of cleanliness of the body entering the coating line, the legislation demands with respect to cascade rinsing, and the existence of a postrinsing stage, the pretreatment line can have up to 14 stages for a safe and reproducible process [4].

3.3.2
Degreasing

The task of degreasing is to remove all kind of contaminations from the metal surface, to achieve a water-break free surface, that is, a continuous water film on the surface after rinsing off excessive degreasing chemicals with water and to obtain a reactive surface which is able to buildup the phosphate coating that is, the conversion layer, within a reasonable period of time.

In car body phosphate lines, liquid- or powder-type alkaline degreasers are used to feed the tanks. Liquid cleaners are typically two-pack products with the builder and the surfactant separate, whereas powder products are typically one-pack products.

Alkaline degreasers, which are the standards for car body cleaning, are composed of inorganic salts, the 'builder', and organic compounds, the 'surfactants'. The predominant task of the builder is to remove inorganic and pigment contaminants like metal grains and welding pearls. The task of the surfactant is to remove oils, lubricants, soaps, and other organic contaminants.

Typical builders used in alkaline cleaners are the following:
- $NaOH$, KOH, Na_2CO_3, K_2CO_3 → maintain alkalinity
- silicates → particle removal, inhibitor, buffer
- orthophosphates → degreasing
- condensed phosphates → degreasing, complexing
- complexing agents → complexing.

Surfactants contain a hydrophilic group, that is, a long chain of ethoxy (EO) and/or chain of propoxy (PO) molecules, and a hydrophobic group, which is typically a long chain alkyl. They are classified into anionic, cationic, nonionic, and

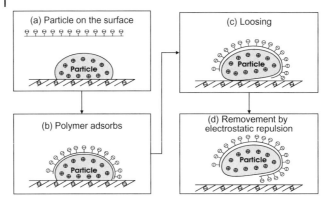

Fig. 3.1 Mechanism of particle removal from surfaces.

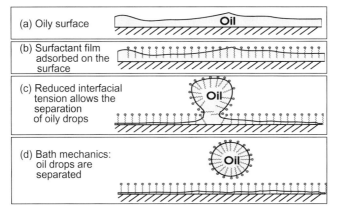

Fig. 3.2 Mechanism of oil removal from surfaces.

amphoteric surfactants. In car body degreasing, typically nonionic surfactants are in use today owing to their better environmental compliance. The mechanism of particle removal from surfaces is shown in Figure 3.1. Owing to their amphiphilic characteristics, surfactants in solutions reduce interfacial tensions, absorb at solid interfaces, and form micelles. The mechanism of oil removal from surfaces is shown in Figure 3.2. The surfactants first adsorb on the oil surface, and then try to reduce surface tensions and remove separated droplets containing the oil from the surface.

Nonionic surfactants are solubilized through hydrogen bonds between water molecules and the ether group. As the hydrogen bond breaks at increasing temperature, the surfactant becomes insoluble and forms a second phase. This is called the *cloud point* and the corresponding temperature is specific for each surfactant, depending on the number of EO and PO groups. Nonionic surfactants above the cloud point have a low tendency to build foam, but are still surface active.

Tests/methods used to analyze the result of degreasing, that is, the cleanliness of surfaces are as follows:
- water-break test→ for residual oil, very sensitive up to <10 mg m^{-2}
- wiping test with white cloth→ for oil and particles
- adhesive tape test → residual particles
- electron scanning for chemical analysis (ESCA), secondary Ion mass spectrometry (SIMS), auger electron spectroscopy (AES) → sensitive, but expensive and time consuming
- test inks → meaningful, easy to handle
- contact angle measurement → relatively easy to handle.

Degreasing solutions are applied both by spray and immersion applications. The advantages of spray application are as follows:
- short treating times
- used for parts with simple geometry
- excellent for particle removal
- restricted in terms of low-temperature application (surfactants tend to foam at low temperatures)
- requires only small bath volume for less investment, less space, low capital cost
- can be combined with brushing operations, and so on.

The advantages of immersion application are as follows:
- excellent cleaning of areas difficult to access (box sections, hem flanges)
- requires higher concentration and treatment times compared to spray application
- higher stability due to greater bath volume.

In practice, degreasers are typically applied by spray process followed by immersion application in order to make use of advantages of both technologies. In addition, immersion tanks are equipped with entry and exit sprays.

The bath is kept level either by addition of water from the city supply, or by addition of deionized (d.i.) water, the latter being preferred in order to prevent precipitation of calcium and magnesium salts.

Control of degreasing baths is accomplished mostly by titration methods:
- total alkalinity
- free alkalinity (optional)
- conductivity
- pH (optional).

Total and free alkalinity are defined by different titration methods, which are developed by the suppliers of the pretreatment chemicals. In practical terms, degreasing baths pick up oil and other contaminants that have to be either dumped at regular intervals, or need to be continuously passed through filtration systems in which oil and other contaminants are removed. Degreasers with emulsifying properties may take up >10 g l^{-1} of oil. As the risk of cathodic electrocoat tank oil contamination and hence the risk of craters being produced in the paint film increases with the steady-state oil concentration of the degreasing bath, special emphasis is placed on the reduction of the oil contents of the degreasing bath. Oil removal can be accomplished by ultrafiltration, decanter, thermal separators (mostly installed in East Asia), coalescence separators, or static separators.

Ultrafiltration devices are effective on the one hand, but expensive in terms of investment and operating/maintenance cost. In order to avoid blocking of the membranes by the precipitation of silicates, cleaners should be operated at pH > 11.0 or should be silicate free. As surfactants are removed in ultrafiltration as well, two-pack cleaners must be used to compensate the excessive surfactant consumption compared to other processes.

Most recently, static oil separators have become more popular in combination with the use of demulsifying degreasers. Compared to ultrafiltration, both the investment and operating costs are significantly lower. A flow scheme is shown in Figure 3.3.

Removal of dirt particles may be accomplished by a set of microhydrocyclons combined with pressurized paper band filters and magnetic separators.

Amphoteric substrates like zinc-coated steel and aluminum are etched by alkaline degreasers. Inorganic inhibitors like borates and silicates are used to protect the

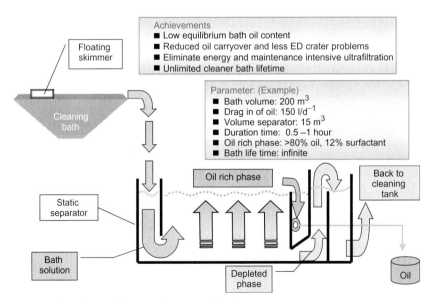

Fig. 3.3 Flow scheme of static oil separator for effective removal of oil from degreasing tanks in connection with demulsifying surfactant.

metal surface against excessive etching on these substrates. Inadequate formulation of the degreaser, for example, with lack of inhibitors like borates or silicates, may lead to excessive local etching and formation of so-called *white spots*. After cleaning and rinsing, defects that are several millimeters in diameter would result on the zinc surface, which would interfere with the painting process.

For degreasing of prephosphated zinc-coated steel sheet, mildly alkaline cleaners are that do not greatly affect the prephosphate layer are used. If the prephosphate layer is affected substantially, there is a significant risk that a zinc phosphate coating morphology is changed in the phosphate process, which greatly differs from the prephosphate layer and which results in a nonuniform cathodic electrocoat film.

On aluminum substrates, excessive etching may lead to the formation of deposits, that is, sludge on the surface, especially for aluminum alloys with high silicon content.

Legal regulations on cleaners vary considerably on a worldwide scale and, to some extent, also within Europe. With effect from October 2005, surfactants need to have a primary degradation of $>60\%$ according to OECD 301. Alkylphenolethoxylates (APEOs), which have been in use for long, may break down to form products that are more toxic than the APEO itself, and so this class of surfactants has been banned in Europe since April 2005. In Germany these were already banned several years ago.

Limitations already exist in European countries for the use of complexing agents and certain salts that are normally used to formulate the inorganic part of cleaners. In Denmark and in Italy, for example, the use of borates is prohibited in metal cleaning formulations. Throughout Europe, the use of the very active complexing agent ethylenediaminetetraacetic acid (EDTA) is forbidden.

In Japan, degreasers free of nitrogen and phosphorus compounds ('P&N free cleaner'), that is, aminocarboxylic acids and phosphates, have been introduced more recently.

In North America, cleaners mostly need to fulfill local regulations. The use of the complexing agent nitrilo-triaceticacid (NTA) is prohibited throughout the United States.

3.3.3
Activation

Activation increases the number of crystallization nuclei on the metal surface. This results in an increased number of phosphate crystals per unit surface area and a reduced coating weight for the applied conversion layer. As the surface will be uniformly covered with crystals in a shorter time, activation treatment also has an accelerating effect in the phosphate process.

For activation prior to zinc phosphating, aqueous dispersions of titanium ortho phosphates with a pH between 7 and 11 are typically used. The manufacturing process determines the performance of these activators and is proprietary know-how of pretreatment manufacturing companies. Titanium phosphate activators are available on the market both in powder and liquid form, the latter having a shorter shelf lifetime and being sensitive toward freezing temperatures during transportation and storage.

Most recently, liquid activators based on zinc phosphate have become available and have been introduced in the Japanese automotive industry (Nipppon Paint, Private information, 2004).

The performance of activating baths diminishes with time, independent of the throughput of parts. This phenomenon varies to a great extent and depends on the specific product formulation. This can be explained in terms of titanium phosphate colloids being negatively charged and being precipitated out by divalent or trivalent cations, especially Ca(II) and Mg(II) ions, which are contained in hard water. In order to reduce degradation by divalent ions, most activator products contain condensed polyphosphates to form complexes with the aforementioned cations.

It is strongly recommended that activator baths be made up using demineralized water and losses be compensated with water of the same quality. Depending on the activator product, the level of contamination in the bath, and the specifications with regard to crystal size and coating weight, activating baths are dumped at fixed intervals. These time intervals may range from about 1 week to about 6 months.

Liquid activator products are normally fed directly into the supply tank, although special pumps are required to cope with the high viscosity of these products. When powder products are used, a 5% slurry is typically prepared in a premixing tank and used for product dosage in the working tank.

It is recommended that activator baths are stirred continuously in order to prevent settling of the titanium phosphate dispersion.

3.3.4
Zinc Phosphating

Low-zinc phosphate is the standard phosphate conversion layer worldwide for processing multimetal car bodies. Other types of pretreatment like iron phosphate or conventional zinc phosphate, which are used for steel surfaces, and chromating or chrome-free processes on basis titanium/zirconium compounds for aluminum surfaces have been tried in the past, but neither provides the required quality on all substrates, nor are they suitable for multimetal treatment.

The initial step in all conversion treatments is a pickling attack of the metal surface by free phosphoric acid. The metal loss on cold-rolled steel sheet, zinc-coated steel sheet, and aluminum is typically in the range of 0.5 to 2 g m^{-2} (see Figure 3.4).

The consumption of hydrogen ions leads to a shift of pH in the diffusion layer adjacent to the metal surface, exceeding the solubility limits, and consequent precipitation of zinc phosphate. Accelerators are therefore added to the zinc phosphate chemicals to speed up the pickling reaction by replacing the unwanted H_2 evolution by chemical reactions that take place more easily, like reduction of the accelerators themselves.

Zinc phosphate solutions typically contain dihydrogen phosphates of zinc, nickel, manganese, free phosphoric acid, sodium nitrate, fluorosilicic acid, one or several oxidizing compounds like sodium nitrite, hydrogen peroxide, hydroxylamine,

Pickling reaction

$$Fe + 2H_3PO_4 \rightarrow Fe(H_2PO_4)_2 + H_{2(g)} \uparrow \text{ or}$$

$$Fe + 2H^+ \rightarrow Fe^{2+} + H_{2(g)} \uparrow$$

Coating formation

$$3Zn^{2+} + 2H_2PO^-_4 + 4H_2O \rightarrow \underline{Zn_3(PO_4)_2 \cdot 4H_2O \downarrow + 4H^+}$$
$$\text{Hopeite}$$

$$2Zn^{2+} + Fe^{2+} + 2H_2PO^-_4 + 4H_2O \rightarrow \underline{Zn_2Fe(PO_4)_2 \cdot 4H_2O + 4H^+}$$
$$\text{Phosphophyllite}$$

$$2Mn^{2+} + Zn^{2+} + 2H_2PO^-_4 + 4H_2O \rightarrow \underline{Mn_2Zn(PO_4)_2 \cdot 4H_2O + 4H^+}$$
$$\text{ZnMn-phosphate}$$

Sludge formation

$$Fe^{2+} + H^+ + Ox \rightarrow Fe^{3+} + HOx$$
$$Fe^{3+} + H_2PO^-_4 \rightarrow FePO_4 + 2H^+$$

Fig. 3.4 Phosphate reactions on steel surfaces.

sodium chlorate, nitroguanidine (CN4), N-methylmorpholine-N-oxide (NMMO), acetaldoxime, and sodium nitrobenzenesulfonate (SNIBS). The Zn, Ni, Mn compounds and the phosphoric acid along with the ferrous ion (FeII) from the steel surface are the layer forming compounds. All other chemicals have supporting functions like acceleration, oxidation, etching, and stabilization of the bath and the film.

Processes for phosphating steel and aluminum surfaces at the same time additionally contain fluoride (hydrofluoric acid, alkali fluoride, or alkali bifluoride) to generate amounts of about 50–250 ppm free fluoride, depending on the type of process and application, that is, spraying or dipping. In patent literature, low-zinc phosphate processes are defined by the low Zn : PO_4 ratio in the working solution. This ratio had been set to 1:12 to 1:110 originally, but has been changed to 1:20 to 1:100 as of today.

The reactions taking place during phosphating of steel, zinc and aluminum surfaces are summarized in the Figures 3.4, 3.5, and 3.6.

The main crystal phases produced in low-zinc phosphates are hopeite and phosphophyllite. Phosphate crystals are electric insulators, but contain pores constituting approximately 1% of the surface. This is a very important precondition for the deposition of electrocoatings.

Typical compositions of the conversion layers are different on steel, zinc, and aluminum substrates and are shown in Table 3.2.

Scanning electronic micrographs (SEMs) of the zinc phosphate coatings on steel, zinc, and aluminum surfaces are shown in Figure 3.7. The morphology depends on the type of substrate; however, the results in adhesion of the electrocoat primer are similar [5].

Pickling reaction

$$Zn + 2H_3PO_4 \rightarrow Zn(H_2PO_4)_2 + H_{2(g)} \uparrow \text{ or}$$

$$Zn + 2H^+ \rightarrow Zn^{2+} + H_{2(g)} \uparrow$$

Coating formation

$$3Zn^{2+} + 2H_2PO_4^- + 4H_2O \rightarrow \underline{Zn_3(PO_4)_2 \cdot 4H_2O + 4H^+}$$
$$\text{Hopeite}$$

$$2Mn^{2+} + Zn^{2+} + 2H_2PO_4^- + 4H_2O \rightarrow \underline{Mn_2Zn(PO_4)_2 \cdot 4H_2O + 4H^+}$$
$$\text{ZnMn-phosphate}$$

Fig. 3.5 Phosphate reactions on zinc-coated steel sheet.

Pickling reaction

$$Al_2O_3 + 6H^+ \rightarrow 2Al^{3+} + 3H_2O$$

$$Al + 3H^+ \rightarrow Al^3 + 1\tfrac{1}{2} H_{2(g)} \uparrow$$

Complex formation

$$Al^{3+} + 6F^- \rightarrow AlF_6^{3-}$$

Coating formation

$$3Zn^{2+} + 2H_2PO_4^- + 4H_2O \rightarrow \underline{Zn_3(PO_4)_2 \cdot 4H_2O + 4H^+}$$
$$\text{Hopeite}$$

$$2Mn^{2+} + Zn^{2+} + 2H_2PO_4^- + 4H_2O \rightarrow Mn_2Zn(PO_4)_2 \cdot 4H_2O + 4H^+$$
$$\text{ZnMn-phosphate}$$

Sludge formation

$$Al^3 + 6F^- + 3Na^+ \rightarrow Na_3AlF_6 \downarrow$$
$$\text{Cryolite (sludge)}$$

$$Al^3 + 6F^- + 2K^+ + Na^+ \rightarrow K_2NaAlF_6 \downarrow$$
$$\text{Elpasolite (sludge)}$$

Fig. 3.6 Phosphate reactions on aluminum surfaces.

The sludge generated during the phosphate process has to be continuously removed by filtration techniques in order to maintain constant performance of the conversion layer.

At present, most trication phosphate processes are nitrite accelerated. On the one hand, from a technical point of view, sodium nitrite can be considered as an ideal accelerator. On the other hand, however, sodium nitrite is classified as toxic, fire-stimulating, and hazardous to the aquatic environment. It develops fumes (NO_x) upon contact with the acidic phosphating bath and interferes with effluents.

3.3 Pretreatment Process

Table 3.2 Zinc phosphate coatings and compositions on steel, zinc, and aluminum surfaces

	% Zn	% Ni	% Mn	% Fe	% P_2O_5
Cold-rolledsteel	31	0.9	2.2	6.5	41
Electrogalvanized steel	45	0.9	5	–	40
Aluminum	44	0.9	9	–	42

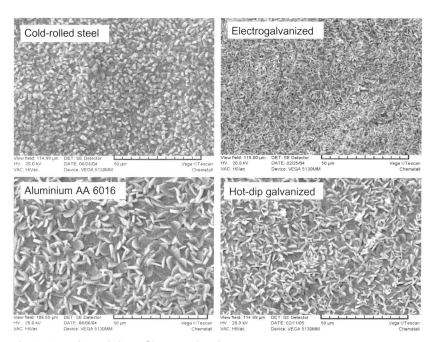

Fig. 3.7 Crystal morphology of low-zinc phosphate coatings on cold-rolled steel, electrogalvanized steel, hot-dip galvanized steel, and aluminum sheet.

Several types of nitrite-free phosphate processes were developed and introduced during the early nineties mainly in Europe. The chemicals replacing nitrite-compounds have already been mentioned earlier in this chapter.

The relevant transportation and safety regulations for these newly developed compounds are shown in Table 3.3.

Phosphate processes in Europe and North America are typically operated at 50–55 °C, whereas, application temperatures of 40–45 °C are more typical in Southeast Asia without any significant detrimental effect on the final performance.

In recent years, so-called low-temperature low-sludge phosphate processes have been introduced in various Japanese car body lines and Japanese plants around the world. The standard operating temperature for this process is 35 °C, and the amount

Table 3.3 Transportation and safety regulations for phosphate accelerators according to German legislation

	Hydrogen peroxide (H_2O_2)	Nitro-guanidine (CN4)	Hydroxyl-amine (HA)	Acetal-doxime (CH_3–CH–NOH)	Acetal-dehyde (from Acet-aldoxim); see also HA	Sodium nitrite ($NaNO_2$)
Supply form	Liquid 10–25%	Liquid 25–50%	Liquid 25–50%	Liquid	Liquid	Powder/granular
Risk symbols	Xi, irritant	None	Xn, health hazard; N, environmental danger	Xi, irritant	F +, extremely flammable; Xn, health danger	T, toxic; N, environmental danger; O, fire promoting
R labels	36, 38	None	2, 37/38, 41, 43, 48/22, 50	10, 36, 37, 38	12, 36, 37, 40	8, 25, 50
Details and transportation	5.1 oxidizing	None	9	–	–	5.1 oxidizing
Water hazard class	1 = Low hazardous	1 = Low hazardous	3 = Severe hazardous	3 = Severe hazardous	1 = Low hazardous	3 = Severe hazardous
Others	–	–	Class 3 carcinogenic	–	Class 3 carcinogenic	–

of sludge generated on steel and galvanneal surfaces is said to be reduced by about 30% as compared to the Japanese standard phosphate process operated at 43 °C.

Phosphating of aluminum–steel structures by immersion is crucial. In order to obtain uniform phosphate layers without precipitation of cryolite particles on the surface, certain preconditions in terms of flow, composition, and replenishment of the phosphate bath are required. Systematic studies have shown that a flow of >0.4 m s^{-1} parallel to the surface is optimal. As cryolite precipitation in the phosphate film is favored by both higher sodium and fluoride concentrations, both sodium and free fluoride levels need to be operated below certain limits. This is only possible if optimal flow conditions are maintained.

Corrosion tests have shown that the paint creepage in both copper accelerated salt spray test (CASS) and outdoor exposure test is much lower for aluminum sheet that received a continuous phosphate layer at a high flow and low free fluoride levels of 70–100 ppm as compared to a significantly higher free fluoride level of about 200 ppm (Chemetall, Private Information, 2005).

3.3.5
Passivation

In order to improve the corrosion resistance of phosphated and coated metal sheet, the conversion layer can be given a passivation as a postrinse with chrome(VI), Cr(III), or chrome-free solutions. Today, hexavalent chrome has been replaced by zirconium-based solution due to the toxic danger of chromiumVI-compounds.

Although the mechanism of passivation by zirconium-based solutions is not fully understood, it is generally accepted that the effect of improved corrosion protection is mostly associated with reduction of the pore size by precipitation of insoluble compounds and removal of secondary phosphate crystals from the surface of the phosphate layer by the acidic solution of the passivating agent.

Zirconium-based postrinses are typically operated under the following conditions:

- zirconium concentration   50–150 ppm
- fluoride concentration   50–200 ppm
- pH   3.5–5.0
- conductivity(20 °C) $<600\ \mu S\ cm^{-1}$
- controls   pH, conductivity, Zr content and/or total acid titration
- preparation   in demineralized water
- dump frequency   1–4 weeks, depending on contamination level and quality of preceding rinsing bath.

Historically, Japanese and Korean automotive manufacturers do not use passivation treatment. Instead, demineralized water rinsing cascades have been installed with more than three stages. In recent years, European manufacturers have started to adopt this production philosophy and are eliminating the passivation step.

As zinc phosphate coatings produced in immersion treatment are more dense than those precipitated in spray phosphating, immersion phosphating is generally accepted to be an important precondition for the elimination of the postrinsing step. Zinc-coated steel sheet shows very little sensitivity toward a post rinse, cold-rolled steel sheet, while that of aluminum is significant. Furthermore, the quality profiles depend significantly on the paint system and test conditions.

3.3.6
Pretreatment of Aluminum – Steel Structures

Pretreatment of aluminum requires the addition of fluoride ions in order to obtain sufficient etching of the surface for the phosphate film deposition and to complex the aluminum ions dissolved in the etching reaction. Free aluminum ions interfere heavily with zinc phosphating and may inhibit the formation of coatings at very low concentration levels (<5 ppm Al^{3+}).

The free fluoride levels typically used to form a uniform and crystalline phosphate coating are approximately 70–180 ppm in spray application and 70–250 ppm in dip application. Important factors to achieve a uniform and crystalline phosphate coating by dip are as follows:
- high activation by appropriate chemicals and their concentration in the formulations;
- continuous control of free fluoride level, preferably by means of a fluoride sensitive electrode;
- adequate movement of the phosphating bath (>0.4 m s^{-1});
- adequate balance of alkali ions (<3 g l^{-1} Na $\pm$ and K $\pm$ ions) in the phosphate bath;
- limitation of the level of free fluoride ions to avoid inclusion of cryolite particles in the phosphate coating;
- keeping Ti and/or Zr contaminants <5 ppm in the phosphate bath;
- efficient sludge removal of the phosphate bath, for example by filter presses.

Typically, a metal loss of 0.8–1.2 g m^{-2} is observed, which results in 6–10 g m^{-2} dry sludge. This is in contrast to 3–5 g m^{-2} of dry sludge in phosphating of steel surfaces and virtually no sludge in phosphating of zinc-coated steel sheet.

In recent years, a process modification has been developed that suppresses the coating formation in the phosphating stage, but provides the aluminum surface with a very thin passivation layer in a zirconium passivation stage. As most automotive manufacturers have specified uniform and continuous phosphate layers on aluminum, the latter process has not be introduced on a broader basis so far. A continuous phosphate layer is accepted as a protection against corrosion of any surface defects on aluminum supported by sanding in the body shop for best adhesive bonding results [6].

3.3.7
Pretreatment of Magnesium

Weight reduction and fuel saving requirements led to the investigation of magnesium alloys for car body construction. Cast magnesium parts of high purity alloys have predominantly been investigated, and more recently cast sheet materials have also been considered [7].

It has been concluded that magnesium needs appropriate pretreatment and painting when used in exposed parts of the vehicle. Passing the parts together with the car body through the actual zinc phosphate lines is not an appropriate pretreatment and does not provide the proper corrosion protection. Magnesium needs to be treated and painted separately and then attached to the car body either in the body shop or in the paint shop.

Proper pretreatment for magnesium parts comprises cleaning and etching, followed by a Cr-containing conversion treatment or a Cr-free conversion treatment based on titanium/zirconium fluorides. In order to achieve optimal corrosion protection, the Cr-free pretreated parts require drying at approximately 120 °C prior to painting.

With regard to painting, magnesium parts need to receive a minimum of 60 μm of coating, either cathodic electrocoat plus top coat or a powder coating in order to be protected sufficiently against corrosion.

Magnesium parts that are not exposed to weather conditions and do not need to receive an anticorrosion pretreatment and painting can be processed through a car body pretreatment line without interfering with the phosphate process. As the phosphate solutions typically contain free fluoride also, magnesium will be partly dissolved and precipitated in the form of magnesium fluoride.

3.3.8
Pretreatment of Plastic Parts

Plastic parts are either pretreated and painted off-line like fascia, trim parts, and so on, or on-line like plastic hang-on parts, that is, fender, wheel-house for SUV, trunk lids, and so on.(see Section 9.1).

Plastic parts are mostly produced from thermoplastic materials by injection molding, either without release agents or when necessary, with either internal or external release agents. The task of plastic cleaning is to provide a clean surface for the painting process with the proper surface tension and surface energy to guarantee wetting of the paint and adhesion of the coating.

For plastic cleaning, either mildly alkaline cleaners or acidic cleaners based on phosphoric acid or organic acids are used. They typically contain sequestering agents, hydrotropic agents, surfactants, and alkanole amines in case of alkaline cleaners.

Parts are treated in conveyor lines by a so-called *power wash* process (Section 9.1.4) which may comprise about four stages, consisting of two cleaning stages at 40–50 °C, and two stages of city water rinsing and an exit demineralized water rinsing (Table 3.4).

As painting with waterborne paints requires a high surface tension, which cannot be obtained by simple cleaning of low surface tension plastics, additional steps like plasma treatment and flame treatment [8] are mandatory to achieve the required coating adhesion.

3.4
Car Body Pretreatment Lines

Car bodies are pretreated mostly in lines with 8–12 stages, made up of 2–3 degreasing stages, an activation stage, a phosphating stage, and, in many cases, a passivation stage.

The remainder are rinsing stages between the degreasing and activation stage, after the phosphating stage, and after the passivation stage.

Table 3.4 Sequence of treatment for continuous conveyorized line for plastic cleaning[a]

Stage Number	Task	Tank (m³)	Temperature (°C)	Pressure (bar)	Treatment time (min)	Concentration/ Conductivity
1	Cleaning	5	50	1.5	2	10–15 g l^{-1}
2	Cleaning	3	40	1.5	1.5	3–5 g l^{-1}
3	Rinsing	3	35	1.0	1.5	350 µS cm^{-1}
4	Rinsing	3	30	1.0	1.5	<50 µS cm^{-1}
–	Final rinsing	–	RT	0.8	0.5	<20 µS cm^{-1}

a Data apply for:
  Materials: R-TPU/R-RIM/PC-PBT/PP-EPDM
  Throughput: 1500 m² per day
  Cleaner: Acid cleaner.

Table 3.5 Typical sequence of treatment in a spray/dip pretreatment line and the respective temperature and time conditions

Number of stages	Stage/Step	Temperature (°C)	Time (s)
1	Flood wash	Ambient-40	60
2	Spray degrease	50	90
3	Dip degrease	50	15–120–15[a]
4	Spray rinse	w/o heating	30
5	Spray rinse	w/o heating	30
6	Dip activate	<50	In and out
7	Dip phosphate	50	135–15[a]
8	Spray rinse	w/o heating	30
9	Dip rinse	w/o heating	In and out
10	Spray passivate	w/o heating	30
11	Dip DI rinse	w/o heating	In and out
12	DI water spray	w/o heating	30

a 15–120–15 = 15″ entrance spray, 120″ roof under, 15″ exit spray; 135-15 = 135″ roof under, 15″ exit spray.

Today, most lines worldwide are designed as horizontal spray-dip lines, with a combination of spray and dip degreases, spray and dip rinses, spray or dip activate, dip phosphate, spray and dip rinses, spray or dip passivation followed by demineralized water rinse. A typical sequence of the stages is shown in Table 3.5.

The schematic outline of a pretreatment line in Figure 3.8 demonstrates the basic material flow and handling operation of the tanks for each stage of the process.

The exact sequence of the pretreatment process, including number of stages, spray or dip application, type of dip application, array of entrance and exit sprays,

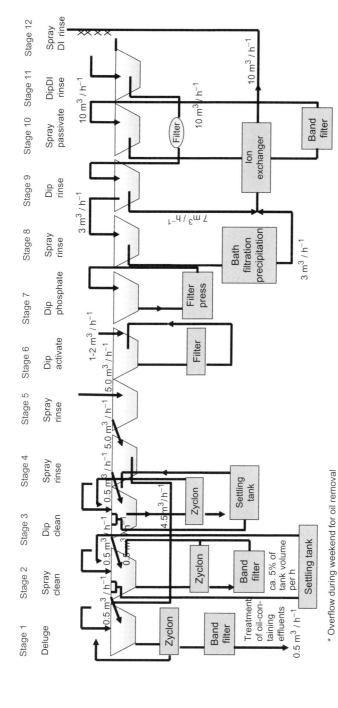

Fig. 3.8 Schematic outline of a car body pretreatment line.

treatment times, length of transfer zones ('neutral stages'), application temperatures, construction materials, and so on, is part of the specifications issued on this subject by the automotive manufacturers and is mostly outlined in a 'bill of process' or 'standard of operation (SOP)'.

The following types of lines for the standard continuous manufacturing process exist:
- spray line
- horizontal spray/dip line with dip phosphate, in some stages spray application
- RoDip3
- Vario shuttle

Besides these continuous lines, in the automotive industry there exist a small number of cycle lines, especially for truck and special car manufacturing.

3.4.1
Spray Lines

Spray lines have a considerably smaller space demand than horizontal dip lines, but have the disadvantage that box sections and possibly other interior parts of the car body are not properly degreased and covered by the conversion layer. The carryover of oil through the pretreatment line and subsequent contamination of the cathodic electrocoat tank is considerably higher than for spray/dip or dip lines. This is caused by the fact that the box sections are normally difficult to access by spraying, and are therefore not very well cleaned, so the remaining oil and contaminants are finally washed out for the first time in the electrocoat tank.

3.4.2
Continuous Horizontal Spray/Dip Line

This is by far the most common type of equipment around the world. The advantages are that the box sections can be cleaned and phosphated. A dense phosphate layer is deposited by the dip process, which might not require a passivation step. The disadvantages are the high investment cost and large space requirement (length of 170–230 m), depending on the bill of process.

Most recently, several spray/dip lines have been designed with a spray phosphate stage for treating vehicles comprising 100% zinc-coated steel sheet. It is assumed that proper degreasing of the box sections is sufficient and that no conversion coating will be required.

3.4.3
RoDip3 Line

In RoDip3 lines, the bodies are carried through the line by a conveyor in a horizontal line, but are immersed into the tank with a 180° turn, that is, they are turned upside

3.4 Car Body Pretreatment Lines

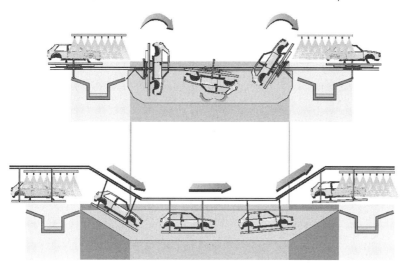

Fig. 3.9 RoDip3 line (source : Dürr).

down for treatment in the tank, then turned again through 180° at the exit of the tank to complete a 360° turn, and become upright again [9].

The advantages of the system are low carryover, efficient rinsing, and fewer immersion marks compared to current horizontal lines. As the phosphate process of aluminum results in enormous generation of sludge, there is a risk of sludge sedimentation on horizontal parts of the car body; this equipment minimizes this phenomenon on visible defects on horizontal areas. Furthermore, the space occupied and the tank volume are lower compared to standard horizontal lines (Figure 3.9).

3.4.4
Vario Shuttle Line

Vario Shuttle lines do not use the conveyor system; instead, they use a motorized transport and immersion system ('Vario Shuttle') to take the individual car body through the pretreatment and, in most processes, also the electrocoat line (Figure 3.10). There exist single-loop and double-loop systems, depending on whether the same Vario Shuttle takes the car body through the pretreatment and the electrocoat stages or not. The advantages described for the RoDip3 system also apply to the Vario Shuttle system [10–12].

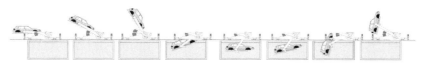

Fig. 3.10 Vario Shuttle line (source : Eisenmann).

3.4.5
Other Types of Lines

In addition to the above equipment types, two other process lines are the cycle box spray lines and the cycle immersion lines. Both have a fairly low capacity of up to 25 units h^{-1} and are not commonly used for passenger car body pretreatment.

Cycle dip lines normally consist of the same number of stages as continuous immersion lines. Adequate bath agitation and suitable measures to prevent the surface from drying during the intermediate stages are necessary for achieving a uniform and dense phosphate coating.

Cycle box spray lines are typically designed to have three boxes. The first box serves for degreasing and rinsing, the second box for phosphating and rinsing, and the third one for passivation and d.i. water rinsing. Like cycle immersion lines, cycle box spray lines are designed for low production volumes and have several major disadvantages. These are mixing/dissolution phenomenon of process solutions, high material consumption, difficult bath operation due to significant drag off, and drying of the body in intermediate stages. Quite frequently, drying of the body can cause electrocoat marks.

3.4.6
Construction Materials

Construction materials for the pretreatment line are typically acid-resistant steels, that is, according to DIN 1.4541 (BS 304) or DIN 1.4571 (BS 316). For the degreasing stages through activation stage, materials like DIN 1.4301 or mild steel are also appropriate and are being used. Supporting pipes for the spray nozzles (risers) and heat exchangers in the phosphate stage are of stainless steel DIN 1.4571. Spray nozzles also should be made of stainless steel or, in case of less abrasion, of plastic.

For lines applying free fluoride in the phosphate stage, heat exchanger made from DIN 1.4539 and dosing tubes made from acid-resistant plastic are recommended. For more detailed information, the reader is referred to the technical data sheets of the pretreatment suppliers.

3.4.7
Details on Process Stages

3.4.7.1 Precleaning Stage
Frequently, the body-in-white is passed through a precleaning stage where part of the contaminants and dirt are removed before the vehicle enters the pretreatment line. Manual precleaning decks, although popular in the past, have almost disappeared and have been replaced by automatic equipment. There are several types of equipment based on different engineering favorites, which are described below:

Bodywasher:
- separate spray equipment (1–2 stages) followed by drying;

- in some cases operated at high pressure or with contour spray;
- typically operated with a so-called neutral cleaner providing bare corrosion protection for storage;
- usually part of the body shop;
- in few cases combined with 'Hilite' operation for visual inspection and repair of body-in-white defects.

Flushing ('Deluge'):
- high volume spray-flooding;
- typically integrated in pretreatment line.

Hot-water soak:
- spray/half-dip at phosphate line entrance;
- fed from rinsing after degreasing;
- common practice in Japanese assembly plants.

Brush wash:
- spray predegrease combined with brush washing by machines similar to those used in car washes.

Spray predegrease:
- combined with rotation of the body-in-white to increase Effectiveness.

3.4.7.2 Spray degrease

The main task of spray degreasing – in a spray/dip line – is to clean the exterior of the car body. The characteristic data are as follows:
- controls: free alkalinity (FA), total alkalinity (TA), conductivity, pH
- dump frequency: daily or weekly
- heating: tube bundle exchangers or plate heat exchangers
- time: about 60 seconds
- temperature: 40–60 °C
- oil removal: ultrafiltration, centrifuge, static oil separators or thermal oil separator.

3.4.7.3 Dip Degrease

The main task of dip degreasing is to clean the interior and box sections of the car body. Characteristic data are as follows:
- controls: FA, TA, conductivity, pH
- dump frequency: monthly, annually or never dump
- heating: tube bundle exchangers or plate heat exchangers
- time: app. 120–180 seconds

- temperature: 40–60 °C
- oil removal: Ultrafiltration, centrifuge, static oil separators, or thermal oil separator
- spray application: at entrance and exit
- consumption: 0.2–0.4 kg of powder cleaner for 100 m^2 body-in-white; in few cases up to 1.0 kg; surfactant consumption typically 10% relative to the builder that is, 0.02–0.04 kg per 100 m^2 body.

3.4.7.4 Rinsing

The main task of rinsing is to remove excessive chemical from the metal surface and hence to avoid contamination of the following stages, which may cause severe problems in the chemistry of the process.

3.4.7.5 Activation

The main task of activation is to deposit Ti phosphate colloids on the cleaned metal surface, providing a denser and faster growth of the conversion layer. Characteristic data are as follows:

- controls: conductivity, pH
- dump frequency: weekly to monthly; in some cases biannual
- heating: no heating
- agitation: continuous, even when production stops to avoid settling and agglomeration of Ti phosphate colloids
- time: in/out to 60 seconds
- temperature: <40 °C, for selected products <50 °C
- spray application: at entrance and exit
- consumption: 0.03–0.1 kg of powder activator for 100 m^2 body-in-white.

3.4.7.6 Phosphating

The main task of phosphating is to deposit a thin, dense, and uniform conversion layer on the cleaned and prepared metal surface. Characteristic data are as follows:

- controls: free acid (FA), total acid (TA), Zn concentration, accelerator free fluoride (where applicable)
- dump frequency: never dump
- heating: plate heat exchanger; $\Delta T < 10$ °C
- time: approximately 180 seconds (120 seconds in Japan, for trials with optimized process 90 seconds)
- temperature: 35–55 °C
- sludge removal: filter press, pressure band filter, tilted plate clarifier
- spray application: none
- consumption: depends on substrate
  - ~15 g m^{-2} for cold-rolled steel

- $\sim$8 g m^{-2} for zinc-coated steel
- $\sim$6 g m^{-2} for prephosphated zinc-coated steel
- $\sim$20 g m^{-2} for aluminum (in case of crystalline coating)

this results in approximately 1.0 kg replenisher for 100 m^2 body with 80% zinc-coated steel and 20% cold-rolled steel sheet for a high production car body line of >200 000 vehicles per annum.

3.4.7.7 Rinsing
The main task of rinsing is to remove excessive chemical from the metal surface and hence to avoid contamination in the following stages, which would eliminate the effectiveness of the passivation step.

3.4.7.8 Passivation
Zr-based passivation was introduced during the late 1980s, and should replace the CrVI-based passivations completely owing to environmental constraints and the End-of-Life Vehicles (ELV) directive.

3.4.7.9 Deionized water rinsing
Requirement for conductivity of water varies slightly from one automotive manufacturer to another and is typically in the range of 15–50 μS cm^{-1} at 20 °C.

3.4.7.10 Entering the Electrocoat Line
In order to avoid structural defects in the electrocoat, the pretreated car body should enter the electrocoat tank either completely dry or completely wet. If the car body is dry and needs to be wet, it is usually wetted by either d.i. water or ultrafiltrate of the electrocoat tank prior to entering the tank. Addition of surfactant to the d.i. water occasionally helps to improve the situation and reduces entry marks completely.

3.5 Properties and Specifications of Zinc Phosphate Conversion Layers

The requirements and specifications for zinc phosphate conversion layers vary form one vehicle supplier to another and are somewhat different for European, North American, Japanese, and Korean car manufacturers.

Table 3.6 gives an overlook of the three important technical criteria, that is, crystal size, coating weight, and P ratio (crystal phase information, that is, phosphophyllite $Zn_2Fe(PO_4)_2 \cdot 4H_2O$ content in phosphate coating; remainder is hopeite $Zn_2Fe(PO_4)_2 \cdot 4H_2O$) and their specifications on different substrates. The latter is only used by some of the Japanese car manufacturers. Most variation between the car manufacturers can be seen in the Nafta region and Japan. The tightest specifications can be found at some of the European manufacturers. Not all of them have issued specifications for crystal size and coating weights, however.

Table 3.6 Specifications of worldwide vehicle manufacturers on zinc phosphate conversion coatings (immersion application)

	Crystal size (μm)			Coating weight (g m^{-2})			P ratio (%)
	Fe	Zn	Al	Fe	Zn	Al	Fe
Europe– A	~10	~10	~10	1.5–2.5	1.5–2.5	1.5–3.5	
– B	<10	<10	<10	2.0–3.5	2.0–3.5	2.5–3.5	
Nafta– A	2–10	2–10	2–10	1.6–3.3	2.2–3.7	1.1–2.2	
– B	<25	<25	<25	1.3–3.8	1.6–4.3	1.6–4.5	
Japan– A	<20	<20	<20	2.0–3.0	2.0–3.0	2.0–3.0	
– B	3–7	<20	–	2.0–2.5	3.0–5.0	>1.0	>85
Korea– A	2–10	2–10	–	1.8–3.0	2.0–4.0	–	

a A,B = Different car manufacturers.

3.6
Environmental Legislations

Environmental legislations for metal surface treatment chemicals vary for different parts of the world and may be summarized as follows:

Surfactants: OECD 301, effective October 2005; 60% total degradation within 28 days in water treatment plants.

Borates: not permitted in cleaners for metal working industry in Denmark and Italy.

Nickel: limit for effluents or waste in Germany (40. Abwasserverwaltungs – Vorschrift, Anhang 40) and many other countries: <0.5 ppm; mostly recently, the legislation has changed in Germany and elsewhere regarding the water hazard class (WGK); phosphate bath solutions have been classified in WGK 2 (formerly WGK 1) and phosphate concentrates in WGK 3 (formerly WGK 2); the World Health Organization (WHO) intends to introduce 'zero emission of nickel' by the year 2016.

ChromeVI: according to the ELV directive of the European Union, only 2g of hexavalent chrome per total vehicles are permitted until 2007.

3.7
Outlook

The requests of the automotive industry to reduce investment and operating cost, to reduce complexity, and to improve environmental compliance of car body pretreatment has stimulated R&D to develop new techniques for substituting the zinc phosphate process.

Major progress has been made over the last one or two years so that introduction of the new techniques into the automotive industry may be feasible as early as year 2007.

These new zinc phosphate replacement processes are based on inorganic-organic components, that is, transition metal fluorides combined with polymers and/or silanes and have the following advantages [13, 14]:

- free of hazardous metals
- operation at ambient temperature (no heating required)
- virtually no sludge production
- fast reaction on all substrates
- smaller footprint (no activation stage or passivation step necessary) compared to zinc phosphating line
- high flexibility toward treating mixtures of cold-rolled, zinc-coated or aluminum
- easy bath make-up and control
- closed-loop operation feasible
- drop-in replacement.

Currently, line trials are conducted with the new processes in the general industry and the automotive components Industry.

References

1 *AutomobilProd* (**2001**), 126.
2 Alsmann, M., Reier, T., Weiß, V. **2000** *Journal für Oberflächentechnik*, **40** (9), 64.
3 Rausch, W. (**1990**) *Phosphating of Metals*, SAM International Finishing Publications.
4 Depnath, N.C., Bhar, G.N. (**2002**) *European Coatings Journal*, 46.
5 Rausch, W. (**1988**) *Proceedings Galvatec*, Tokyo, September 1988.
6 Kiene, J. (**2004**) *Metalloberflache*, **58** (4), 52.
7 de Brondt, T. (**2004**) *Metalloberflache*, **58** (4), 42.
8 Hauser, W. (**2002**) *Journal für Oberflächentechnik*, **42** (12), 38.
9 Milojevic, K., Kreuzer, B., Koerbel, U., Schmidt, D. (**1999** *Proceedings Surcar XIX* Cannes.
10 Klocke, C. (**2003**) *Industrial Paint & Powder*.
11 Weinand, J. (**2003** *Proceedings Surcar XXI*, Cannes.
12 Kochan, A. (**2004**) *Automotive News Europe*, 20.
13 Hosono, H., Nakayama, T., Huys, P. (**2005**) *Proceedings Surcar XXII* Cannes.
14 Gehmecker, H., Hueber, M. (**2005**) *Proceedings Eurocar* Barcelona.

4
Electrodeposition Coatings

Hans-Joachim Streitberger

4.1
History and Introduction

The automotive industry in the 1960s initiated the commercialization and introduction of this coating technology driven by many factors. Under them safety, environmental, and processing aspects were the most important ones. The waterborne or solvent-borne dip tanks that were used till then were either dangerous in terms of fire hazards owing to high solvent emission even in the case of waterborne products or had a lot of processing problems e.g. cavities caused by boiling out and accuracy issues due to film thickness.

One of the driving companies was Ford which supported the development of paints and the coatings process heavily [1]. The first tanks were filled in USA followed by Europe in the mid 1960s. Until 1977 the anodic process had been used widely in automotive body priming throughout the world, which was then almost completely changed to the cathodic process, further implemented to nearly 100% today. Reasons for this complete change were improved corrosion protection by better resin chemistry, passivation of the substrates instead of dissolution during the deposition process, and a more robust application process provided by a predispersed 2-component feed technology.

This impressive penetration of the electrocoating process into the market was boosted further by the introduction of the ultrafiltration process, which enhanced the material usage to nearly 100% [2].

Today the film thickness and material usage per body has been optimized to 20–22 µm on the outside and about 3 kg solid content per 100 m^2 surface area, respectively. The solvent level in the electrocoat tanks is below 0.5%, so that the final rinsing with d.i. (deionized) water in the multistage cleaning step is not necessary any more. The reliability of the process has led to the fact that in most paint shops of the automotive industry only one electrocoat tank feeds all top-coat lines, coating sometimes more than 1500 units a day.

Automotive Paints and Coatings. Edited by H.-J. Streitberger and K.-F. Dössel
Copyright © 2008 WILEY-VCH Verlag GmbH & Co. KGaA, Weinheim
ISBN: 978-3-527-30971-9

Table 4.1 History of dominant electrodeposition technology for the automotive industry

Time frame	Technology	Chemistry	Film thickness[a] (μm)
1964–1972	Anodic	Maleinized natural oils	25
1972–1976	Anodic	Maleinized oligobutadienes	25
1976–1984	Cathodic	Epoxy-polyurethanes	18
1984–1992	Cathodic	Epoxy-polyurethanes	35
1992–today	Cathodic	Epoxy-polyurethanes	20

a Cured film.

4.2
Physico-chemical Basics of the Deposition Process

Electrocoat paints are aqueous dispersions consisting of typical paint ingredients like film forming agents (resins), pigments, extenders, additives, and some solvents. The dispersion has to be stabilized by electrostatic forces. Negatively charged paints, normally called anodic electrocoatings or anodic electrodeposition (AED) coatings are deposited at the anode, positively charged paints, called cathodic electrocoatings or cathodic electrodeposition (CED) coatings at the cathode. To enable the paint to be deposited, the object must be immersed in a tank filled with the electrocoat and connected to a rectifier as the corresponding electrode. The counterelectrode must be immersed at the same time and a direct charge must be provided by applying sufficient voltage of more than 300 V on the technical scale.

Under these suitable conditions the water is split at the electrodes into hydrogen and oxygen according to Equation (1):

$$H_2O = H_2 + 1/2\, O_2. \qquad (1)$$

The particular reactions at the anode and cathode are as follows:

$$\text{cathode reaction:}\ 2H_2O + 2e^- = H_2 + 2OH^-, \qquad (2)$$

$$\text{anode reaction:}\ H_2O - 2e^- = 1/2\, O_2 + 2\,H^+. \qquad (3)$$

From Equations (2) and (3) it becomes evident that double the volume of gas is being generated at the cathode in comparison to the anode. But the most important factor in the electrocoating process is the change of pH in the diffusion-controlled layer at the electrode. The thickness of this layer can vary between 0–80 μm depending on the speed of application of the electrocoat material at the surface of the electrode relative to the specimen to be coated. This has been explored by Beck [3] and has been established as the basic reaction for the electrocoat process at the anode (see Figure 4.1) and the cathode.

The pH or C_H^+- and C_{OH^-}-concentration at the electrodes can be calculated based on Faraday- and Fick-law respectively. According to the Sand or Cottrell [4]

4.2 Physico-chemical Basics of the Deposition Process

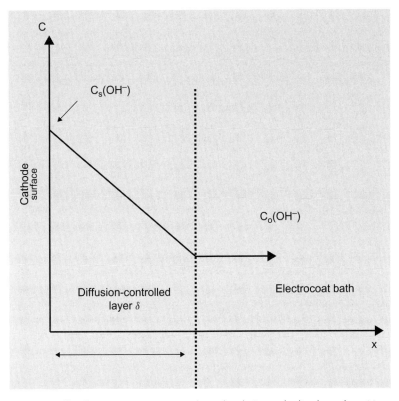

Fig. 4.1 Profile of OH⁻-concentration at electrodes during cathodic electrodeposition.

Equation (4) for OH⁻ and potentiostatic application conditions which uses the diffusion constants of H⁺ or OH⁻ and the electrical current density:

$$C_{OH^-} = 2j/F \cdot (t/\pi D)^{-1/2}, \tag{4}$$

where j is the current density in A cm^{-2}, D the diffusion constant of OH⁻-ion in cm² s⁻¹ and F the Faraday constant. Without considering the influence of the electrocoat migrating and diffusing into the diffusion-controlled layer, this results in pH of 12 at the cathodes and 2 at the anodes. In the presence of the electrocoat the pH-profile in the diffusion layer differs from the ideal conditions of pure water decomposition. The reason for this is the fact that the electrocoat material coagulates according to the pH-depending equilibrium of the carboxylate-ions in the case of an anodic depositable electrocoat in the diffusion layer according to Equation (5) and for a cathodic one according to Equation (6):

$$\text{at the anode: R--COO}^- + \text{H}^+ \rightarrow \text{R--COOH}, \tag{5}$$
(dispersible)(insoluable)

$$\text{at the cathode: } R_3N^+H + OH^- \rightarrow R_3N + H_2O. \tag{6}$$
(dispersible)(insoluable)

That the diffusion and migration of the dispersion particle into the diffusion layer plays an important role in the deposition process has been demonstrated in experiments by Beck [5]; he used rotating disc electrodes, at a certain speed of the discs there is no build up of the pH-profile and no deposition occurs. In those experiments it was also found that building up the pH-profile takes some time so that deposition starts after a certain induction time τ, when the OH^- or H^+-ion concentration has reached the level to start coagulation of the electrocoating. This critical ion-concentration is dependent on the applied voltage or the current density at the electrode and the charge of the electrocoating material. Under specific conditions the critical H^+ or OH^- ion concentration is related to the current density which is called the minimum current density j_m. This value is about 0.15 mA cm^{-2}. It could be shown that τ will be reduced by increasing j^2. The product of $j\tau^{\frac{1}{2}}$ is a characteristic value for each coating material [6]. The necessary time for starting the deposition process is less than a second under industrial application conditions [7].

The following film growth after the start of the deposition process takes about 2–4 min under industrial application conditions. It defines the final thickness of the film and the throwing power. The throwing power for the coating of inner segments is a result of the dynamics and resistivity build up of the film during deposition. This dynamics of film growth can be described by potentiostatic experiments [8, 9] which reflect the normal industrial electrodeposition conditions.

Applying voltage to a standardized electrodeposition tank, with defined electrode areas and electrode distances, increases the current A or current density A cm^{-2} to a certain level after which the current decreases. The increase is connected to the build up to the critical level of ion-concentration, the decrease to the resistivity of the film. The current will not go down to zero (see Figure 4.2.). During the applied

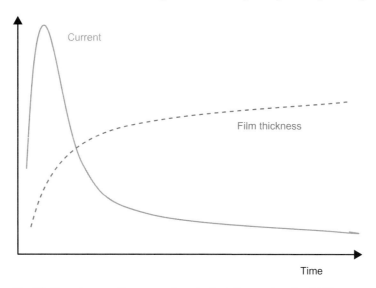

Fig. 4.2 Current versus time curves in potentiostatic experiments of CED.

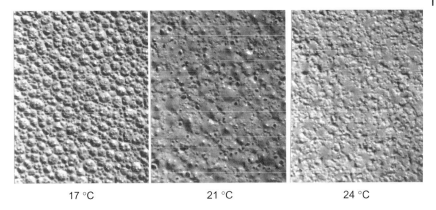

17 °C   21 °C   24 °C

Fig. 4.3 Morphology of cathodical deposited coatings at different bath temperatures.

voltage the dispersion particles of the electrocoat paint migrate as charge carriers to the electrical field of the diffusion-controlled layer as a result of the interaction of the forces of electrical attraction and retarding friction at a constant rate of about 10^{-4}–10^{-5} m s^{-1} [10]. This generates up to 20 µm film thickness in one minute. It has also been shown that the build up of the thickness of the film is correlated to the viscous behavior during the deposition process [8] and to the T_g of the film [11]. The viscosity again is dependent on the type of resin and the temperature of the bath.

Inside the deposited film the increased electrical field up to approximately 10^5 V cm^{-1} generates electro-osmotic pressure to remove water from the electrode and makes the deposited film more dense. After 2–4 minutes the film sticks to the substrate and has only 5–10% water and cannot be washed off. The film has a characteristic morphology due to the fact that hydrogen and oxygen also develop at the electrodes (see Figure 4.3). The gases form bubble-type structures in the high viscous deposited film depending on the bath temperature which needs to be flown out during baking and cross-linking [8].

The film growth Δd can approximately be described according to Equation (7) [8]:

$$\Delta d = 1/E\, c\, (j_t - j_m) \Delta t, \tag{7}$$

where d is the film thickness after curing of a given system, t is the time, c is a paint specific constant and j_t the current density in Acm^{-2} at the time t, j_m the minimum current density for depositing the electrocoat dispersion and E the electrochemical equivalent in C cm^{-3}. If the current density j_t falls below j_m due to the increasing resistance of the deposited film (see above), no further deposition process takes place (Equation 8).

$$\Delta d/\Delta t = 0 \text{ at } j_t < j_m \tag{8}$$

The current density at the time t also can be expressed according to the Ohm's law:

$$jt = (kF)_t \qquad (9)$$

where k is the specific conductivity of the system and F the field strength in V cm^{-1}. Combining Equations (7) and (9) we obtain the general formula for film growth, Equation (10):

$$d/t = 1/E\,c\,((kF)_t - j_m) \qquad (10)$$

Two distinct factors contribute to k:

$$k = k_{wet} + k_{bath} \qquad (11)$$

k_{wet} is the conductivity of the freshly deposited wet film and k_{bath} is the conductivity of the tank. Due to the fact that k_{wet} decreases during the application process Equation (10) gives only the first approximation for film growth. The dynamics of k_{wet} can be measured by experiments.

Another important factor for controlling the electrodeposition process is the temperature, T, connected with the fact that the voltage increases in the film during the deposition, which builds up the resistivity significantly. The consequence is a rise in temperature [12] in the film which needs to be brought down by convection in the tank, for example.

Of great practical importance is the Coulomb-efficiency or electrochemical equivalence of the deposition process. This can be easily measured by deposition experiments. In a given laboratory experiment the total current I, the time t and the volume in cubic centimeter of the deposited and cured film need to be monitored. With those data the electrochemical equivalent E can be calculated according to Equation (12):

$$E = Q\,\text{cm}^{-3}. \qquad (12)$$

In modern laboratory equipment these data are registered automatically in each deposition experiment. The final resistance, the Coulombs, and the bath conductivity are the data required to calculate the equivalent and specific wet film conductivity on the basis of the cured film thickness. E is normally in the range of 40–70 C cm^{-3} and varies from paint to paint. It is also reported in coulomb per gram which can be measured more easily or with the reverse quotient that is cubic centimeter per coulomb or gram per coulomb.

Based on the well-understood process of electrodeposition of paints, many attempts have been made to calculate the film thickness of complex parts like the automotive body prior to coating to improve the throwing power of constructive elements for better corrosion protection and less testing demands [13].

The basic electrical data of a commercial CED tank are shown in Table 4.2.

Table 4.2 Typical electrical and physical application data for CED tanks

Parameter	Voltage (Direct current)	Minimum current density	Bath temperature	Conductivity of the bath (20 °C)	Conductivity of the wet film
Measure	V	A m^{-2}	°C	µS cm^{-2}	nS cm^{-2}
Value	300–450	0.8	28–33	1200–1800	1.5–3.0

4.3
Data for Quality Control

The electrocoat paints are controlled in their application performance by many physical, technical, and analytical data (see Figure 4.3). The most important factors for proper application are voltage and current density, bath temperature, pH of paint, paint conductivity in combination with solid content, and electrolyte concentration.

The chemical as well as the physical composition of the modern cathodic electrocoating defines the technical performance for the industrial application [14]. Because of the complex nature of the formulated coating the analytical data are difficult to obtain in terms of cost and time. In practice the solid content, ash content, pH, solvent content, and conductivity data are easy to measure, but the information is not enough for the physical application data to be evaluated at the same time. Before introducing a new generation of electrocoat the application profile will be evaluated on the laboratory scale for defining the application window. The complex nature of the interaction of each parameter makes it necessary to control running tanks with both analytical data and application tests in a specific laboratory equipment to ensure correct application behavior.

Usually daily measurements are taken for solid content, conductivity, and pH, while weekly measurements are for analytical data like acid and base values, solvent content, ash content, and confirmation of the application parameters according to a standard setting. The frequency depends on the speed of replenishment, the so-called *material turnover* or *throughput rate*.

Most tanks are designed for a theoretical turnover of two weeks to six months, which is the time for theoretical replacement of the solid content of the tank by the feed material. Biweekly controls are sufficient for tanks with more than two months of turnover, four-weekly for those with higher turnovers. Stability of CED can be more than six months.

The new generation of electrocoatings may run into problems like bacteria contamination due to their low solvent level and toxic free composition. In such cases it is advisable to control the d.i. water and other potential sources of contamination as well as run weekly bacteria counts or install an equipment for continuous measurement [15].

Table 4.3 Basic analytical bath data of a CED tank

Parameter	pH	Solid content (2h/180 °C)	Solvent content
Measure		%	%
Value	5.8–6.2	18–20	0.5–1.5

4.3.1
Voltage, Current Density, Bath Temperature, and Bath Conductivity

Voltage and time are the basic parameter for electrocoatings with respect to film thickness and throwing power. The composition of the paints is designed to show increased film build with increased voltage and to resist very high voltages thereby avoiding too much thickness and rupture defects in films. In a given system a voltage limit always exists beyond which rupture occurs and the deposited film deteriorates by cross-linking due to high temperatures either in the already deposited film or in the bath itself. This limit is calculated from laboratory tests and called the 'rupture voltage'. A basic rule says that for reasons of safety, the applied voltage must be at least 10% below the rupture voltage. Sometimes microcoagulation occurs close to the rupture voltage.

At the start of the deposition process, when the resistivity of the layer is still low, the high current flow or current density can also create defects during the immersion process. This is mostly related to wetting properties in combination with limited heat transfer so that marks and structures are preformed due to increased film and bath temperature in the diffusion layer.

All these effects are supported by the conductivity of the bath with respect to the electrocoating material and the bath temperature which raises the conductivity with increased temperature. If the conductivity is high the current density increases correspondingly at a given voltage and tends to create surface defects. Small electrolyte ions like sodium chloride or similar products, mostly contaminants of the pretreatment step, also give rise to side reactions at the electrodes and disturb the smooth build up of the electrocoat films.

4.3.2
Wet Film Conductivity

The wet film conductivity as measured in laboratory equipments is usually defined as the final k_{wet} at the end of the deposition process and describes therefore the resistivity of the wet film calculated on the basis of the dry film thickness. This is reasonable because the wet film cannot be measured easily and the ratio between the wet film thickness and the dry film thickness in a given electrocoat material is nearly constant. In any case this value provides information of the exact behavior of the coating with regard to the throwing power and film build. In the wet film conductivity exceeding the limit, high film build with low throwing power is the result.

4.3.3
Solid Content, Solvent Content, and pH

For controlling tanks the solid content, solvent content, and pH are relatively easy to measure but all data represent a complex mixture of different impacts to the performance of an electrocoat process. For example, the variation in solid content influences the bath conductivity, the application behavior affects applicable voltage, the film build, and the throwing power. Therefore it is mandatory to run a tank with certain solid content limits to keep the parameter set of the application as constant as possible. Using a predispersed 2-component feed system makes this task easier than the traditional underneutralized 1-component system, which needs to be dispersed batch-wise with the help of the excess neutralizing agent of the electrocoat bath. This step is difficult to be controlled and so the 2-component feed technology has been established in automotive coating tanks worldwide. The feed can be automated to be continuously replenished by controlling current consumption.

The solvent content has a very sensitive impact on the film build of the CED materials. Not every solvent has the same effect. It depends largely on the equilibrium distribution of the specific solvent between the continuous phase and the dispersed particle phase. The more the solvent is in the continuous phase the less the impact on increased film build is to be seen. The new CED generations avoid this sensitivity by carefully selecting the solvents for manufacturing the resins and reducing the lead in the dispersions as much as possible, if necessary by evaporating or stripping techniques after the dispersing step (see Section 4.4.2).

The pH of the solvent in the tank is a measure of the balance of acidic and caustic components in the tank and can easily be determined. The balance is dominated by the functional groups in the resin and the neutralizing agents. Most of the systems are buffer systems so that fast reaction time is needed when the value is outside the specification. Measuring the trend is the best way to overcome this problem. The pH can be influenced by the pretreatment chemicals in a detrimental way in the case of improper washing and rinsing of the complex bodies. In such a case conductivity and pH drifts to values outside the specification and corrections have to be made by replenishing ultrafiltrate with d.i. water. Typically decreased pH in a CED bath lowers film thickness so that higher voltages need to be applied with the danger of rupture effects.

Depending on the tank size and its material turn over, that is the speed of replenishment in terms of theoretical replacement of the solid content in days or months, the frequency of the bath control data are set. Generally at high turnovers between 3 weeks and 8 weeks it is useful to have a daily measurement of solid content, pH, and conductivity, a weekly check of ash content, acid, and base numbers, and the

4.4
Resins and Formulation Principles

4.4.1
General Remarks

Electrocoatings contain the same paint ingredients that most waterborne paints have. This includes a waterborne main resin, very often a special pigment grinding and stabilizing resin, cross-linkers, pigments, and extenders and additives for dispersing, stabilizing, and as catalysts for the cross-linking reaction. The main resin, including the cross-linker, makes up 70% of the solid content and defines therefore most of the properties of the paint and final coating film. The next important components are the grinding resin, the pigments, and extenders.

The nature of the resins for the electrocoating process can vary from natural oils to polybutadiene and polyacrylic to epoxy-, polyester and polyurethane but they all have some functional groups in their backbone which allows them to become ionic in the presence of neutralizing agents and makes them different from those resins being used in solvent-borne paints.

Anodic depositable resins have mostly carboxylic groups, neutralized by amines or potassium hydroxide, cathodic depositable resins have amine or sulfonium groups neutralized by acetic, lactic, or formic acid. The position and the number of these groups in the resin backbone decide if the resin is easily dispersible or not. It is essential that the micelle need not only the electrostatic stabilization in the continuous phase but also significant hydrophobic interaction inside the micelle itself. Therefore it is understandable that randomly distributed functional groups in a resin backbone are not as good as terminated groups or groups in a side chain (see Figure 4.4). The resins for automotive electrodeposition paints are mainly a captive market and are not sold by the respective manufacturer to the whole paint market.

Last but not least, there is the considerable knowledge of each paint manufacturer in selecting the additives for dispersion stability and the manufacturing process. The fact that so-called electrostatically stabilized secondary dispersions have to be produced [16] makes it important to look at the right type and level of functional groups as well as proper incorporation of the hydrophobic cross-linker into the micelle.

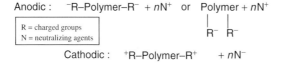

Fig. 4.4 Basic structure elements of anodic and cathodic resins for electrocoatings.

4.4.2
Anodic Electrodeposition Paints

The first important AED paints were formulated with unsaturated oils that were functionalized with maleic anhydride. The reaction of the oil with maleic anhydride was followed by semiesterification of the succinic anhydride derivatives. The remaining carboxylic groups were then neutralized mostly by alkyl–amines for the dispersion process. Those resins were cross-linked by oxidative radical polymerization of residual double bonds in the backbone using catalysts at 160 °C. The film thickness was about 25–28 µm. The surface was very smooth, but the stability of the electrocoating was often critical because oxygen in the spray mist of the washing zones created the danger of precross-linking reactions of the remaining carbon-double-bonds in the resin backbone with an unpredictable change of the coating performance.

Anodic electrodepostion coating based on these resins were first introduced in automotive coating in the 1960s. They were followed in the 1970s by specially developed polybutadiene resins replacing the natural oils before the cathodic electrocoating appeared.

For small parts unsaturated and saturated polyester as well as alkyd resins played some role. These resins could be easily dispersed at a remaining acid index of 40–80 mg KOH/100 g resin solid after polycondensation of the backbone.

Cross-linking can be provided by addition of melamine or phenolic resins.

The paints consist of the resins, pigments, extenders, additives, catalysts, and solvents. Due to the nature of the commonly used 1-component-feed system the main resin is also the dispersant and the solvent content of the feed composition is between 30 and 50%, which leads to a solvent level in the baths of 10% and more.

Recently 2-component feed systems with low solvent contents have also been proposed for industrial coating processes [17].

4.4.3
Cathodic Electrodeposition Paints

For automotive body tanks as well as for small part coating for best corrosion protection CED is mostly in use today. The basic advantages come from the deposition process, that is no dissolution of iron, zinc, or aluminum and the respective inorganic pretreatment and also from the chemical nature of the resins.

As explained the functional groups for AED paints are carboxylic groups, which can be saponified easily under alkaline conditions. The CED paints are attacked much less by alkaline conditions because of the chemical nature of their functional groups and their backbone chemistry. The processing advantage and the chemistry provide the precondition to introduce galvanized steel and aluminum components as integral parts of the body into the body shops (see Chapter 2).

The first generation of commercially available CED of the CATHODIP®-type (see Table 4.1) were designed for good corrosion protection and alkaline resistance because galvanized steel and cold rolled steel are being used side by side in body

Fig. 4.5 Resin backbone of early CED paints.

construction in the automotive industry. The main resin backbone (see Figure 4.5) was a polyester modified epoxy chain with amine modification at the chain ends [18].

Molecular weight M_w was about 3000–5000. The polyester modification was necessary to achieve enough flexibility in the cross-linked film to fulfill the different chip resistance requirements in the automotive industry. The amines were introduced either by simple reaction of epoxy-groups with primary, secondary, or tertiary amines resulting in secondary, tertiary, or quaternary amines. For better solubility it is preferred to use primary amines which could be introduced by the reaction of the epoxy-groups with ketimine groups which were hydrolyzed by water to primary amines later during the dispersing process [18]. Before dispersing the resin was blended with a blocked isocyanate cross-linker with the basic isocyanates of toloylene-diisocyanate (TDI) or diphenyl-methane-diisocyanates (MDI). These cross-linkers were incorporated in the micelle and stabilized by hydrophobic interaction. Blocking agents were primary alcohols deblocking at 160°C providing a safe cross-linking reaction at about 180°C for 20 minutes. The weight loss during baking varies depending on the type of cross-linker, the level of the cross-linker in the paint, and the solvent content. The weight loss can reach up to 20% on the basis of an air dried deposited electrocoat film. The material loss is important in two ways: first it affects the film surface with respect to smoothness and appearance, second high losses represent low material efficiency. The lost compounds do not contribute to the film build and are therefore only transfer media which are normally not accounted for directly when solids are measured at baking conditions. In general low weight losses have better material efficiency because more compounds put into the resin manufacturing vessel will contribute to the electrocoat film performance.

The respective electrocoat formulae were based on chromic pigments, lead, tin-based catalysts, and about 10–15% of solvents, which belong today to the class of hazardous compounds. The development of the following generations in the early 1980s was focused on higher film build up to 30–35 µm, high smoothness of the surface, and less solvents but did not change their basic chemistry. The target of the North American car manufacturers was to replace the primer surfacer by direct application of the newly introduced 2-layer top coat over the electrocoat. Severe problems with the UV-stability of the epoxy-based electrocoat leading, even after a few months to top coat delamination led to the quick reintroduction of a primer surfacer. In many cases this was a powder primer surfacer based on UV-resistant acrylic formulations and due to the VOC restrictions set by the American Environmental Protection Agency (EPA). With the introduction of the so-called high-film-build cathodic-electrocoat material, low solvent resins became standard. The resins were manufactured in the traditional way by building the resin backbone, mixing the cross-linker, and then dispersing the mixture into a separate tank. This dispersion was, in a further step, set under vacuum at elevated temperatures to distil off water and solvent. These were finally replaced by water

Table 4.4 Typical material data of a 2-component cathodic electrocoating as delivered for the automotive industry

Properties	Dimension	Resin-component	Paste-component
Solid content(2h/180 °C)	%	33–40	50–70
Solvent content	%	2–4	0–10
pH		5,5–6,0	5,8–6,5
Pigment to binder ratio		–	2,5–4,0
Spec.conductivity (20 °C)	S/cm	1,0–1,8	1,8–2,8
Feed ratio	by weight	3–7	1
Storage stability (0–30 °C)	months	6–12	3–6

to adjust to the suitable solid content to about 35–40%. This process is also called 'stripping'.

At the same time the most hazardous solvents like ethylcellosolve (ethylene glycol) have been replaced by less harmful ones.

Today's formulations are concentrating on high throw power, ideal film thickness of 20 µm, being free of lead and tin, having lesser solvents and less oven loss (see above) and as low as possible a specific film density for better material efficiency [19].

Commercial products are for example Enviroprime of PPG and CATHOGUARD 500® of BASF Coatings (see Table 4.4). They also provide improved corrosion protection even on Cr-, Ni-, and NO_2-free pretreatment systems (see Chapter 2). The new target of the development, after reducing the total solvent level, was building resistance to microbiological contamination. This gained importance owing to the development of low solvent-containing electrocoating which had less than 5% of organic solvents in their resin and pigment dispersions and so were prone to microbiological attack. In accordance with the newest regulatory requirements contaminant resistance can be achieved only by inherent incorporation or addition of biocides and avoiding bacteria penetrating the coating via d.i. water.

4.5
Film Performance of Cathodic Electrocoatings

4.5.1
Physical Film Data

The baking conditions for achieving best performance of the cathodic electrocoating in the automotive industry may vary somewhat. The baking window, especially, can be different. Most of the products specify a so-called *metal temperature* of 10 minutes at 175 °C as the standard condition. The minimum then can be 10 minutes at 165 °C. Chip resistance and adhesion are the most sensitive to baking conditions. Corrosion protection is less affected.

So-called *high film build electrocoats* of 30–35 µm film thickness were introduced in the mid 1980s to eliminate the primer surfacer in the coating process. Owing

Table 4.5 Physical film properties of CED

Film thickness	Spec. film density	Roughness R_a	Color
μm/μm 20–22 outside/8–20 inside	g/cm³ 1.3–1.5	μm 0.5–0.8	Gray/Black

to insufficient UV-protection this process was not successful, so today's cathodic electrocoats concentrate on their film thickness to achieve the best compromise between cost and quality. Actually the specific film density was another attempt at cost reduction. Early formulations had densities $>1.5\,\text{g cm}^{-3}$ while today's have about $1.3\,\text{g cm}^{-3}$.

The color is mostly in grayish tones. The purpose is to increase the hiding power of the next primer surface, which is usually in different colors. The surface roughness R_a (see Chapter 4 and Section 5.4) is about 0.5–0.8 μm depending on the substrate roughness.

4.5.2
Corrosion Protection

As mentioned earlier the electrocoating for automotive application is designed to give as much protection as possible against corrosion of a completed body construction for cars, trucks, and busses. This was a problem of great economic importance, both from a national point of view and from the car owners' point of view. Before introduction of the CED process and the increased usage of galvanized steel for automotive body construction, the cost of automobile corrosion was reported to be 16 billion dollars in USA alone [20]. Rusting of the steel sheets may start in inner sections that were not protected sufficiently either by less cleaning, pretreatment, or coating performance boosted by collection of salt and humidity in those sections finally resulting in perforation of the steel or may start from the outside, that is by chipping or scratching of the coating. Special focus is still placed on the edges of the steel panels which are weak spots of coating thickness. Even the process provides enough film thickness on the edges of the coating the deposited film withdraws more or less from the edges owing to the flow behavior during the baking and cross-linking process as can be seen from Figure 4.6. The later types of corrosion can be observed by the customer very easily and is called cosmetic corrosion, reducing the value of the car significantly because the appearance of the whole car deteriorates dramatically. The prevention of corrosion can be provided by three different main mechanisms performed by coating layers [21]:

1. electrochemical inhibitors,
2. sacrificial coatings,
3. barrier coatings.

Inhibitive pigments like chromates used in the first generations of CED passivate the metal surface by generating an oxide film. Sacrificial coatings such as pure

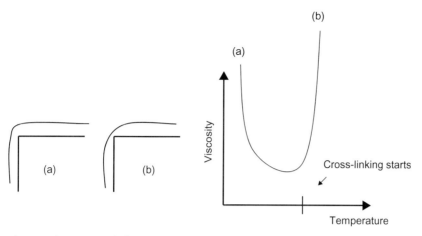

Fig. 4.6 Edge coverage before (a) and after baking (b) of a cathodic electrocoat and a respective viscosity/time-curve during baking.

zinc or zinc-rich primers uses a less passive anode (zinc) and render steel in a continuous electrolyte entirely cathodic and prevent steel from dissolution. Barrier coatings block the permeability of water and oxygen and reduce the fuel for the corrosion process [22]. Furthermore the strength of adhesion plays an additional role for good corrosion protection of all metals [23].

Legal enforcement has banned many inhibitive pigments in the past so that today the basic corrosion protection is provided by using a combination of sacrificial coatings and barrier coatings. At least critical areas of the automobile body or the complete body is made of galvanized or similar steel, which then is protected by the CED as a barrier. In case of a damage where the steel is exposed to the corrosive environment the preferential anodic dissolution of the zinc layer protects the steel as long as zinc is available. Most car manufacturers use hot dip galvanized steel with a thickness of Zn up to 15 μm for inner parts and electrolytically galvanized steel with thicknesses of the Zinc layer of 4–8 μm for the visible areas.

The many edges of an automobile body are critical areas. The electrocoat process provides same film thicknesses at the edges as on flat panels (see Figure 4.6). The task of an electrocoat material is to keep the edges covered during the baking process also when the film undergoes a viscosity drop and flows out to a smooth surface. An appropriate viscosity profile and the surface tension play important roles [24].

Together with an excellent pretreatment the corrosion protection against perforation is today granted for up to 12 or 15 years by many car manufacturers.

All the formulations of the latest generations of cathodic electrocoatings therefore are based on epoxy resins which provide the best adhesion to metallic substrates and corrosion resistance by passive protection owing to the barrier mechanism. Short term tests like the ASTM B117 salt spray test, the climate change test according to VDA (Verband der Deutschen Automobilindustrie) 641, and outdoor exposure tests with weekly salt spray demonstrate the superior performance of

Table 4.6 Typical data for delamination at the scribe of CED versus AED on steel and galvanized steel substrates without top coats in different corrosion tests

	ASTM B117 1000 h	VDA 641 10 cycles	Outdoor exposure[a] 1 Yr (Germany)
	AED/CED	AED/CED	AED/CED
Cold roll steel pretreated[a]	>10 mm/<2 mm	>10 mm/<2 mm	>10 mm/<2 mm
Galvanized or hot dip galvanized	−/−	−/<5 mm	−/<1 mm

a Zn/Mn/Ni-phosphatation (see Chapter 3).

cathodic electrocatings versus anodic ones. The performance improvement can not only be seen in rust development at the scribe but also at the edge, where the coverage after the baking process plays an important factor.

A comprehensive picture of corrosion protection and durability of cars can only be seen when testing 'complete' cars in special facilities, so-called *proving grounds* or observing them in use at different places around the world (see Section 12.3).

Short term tests have been available for many years, they are standardized in ASTM, ISO, DIN (Deutsche IndustrieNorm) and VDA norms. It is very important to judge them as laboratory tests where the corrosion protection performance is evaluated, but depending on substrates they have very little correlation to real environmental conditions [25]. For example, it is very critical to test CED on galvanized steel in ASTM-B 117 (American Society for Testing and Materials-B 117) saltspray test. This test shows heavy delamination and white rust which however, does not correlate to outdoor tests where almost no corrosion is seen after many years.

On the galvanized steel specimen test results in any corrosion test procedure are dependent on the type and especially on the depth of the scribe. Only scratching with zinc gives a completely different picture and a degree of corrosion comparable with a scribe which also engraves the steel basis of the panel.

4.5.3
Chip Resistance

Chip resistance is an important property for CED to avoid any type of cosmetic corrosion. This occurs when chips or gravel on roads delaminate the film from the substrate in a small spot. This spot which is exposed to water, salt, and oxygen will corrode fast. First, white zinc salts will be generated followed by red rust after some time. But even invisible impacts of chipping can result in local reduction of adhesion of the electrocoat and subsequent delamination and corrosion, which can be detected early by modern methods [26].

All car manufacturers have chip or gravel tests in their portfolio for CED testing (see Chapter 12.4).

Paint formulators approach this issue in two ways:
1. improve adhesion to substrate and intercoat adhesion to the primer surfacer,
2. improve energy absorption in the film.

The adhesion to the substrate is influenced mostly by the pretreatment (see Chapter 3). Values can reach, depending on the measurement method, between 10 and 15 N mm^{-2}. This is also the level of adhesion between the different coating layers of an automotive coating system like between base coats and clear coats. The baking process can sometimes influence the adhesion that is depending on the use of direct or indirect fired ovens [27].

To improve the energy absorption, the polymer structure and the cross-linking density are the dominant factors. Polyesters and Polyurethanes are the favorites of the polymers and are being used in almost all cathodic electrocoats. Furthermore rubber domains can be incorporated to reach best levels of energy absorption and best test results in chip and gravel testings.

The performance level, together with an applied primer surfacer, has become very high so that additional applications of stone chip primers, as used in the 1980s and 1990s even on front ends, are not necessary any more.

4.5.4
Surface Smoothness and Appearance

A great deal of work has been focused recently on smooth surfaces to increase the overall top coat appearance of a car. It has been known for many years that the roughness of steel and the damping factor of electrocoating influence the smoothness and brilliance of the top coat.

The roughness can be measured in several ways [28]. A typical and often used number is R_a which describes the average difference of all upper and lower roughness peaks registered over a certain distance of the film. The surface profile can be measured by a profilometer device which tips the surface and monitors the profile. Many typical data can be generated via Fourier transformation for correlation with visual impressions and differences from film to film [29].

The newest generations of electrocoating have been developed to be as smooth as possible by appropriate resin design, paint formulation, and careful selection of additives to provide the best appearance of the total coating system. Synergies have been identified with the target of less material loss during baking. The lower the weight loss is, the higher the potential for smooth surfaces can be seen.

Today's formulations have material losses in the oven of 8–12% while this number was as high as 20% in earlier products (see Section 4.4.3).

The damping factor is also influenced by the weight loss and additionally by the surface tension profile and the viscosity profile during the baking process. The higher the surface tension and the viscosity the better the damping factor. This is on the other hand contradictory to the edge protection which needs a low surface tension profile.

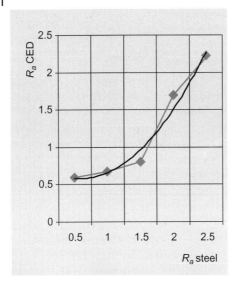

Fig. 4.7 Roughness numbers R_a of smooth electrocoat films depending on steel roughness.

4.6
Design of Cathodic Electrocoating Lines

4.6.1
Integration into the Coating Process of Cars and Trucks

As described in Chapter 1 (see Figure 1.3) the electrodeposition process is integrated as the second surface modification process after pretreatment and before primer surfacer or basecoat application on a car or truck body. In case of underbody protection and sealants being used their application takes place before the primer surfacer using the same baking oven. For higher first-run-ok rates in some paint shops these operations are transferred either to the body shop or after the final top coat application before being sent to the assembly line.

The conveyors are normally overhead conveyors designed for both the dipping processes of the pretreatment and of the electrodeposition.

Before entering the CED tank the body must be free of chemicals like salts in the pretreatment process. Otherwise corrections are necessary to reduce conductivity in the CED tank by replacing certain amounts of ultrafiltrate with d.i. water which involves waste water treatment costs. The body can be wet or dry when it is dipped into the tank. Before entering the next application step there can be a control station and some sanding operators necessary in case of defects. The control of paint and tank parameters are very important (see Section 4.3). After some time of optimization of all parameters including the rinsing operation the surface of the electrocoat after the baking process will be free of defects (see Section 4.7). The line speed as well as the volume of the bodies define the tank size and the tank design (see Section 4.6.3).

The automatization of the electrocoating process is high. Normally only bath, process and quality controls of the feed materials at the paint suppliers are necessary on a regulary basis which let this segment of a paint shop appear to be free of workers.

4.6.2
Pretreatment

The condition of the metal surfaces prior to the phosphating plays an integral part in the formation of a uniform and smooth conversion coating in the pretreatment zone (see Chapter 3). In metal working operations, that is the body shop, lubricants containing many compounds like graphite or molybdenum disulfide should be avoided where possible. Furthermore very often sealers are applied in the body shop. Material that has been carelessly applied or is in excess must be removed even by manual operation if necessary. In any case for defect-free operations a body washer should be installed in the body shop to remove all heavy contaminants as far as possible before entering the pretreatment zone.

The pretreatment zone of a modern paint shop is somewhat longer than the electrocoat area due to the fact that at least two tanks, quite often three tanks, are involved for a perfect pretreatment step. Till some years ago some pretreatment shops consisted mostly of spray steps for the basic processes as cleaning, activating, conversion, and rinsing; all the modern and future zones have dip tanks. For space saving the Ro-Dip-Process [30, 31] is the latest state-of-the-art-technology (see Chapter 3). The ability to penetrate better into interior spaces in connection with improved rinsing of the total body to minimize carry off of chemicals into the electrodeposition tank as well as easier and more efficient tank control are the main reasons for saving space and for making possible investment into three or more tanks.

Due to environmental concerns the last 'sealing' step of a traditional pretreatment zone, where chromates are applied to the conversion layer for improved corrosion protection especially over cold roll steel, has been removed from the process. To minimize the negative effect on corrosion protection a baking zone has been installed where the pretreatment is brought to 140°C for a few minutes. This results in almost no performance reduction in adhesion and corrosion protection as against a chromate rinse.

The final rinse and the water adhering to the body should be almost on the conductivity level of d.i. water which then guarantees that no significant contaminations in the form of acids or electrolytes enter the electrocoat tank.

4.6.3
General Functions and Equipment of an Electrocoat Line

Several units and functions have to be installed for a mostly automated electrocoat process:
- the tank and probably a dumping tank
- the power supply unit
- the heat exchanger and filters

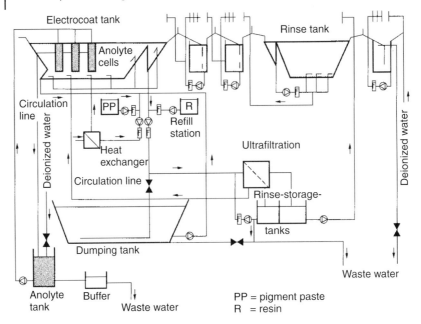

Fig. 4.8 Scheme of an industrial CED application unit for automotive coating process (BASF Coatings).

- the replenishment and anolyte circuit
- the ultrafiltration unit
- the rinsing zones.

All functions contribute to this highly efficient application process. The investment costs vary between €100 and 300 million for a standard line of 1000 units a day.

4.6.4
Tanks, Filters, Heat Exchanger, and Power Supply

Depending on the size and throughput of cars or trucks the tank size varies. The smallest tanks of about 40–60 m^3 can be found in truck painting lines. Those tanks are batch systems where the automotive bodies are dipped in and then transported to the next station, the rinsing station. The throughput can reach about 15 units per hour. The immersion of a complex body is a time consuming process due to buoyancy as a result of flushed cavities. New techniques like RoDip have increased the capacities of these lines by immersing the body vertically and rotating it, so that all cavities are wetted and coated [30].

Continuous lines can have tanks as big as 600 m^3 for high throughput of up to more than 1000 units per day at line speeds of 4–8 m min^{-1}, for example, for light trucks.

Immersing the body is also a critical part of the dipping process in continuous lines. To overcome problems with buoyancy and dirt the conveyor techniques being

Fig. 4.9 Car body entering the electrocoat tank on a skid (BASF Coatings).

used will differ to a large extent. The most common system is an overhead conveyor system, which will be changed to ground based conveyors for the following spray application station. The trends is to use special skid conveyors which fix the body better to the conveyor system and do not have to be changed for the subsequent spray application stations. New developments are already being evaluated on an industrial scale which can provide a better immersing process in the form of rotational moves during the dipping step similar to the technique used for batch tanks [31]. This allows a faster dipping process for a smaller tank design and protects horizontal surfaces against settling because during most of the deposition process the body moves on top through the tank. The paint should not have any foam on top of the liquid surface especially in the entry zone. Otherwise severe markings appear on the hoods entering first into the tank.

The tanks are built as reinforced steel construction and are isolated at least inside with an epoxy coating. The coating is carefully applied in many layers to avoid any leakage and provides a perfect isolation. Bare steel spots are coated (see below) and created by the coagulants and dirt. At the end of the tank is an overflow from where the different circuits start – the heat exchange, the feed to the ultrafiltration units, the feeding system, and sometimes a separate circuit for the heat exchanger. The generation of heat is caused by the many pumps and the deposition process itself. The bath temperature is designed to be as high as possible that is $35\,°C$ so that cooling is significantly provided by the rinsing zones and the temperature in the manufacturing hall.

For cleaning up and repair work often a dumping tank is under the body tank to take the entire volume of the electrocoat.

All the paint circuits are designed to maintain a constant flow of more than 0.3 m s^{-1} but less than 3.0 m s^{-1} of the coating material in the stainless steel or PVC (polyvinylchloride) pipes as well as in the tank. As a basic number these circuits move the total tank volume five times per hour. There are two reasons for the flow limits: first to avoid settling of the material, second to provide enough movement on the surface of the body to establish sufficient heat transfer from the deposition process to the bath material. Too high flow can affect the deposition process, for example, by lowering the film build.

The paint circuits are entering the tank on the ground in a direction against the movement of the bodies and the flow is enforced normally by Venturi nozzles. At the entry zone the flow will be directed back at the top of the tank. This guarantees no settling as well as the necessary movement all over the tank. The body moves in the same direction as the paint at the upper part of the tank but at higher speed.

The common type of pumps used for all paint circuits has double sealed packings. The sealing fluid is an ultrafiltrate of the respective paint and pumped under pressure between the two packings. These types have the advantage of not generating coagulants and foam.

Impurities from various sources can contaminate the electrocoat. Those are
- entrainment of dirt by the body
- dirt from the periphery of the CED system via air,
- abrasion and dirt from the conveyor system,
- dried-on paint from the hanger,
- coagulation and deposits from the electrocoat paint,
- contaminated d.i. water,
- coagulants generated by electrical breakthrough and high peak voltage.

To further avoid any settling on horizontal surfaces by dirt particles or coagulants all paint circuits are normally filtered. Standard filter types are bag filters with particle retention of greater than 25 or 50 µm. The preferred filter material is polypropylene in a needle-felt finish. The filter bags are placed in a special steel basket and these are placed in steel filter vessels holding two to eight such baskets. The electrocoat flows from top to bottom under pressure which is maintained as low as possible for best filtering effect. Every week the filters are flushed and cleaned.

Another paint circuit may feed the heat exchanger. Excess heat based on the deposition process and the corresponding electrical resistance of the tank and film will be detracted and the paint cooled down to the operational temperature. This temperature can be between 28 and 35°C for the actual cathodic electrocoating due to their high thermal stability. Temperature variation should not exceed 0.5 °C, otherwise the film build will vary too much. Higher temperatures normally mean increased film build.

It may be necessary to increase the flow in the entry area to avoid markings on the hood of the body. The so-called entry marks depend on the wetting properties of the electrocoat formulation and on the heat generation and heat transfer in this of the tank. The reason for the high heat generation in this area is reason that most

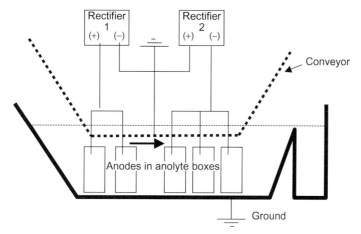

Fig. 4.10 Principles of electric circuits for continuous conveyor cathodic electrocoat tanks with two rectifiers.

current is running in the entry area generating the highest current density (ampere per square meter) during the electrodeposition process.

Power supply is carried out by direct current which is normally generated by thyristor rectifiers from an indirect current. The remaining ripple should not exceed 5% otherwise the smoothness of the deposited electrocoat will deteriorate.

For stationary tanks one rectifier is sufficient which can be programmed in terms of a voltage-versus-time-program. This is necessary to avoid high current densities at the beginning of the deposition process. Too high voltages create rupture effects of the deposited film by heat and gas generation supported by low flow and low heat transfer of the electrocoat bath.

In the case of the continuous conveyer mode at least two rectifiers are mandatory. The first supplies the entry zone, the second the main and exit zone. Often a third rectifier is in place to supply specific voltage to the exit zone alone (see Figure 4.9).

The voltages are between 300 and 450 V depending on the electrode distances, surface area and requested throwing power. The coating time is more than two minutes, for bigger bodies it can reach four minutes. Based on the differences in the electrochemical equivalence per type of paint and the coating time the practical electrical energy for the electrodeposition process of an average car of $70\,m^2$ is between 5 and 8 kW h.

4.6.5
Replenishment and Anode Cells

To keep the process parameter constant the electrocoat bath must be kept in a constant composition. In other words, the deposited material and the removed parts (i.e. solid content) must be replaced appropriately. It must be remembered that not only the solid paint but also the neutralizing agent will be discharged during current flows. There are two principle ways of replenishing electrocoat tanks. They are connected with the equilibrium of the deposited and removed material and the respective neutralization agent. The latter cannot be removed from the tank without special equipment.

In the anodic deposition process this excess neutralizing agent generated by the application of the electrocoat between the feeding steps is used to help dispersing the partially, and in any case, underneutralized feeding material (see Figure 4.10a). This was done in a separate tank in which the feed material was mixed and predispersed with the tank material and then after a certain time pumped with special pumps into the body tank. This step is not easy to control for appropriate generation and stability of the dispersion. Furthermore it was not a continuous step resulting in quality fluctuations of the respective electrocoat tank. The advantage is that only one component material with a rather high solid content of more than 60% could be used with the disadvantage of a high solvent content of up to 50% based on solids.

For improved processability and low solvent materials of less than 10% based on solids the second way of using fully neutralized feed material is the better one (see Figure 4.10b) and has gained the most frequent application. The specified and approved dispersions, mostly as two components in the form of a low solvent-containing resin dispersion and a paste dispersion, are simply pumped into the tank. One of the paint circulation lines is normally used for replenishing or feeding these components. The ratio of these components vary for the different kinds of paints and can be between three and five parts of resin dispersion to one part of paste dispersion. The ratio may be used for trouble shooting the various defects for the same paint (see Section 4.7). In very rare cases solvent and acid additions may also be necessary to bring the bath components into the proper operational values.

Using current consumption and the corresponding correlation for material consumption or the daily measurement of car bodies and solid content of the bath as controls the necessary amount of electrocoat feed material is often continuously pumped into the respective feed line [32]. For this feeding technique as mentioned one needs to remove the neutralizing agent from the tank, otherwise an excess will be generated and problems of film build and too high voltages or other not acceptable changes of the application parameter will arise.

For the cathodic electrocoating process, anolyte cells (see Figure 4.12) are used to remove the excess acid. The standard cells consist of a plastic box with the stainless steel anode in it and are covered by an anionic exchange membrane, which allows acid to enter the cell during the deposition process but not to return back to the tank. Anionic exchange membranes have cationic charges, which are fixed to the membrane polymer material so that cations cannot pass. To generate a flow of anions the current must flow between the anode inside the cell and the parts. The acid level in the separate circuit of the cell will continue to increase until the conductivity of the anolyte fluid reaches a preset level. Deionized water is fed into the circuit and an overflow is sent to waste pretreatment.

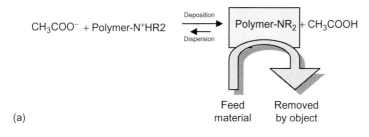

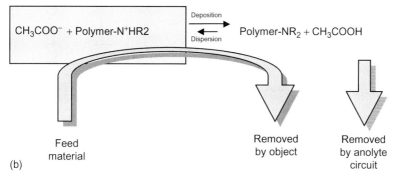

Fig. 4.11 Acid-base-balance of the two types of replenishing processes of CED. (a) underneutralized, (b) fully neutralized.

The anode material is high grade stainless steel with alloys resistant to chlorine anions like 1-4404, 1-4429, or 1-4439 according to the DIN nomenclature. The lifetime is dependent on the current passing through the anode area.

Exceeding the concentration of chlorine-ions of 50 mg l^{-1} in the anolyte causes pitting corrosion and reduces the lifetime drastically. Anodes in the entry area of the electrocoat tank may be changed every half a year every other year. Iridium covered titanium anode panels are also used. Cost versus lifetime were often in favor for these costly anode materials.

For control and monitoring of the acid removal the conductivity of the anolyte is used. The conductivity sensor may be preset between 700 and 1400 µS cm^{-1}. When the upper limit is reached a valve will be opened to let deionized water flow into the system. When the lower value is reached it closes the valve. Depending on the turn over of the tank the limits have to be adjusted.

To keep the system running smoothly the flow of the anolyte circuit must be sufficient at the minimum of 4 l min^{-1} and cell. The flow should be monitored by visual inspections and flow meters. Leakage of the membranes must be avoided, otherwise the inner side of the membrane will be coated and blocked for acid flow. In some cases mold can be generated especially when acetic acid is used as the neutralizing agent. Fungicides have then to be added to the anolyte circuit.

The type of cells has changed from boxes in the early days to tube cells which can be better maintained, more easily replaced and more variably located [33]. Because of the low pH value of the anolyte liquid, the materials of a circuit system are made

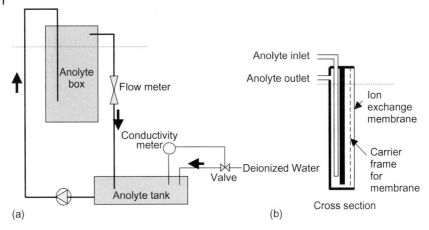

Fig. 4.12 Anolyte cells (b) and anolyte circuit (a).

Table 4.7 Typical data of an anolyte fluid compared to cathodic electrocoat

Property	Units	Anolyte circuit	Cathodic electrocoat
pH		2.2–2.8	5–7
Conductivity (20 °C)	µS cm^{-1}	700–1400	1200–1800
Solid content	%	<0.1	18–22

of plastic, usually PVC. The circulation pump has to be made of acid resistant steel, and the total system must be grounded.

4.6.6
Ultrafiltration and Rinsing Zones

The high material yields of the electrocoating process are achieved by generation of ultrafiltrate out of the electrocoat paint. This feeds the last rinsing zone and flushes back in a cascade system to the electrocoat tank in a closed loop (see Figure 4.12). If sufficient feed is given the material efficiency can reach more than 98% by this process [2].

Ultrafiltration is a separation process used widely, for example, in the dairy business in which low-molecular weight substances are passed through a membrane and separated from the dispersion particles. The membranes have particle size limits between 30 and 300 nm. With cathodic electrocoatings the substances passing the membranes are water, organic solvents in the fluid phase, electrolytes and low-molecular fractions of the resins. This is called ultrafiltrate or permeate.

In big body tanks the ultrafiltrate should be generated prior to initial operation to fill the rinsing tanks. This is better than filling the tanks with deionized water because in the latter case coagulation of the electrocoat by thinning of the material adhered to the surface of the deposited film can occur. Normal feed for a line of 50

Table 4.8 Physical data of ultrafiltrate in comparison to CED

Value	Dimension	CED	Ultrafiltrate
Visual inspection		Opaque	Clear
Solid content (2 h/180 °C)	%	18–22	0.2–0.5
Spec. conductivity (20 °C)	mS cm^{-1}	1.2–1.8	1.0–1.6
pH		5.8–6.2	5.6–6.0

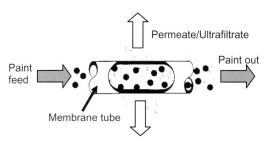

Fig. 4.13 Separation process by ultrafiltration in tubular membranes.

bodies per hour is 1.5–2.5 m^3 h^{-1} to reach high material efficiencies. This accounts for about 0.5–1.0 l per m^2 body surface.

The ultrafiltration process is driven by the concentration gradient from the electrocoating and the permeate across the membranes. It is also pressure sensitive, but high pressures (>4 bar) normally lead to coagulants on the membrane surface and to the reduced life time of the membranes.

The rate at which the permeate is generated is called the flux. There are several factors influencing the flux and the lifetime of the membrane surface, including the

1. flow rate of the electrocoat at the membrane surface,
2. area of membrane surface,
3. fouling of the membrane surface,
4. pressure of the electrocoat,
5. stability of the electrocoat.

The most important factor is the flow rate of the electrocoat material. In combination with the membrane configuration (pipe, flat membrane etc.) it defines the diffusion layer at the membrane surface. At high flow rates the diffusion layer will be thinner compared to low flow rates and the danger of a material build up on the membrane surface will be reduced.

Membrane fouling can occur when between the diffusion layer and the membrane surface either bacteria or other material like dirt, unstable resins and polymers, or other retained components are fixed and neither permeate through the membrane nor diffuse back into the bulk stream of the electrocoat. Build up of a fouling layer can be a lengthy process and soon becomes irreversible. It then can only be removed by mechanical or chemical cleaning processes.

Experience shows that the high pressure of the electrocoat is detrimental to the lifetime of the membranes. The basic formulation in combination with the chemical and physical stability of the distinct electrocoat materials have also some influence on the flux and lifetime of the membranes.

The semipermeable membranes mostly used for this process must be resistant to the constituents of the electrocoat that is solvents, acids, and electrolytes. The preparation of the membrane material is often carried out by flushing the membranes with a special solution. This solution cleans the surface of the membranes and flushes out all deposited materials but also may fix cationic charges at the membrane surface. These charges help avoid the dispersion from concentrating and coagulating on the membrane surface.

There are two main configurations of ultrafiltration membranes, which currently find application in the electrocoating industry. These are spiral wound and plate and frame modules. These types of modules have outperformed the old tubular and hollow fiber modules due to the fact that high membrane areas can be constructed on the smallest space contributing to the fact that high flux rates are mostly determined by the membrane area.

The spiral wound configuration is structured as a flat sheet of membrane folded on both sides of a sheet of porous material. The membrane is sealed along two edges to form an envelope with the porous media protruding from the opening. The porous media is then attached to a perforated tube and the entire envelope assembly with feed spacer material is wrapped around the perforated tube. The assembly is covered with a fiberglass outer layer (see Figure 4.14).

The electrocoat is pumped to one end of the spiral wound assembly and flows into the open space created by the feed spacer, flows through the entire membrane and exits at the other end. Ultrafiltrate is collected in the porous media inside the envelope and flows in a spiral pattern toward the perforated tube and then into the storage tank. As many tubes as necessary for the required flux are connected to a final ultrafiltration unit.

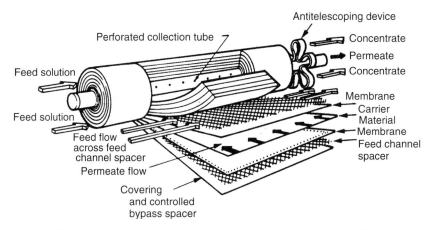

Fig. 4.14 Configuration of spiral wound membrane for ultrafiltration units of CED.

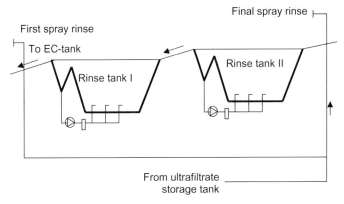

Fig. 4.15 Rinsing zones with two dipping tanks.

This type of membrane configuration provides a rather high membrane surface area at low space requirements.

This also accounts for the second type of the plate and frame configuration. Paint flows along the membranes fixed in frames and stacked to packages of different sizes. Standard units have measures of $1 \times 1.8 \times 1.0$ m for size and generate $700 \, l \, h^{-1}$ at input paint pressure of about three bars.

For some time reverse osmosis was used as an additional purification step to generate from the ultrafiltrate rather pure d.i. water to further increase the material efficiency. Due to the fact that energy cost has become too high and the modern paint formulations have much lower solvents and less low-molecular fractions of the resins, those units have become obsolete.

As seen in Figure 4.8 and already mentioned, the permeate generated by the ultrafiltration units is feeding the closed-loop rinsing zones which are necessary to purge the deposited electrocoat surface from the adhering electrocoat material. Noncleaned surfaces will not be smooth after the baking process. After exiting the electrocoat tank the body must be kept wet by a small rinsing station 50 seconds later followed by a first rinsing zone and a second one. Final rinse can be d.i. water or pure permeate.

Today most of the rinsing zones consist of two dipping tanks (see Figure 4.15). The former rinsing stations with many crowns were cheaper by way of investment but less efficient in rinsing the inner parts and cavities of the car body. Furthermore the large surface of the recirculated spray dust was detrimental to many electrocoat products due to increased oxygen attack of the resin backbone. The solid contents of the two tanks vary between 1–2% for the first tank after exit from the CED tank up to 0.5–1% in the second tank.

The final d.i. water rinse can improve the smoothness of the film, but in modern cathodic electrocoatings the final ultrafiltrate rinse is sufficient, so that the waste generated by a final d.i. water rinse can be eliminated. Under these circumstances the CED process is almost free of emission considering that the low solvent emission of the tank will be conducted to the baking oven and incinerated there.

4.6.7
Baking Oven

Almost directly after the final rinsing step in the rinsing zone the body enters the baking oven. Some ovens use multistages in continuous lines. They consist of an IR (Infra Red)-heating zone and two or three stages of convection zones with circulating air. Some only have circulating air stages. Even the convection ovens in general are not very energy efficient compared to IR-ovens, but are mandatory for the baking process because of the complex shape of a car body.

The state-of-the-art ovens are called "A"-ovens characterizing the shape of the oven in which the body is moved up to a certain height in the entry area, conducted through the oven and moved down back to the starting level at the exit of the oven. The advantage of this shape is significantly reduced loss of energy due to the fact that warm air moves to the top of the oven, in other words it cannot leave the oven easily. The IR-zone also helps to save energy because it heats up the outside of the body very rapidly, so that the circulation zones can concentrate their air flow specifically on the critical parts of the body. Those are the inner parts and the parts with high masses of steel like rocker panels or the B-columns. This is provided by air fans in conjunction with flow apertures and corresponding air speeds of $2-5$ m s^{-1} or higher for special tasks. Nevertheless measurements from the temperature/time curves show differences depending on the location of the bodies (see Figure 4.16).

To guarantee the specified film performance it is mandatory to control these data on a regular basis. The minimum metal temperature has to be reached at all points for example 10 minutes over 170 °C. Sensitivity to overbake has also to be considered while this is not such a critical factor compared to underbake. Owing to the already mentioned fact that cathodic electrocoats based on their chemistry split chemical components the air flow also has to manage this material load. Depending on the zone the share of recirculating air changes in the different stages.

The last stage can have higher recirculating air compared to the first and middle ones in case of a three stage oven. Considering the weight loss of between 8 and

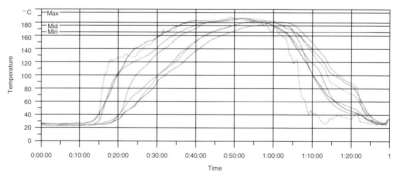

Fig. 4.16 Temperature/time curves on different locations of a body in a convection oven during electrocoat curing.

18% for most electrocoat materials based on the type of coatings and air dried deposited films this results in the concentration of organic emissions in the oven air of several grams per cubic meter. This is well above the value limits specified in the regulations of most countries in the world. Therefore the waste air, that is replaced by preheated fresh air, normally goes into an incineration unit to fulfill the VOC-emission regulations. Other methods like adsorption on activated carbon beds are used rarely. Because the thermal energy generated in the incinerators can also be used for heating air, thermal incineration systems are often an integral part of the oven system.

Gas- or oil-fired-boilers are normally used for heat generation, which is then transferred by a heat exchanger to the recirculating oven air. If natural gas is used the heat exchanger can be dispensed and the hot fuel gases can go directly into the oven. In some cases this can lead to film performance deficiencies like adhesion failures to the primer surfacer [27].

4.7
Defects During Application and their Prevention

As is usual in the coating technology processes, paints have to blend together to provide a stress- and defect-free coating result. If defects occur it is not always easy to identify the source of the problem immediately as related to the process or to the paint. This especially accounts for dirt, craters, throwing power, and film thickness of electrocoatings. In case of defects or insufficient coating result it is mandatory to collect all the data of the paint batches and the actual data of the process parameter of the pretreatment step, the electrocoat application, and the baking conditions. Reviewing all the data may help the operator to make a better approach for the trouble shooting.

4.7.1
Dirt

Dirt is one of the most common recurring problems of coating processes, however this is not the case for electrocoats. As mentioned already the 100% filtering of all circuits and a careful handling of the conveyor systems as well as the cleanliness of the body contributes to this fact.

If dirt appears it may be caused by insufficient cleaning cycles of the conveyor, dirt or other particles in the car body, broken filters, or paint stability.

Conveyors for the electrocoat process are normally so constructed that dirt from abrasion or other sources cannot fall directly into the tank. But in cases of bad maintenance, excess dirt can contaminate the liquid electrocoat material as well as the freshly cured film in the oven area. The highest danger is generated here by the accumulated cured electrocoat films reaching so high a film build that they easily break and fall onto the body from the overhead conveyors.

Typical particles from the body shop are welding pearls when they are not completely removed from the body by the cleaning and pretreatment processes. They easily can be identified by microscopic evaluation like most of the other dirt particles that is mainly paint splits or rust.

Broken filters can be identified by monitoring the pressure differences between the input and output of the filter cartridges. In these cases the dirt will decrease, but the cartridge has to be switched off and the filter bags completely replaced. In cases of heavy dirt load the filter bags should be replaced to coarser mesh-size that is $>100\,\mu m$ (see Section 4.6.3).

4.7.2
Craters

Craters in amorphous films are caused by surface tension differences of inhomogenic particles with very low surface tensions ($<20\,mN\,m^{-1}$) supported by a low critical micelle concentration in the paint dispersion. After deposition these particles are randomly distributed in the films. The poor wettability of the electrocoat film generates a crater by not wetting those particles during film formation. Typical examples of materials with very low surface tensions are oils, silicon oils, and perfluorinated oils or compounds. These compounds can be found everywhere as lubricants, sealing materials, or for other purposes in technical manufacturing processes.

They have two important properties which make them dangerous for waterborne dispersions and coating films: one is the already mentioned low surface tension which is lowest with the perfluorinated compounds, and their hydrophobic nature characterized by a very low hypophilic-lipophilic-balance(HLB-)-value. The latter is the reason that when surfactants are used in the electrocoat formulation, the particles very soon reach the critical micelle concentration for building their own micelles or dispersion particles apart from the electrocoat dispersion. This leaves them inhomogenic compared to conventional coatings where these compounds are more easily dissoluted by the organic solvents. The concentration of the contaminants can be as low as ppb (parts per billion) for special perfluorinated

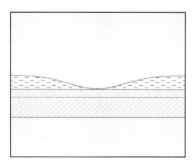
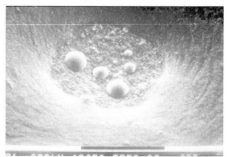

Fig. 4.17 Schematic view and a microscopic picture of a crater in electrocoating films (BASF Coatings).

compounds and ppm (parts per million) for silicon type compounds. Oils on a natural or synthetic basis have to have a somewhat higher concentration.

Deposited in the film, these micelles tend to migrate to the surface during the low viscous phase of the film forming process and then generate craters because of the difference in the surface tension (see Figure 4.17). At that time the surface tension of the electrocoat film is about 30–40 mN m^{-1}. In some cases it was observed that contaminants in very low levels reach the surface of the electrocoat film late, so that craters only occur in the next applied film, the primer surfacer, in the form of wetting defects.

Avoiding craters means careful housekeeping to protect the electrocoat tank from any type of contaminants and selecting approved, that is auxiliary materials like lubricants, working clothes, gloves, cleaning solutions and so on compatible with the paint by testing. If craters occur measure the level and locations on a continuous basis and control material input into the paint shop. As a countermeasure the increase of the pigment-to-binder ratio by changing the feed can help by increasing the viscosity level during baking, but almost at the same time the smoothness of the film deteriorates. Additionally using hydrophobic filter bags filled with polylprolylene fibers and used without pressure also help to filter out the contaminants. The latter procedure is most effective in the case of a one-time-contamination due to the small effect for a tank of the size of many hundred cubic-meters.

It is possible with modern analytical methods to identify the contaminants in the craters even when the overall concentration is very low as described [34]. Especially the time-of-flight-secondary ion mass spectrometry (ToF-SIMS) has been the established routine method for chemical identification of the nature of the contaminants in electrocoat or other paint films. The disadvantage is that even now the preparation of the specimen has to be cut off from the object for sending for evaluation. The required time for this is about a day.

The identification of the chemical nature then helps to locate the potential source in the process faster.

In any case craters can be very costly when they appear because several efforts to analyze and find the source are required, the production process has to be run by either manual repairing or spraying a cover coat. The elapsed time until detection and elimination is then the critical cost factor.

4.7.3
Surface Roughness

Unusual surface roughness exceeding the specification may have many different causes. First rough steel surfaces result in rough and structured electrocoat surfaces. The countermeasure in these cases can only be replacement of the respective coils.

The residual ripple of the rectifiers being too high may be another reason for rough surfaces. Normally this can be corrected by adjustment of the rectifier.

An application voltage that is too high and close to the rupture voltage [35] starts the general deterioration of the surface smoothness of the electrocoat before heavy local defects and ruptures occur. Correction of bath temperature and voltage to lower values are often appropriate countermeasures. If rough surfaces are to be seen on horizontal surfaces like the hood and roof, the pigment level may be too high or the paint may have started to coagulate. The pigment level can be adjusted immediately by the addition of resin dispersion. The tendency to coagulate may be reduced by adding neutralizing agents like acetic acid or the corresponding products recommended by the paint supplier. In this case the voltage or the bath temperature have to be increased at the same time to keep the film thickness within the specified limits. In some cases the flow in the tank has to be increased to reduce pigment settling.

The unusual but severe problem of bacteria contamination starts very often with increasingly rough surfaces followed by higher voltage sensitivity that is increase of film thickness under the same application conditions. Bactericides should be added as soon as the phenomenon is identified by tests [36]. The source should be found at the same time and should be eliminated. Most bacteria sources come from the d.i. water generated by untreated ground water.

The sensitivity of the modern electrocoatings against bacteria is increased when compared to the older formulations because the solvent level in the tanks are far below 5%. Above this level most of the typical solvents like butylcellosolve or aceticesters have bactericide effects. According to the new regulations in Europe and North America the selection of active bactericides has become difficult [37]. On the basis of this situation it is mandatory to avoid any bacteria input into the tanks that is the electrocoat and rinsing tanks including the final pretreatment rinse.

4.7.4
Film Thickness/Throwing Power

The film thickness and the film thickness distribution on a complex body for cars and trucks are the most important qualities defining the property of the electrocoating. The film thickness of the inner segments of the body is characterized by the throw power. It is often not possible to compare the throwing power of the different paints of different suppliers on cars, but in lab tests the potential of an electrocoat material can be judged. On the basis of physical application data of the respective paints and defined body sections calculation programs exist which are designed to forecast the film thickness and material consumption of an automotive body [14].

From a practical point of view, to reach the specified film thickness and the throwing power, the anode to cathode ratio, the appropriate bath temperature, the solid content, and conductivity as well as the voltage programs and sometimes even the feed ratio of the resin and paste dispersions have to be optimized for each specific tank. The performance of the process parameters has to be monitored at least daily, in tanks with short turn over several times a day. This includes the appearance of the applied and cured film as well as film thickness data on defined spots on the body, outside and inside. In cases where the film thickness increases or decreases without changing the process parameters one should carefully analyze the analytical bath parameter. Increasing solid and solvent levels and decreasing degrees of neutralization that is increasing pH may be one of the reasons for film thickness increase. Countermeasures have to be discussed with the paint supplier. The pH is not a sensitive indicator for changes in the degree of neutralization because the electrocoat behaves like a pH buffer system. So the acid and base numbers have to be specified for best operation and have to be checked frequently.

4.7.5
Other Defects

Related to the sensitivity of the paint formula to process parameters there are issues such as hash marks, water spotting, boil outs, and pinholes that may appear causing a lot of manual work and related costs for repairing those defects.

Hash marks can occur on hoods. They are mostly generated by foam in the entry zone, high current density in combination with insufficient flow of the paint, and not removing enough heat. Another property of the paint is also connected with these phenomena: the wetting property of the substrate. Sometimes it helps when the body enters into the tank, in critical situations, wetted by the ultrafiltrate of the respective electrocoat paint. Decreasing current density and avoiding foam as well as increasing paint flow in the entry area are the standard countermeasures for this type of problem. Additionally the geometry and the number of anode cells in the entry area may also be reconsidered and reduced in some cases.

Water spotting can appear when on the air dried electrocoat film water drops from the conveyor or other sources do not wet the film properly and create, in the first oven zones, water marks causing locally severe damage to the film, which has to be completely sanded. In case of a final d.i. water rinse, this water maybe city water or may have too high conductivity levels which also generates these type of defects. Replacing the city water, controlling the d.i. water data, or adding surfactants (i.e. ultrafiltrate) to the final rinse zone helps to eliminate the water droplets.

The same may happen when electrocoat paint drops on a deposited and already dried film. This must be avoided at all costs.

Another very severe disturbance of the electrocoat film is connected with the phenomenon just described: boil outs. A car body has a lot of overlapping panels being welded or clinched together. The electrocoat normally penetrates rather well into these inner spaces of the overlapping panels during the dipping process. Depending on the throwing power of the paint these inner areas are coated. The adherent paint film will not be removed completely during the rinsing processes especially when spray rinsing is applied. But even in the rinsing dip tanks it may happen that the electrocoat is not removed completely. This material will be heated, 'cooked' more or less in the oven, boiled out and will generate a disturbed surface around the overlapping area. The first oven zone has some influence on the film. If the deposited film is heated up somewhat slower the boil outs can flow back to smooth surfaces if the viscosity still drops afterwards. The standard countermeasure is to flush those areas as good as possible.

Pinholes may be found in electrocoatings randomly distributed under certain circumstances. Several factors each one by itself or in combination can cause this problem, which will not necessarily deteriorate the total coating system of the car, but may decrease the film performance of the electrocoat mainly in corrosion protection. Too low a bath temperature, a very high pigment-to-binder ratio and wetting problems of the electrocoat in the liquid or in the solidifying state can lead to pinholes. Corrections are based on the actual situation of the tank and paint parameter. In most cases the additional feed of resin dispersion to decrease the p/b ratio and increase the wettability of the paint overcomes this effect.

4.8
Electrocoating and Similar Processes Used in Automotive Supply Industry

There has been a sharp increase in the volume of components for cars and trucks manufactured by those other than the automotive manufacturers themselves. Wheels, axles, coolers, underbody panels, head lamp systems including the reflectors, hoods, trunks, fenders, and many other hang-on parts are supplied to the automotive industry, in many cases, painted and preassembled. They can be attached to the car in the body shop, paint shop, or final assembly line.

Many of these parts are made of plastic. Steel and aluminum is also used and those parts are mainly electrocoated. For wheels and axles special cathodic-electrocoat formulas are in place providing increased edge protection and/or higher film build up to 35 μm. The principle of these formulae to increase protection of edges can be different. One type uses higher pigment-to-binder(p/b)-ratio to increase the viscosity during baking, the other uses surfactants to decrease surface tension also during the baking process. Both effects can also be used together in one formula. The higher p/b-approach has the disadvantage of rougher surface structures due to less flow during baking.

The application technique is not different from that being used in the automotive industry itself. All units have the same components to guarantee high material usage and high first-run-ok quotes. Differences can be seen in the conveyor systems, which are less sophisticated compared to the car painting process. Also only one rectifier is in place in most cases. To avoid any hash marks or entry marks in this configuration the anode cell configuration is the appropriate way to overcome those problems. It may be necessary to have the first anode cells in the entry area of the tank 1–2 m after the part has been immersed.

For parts, like chair frames, that demand low corrosion protection a technique similar to the electrodeposition technique of paint application, called 'autophoresis', is in limited use [38]. This application is based on a waterborne coating material and applied also in a tank. The parts are immersed and coated by coagulation of the paint on the surface of the part caused by dissolution of iron. This dissolution is forced by acids as part of the paint formula. The advantage of this technology is the missing energy cost for electrical power, while the disadvantage is the incorporation of iron in the film. This provides limited corrosion protection and yellowing of the film. Only black coatings are available in most cases. Only parts made of iron can be coated. The surface must be very clean and smooth to result in a defect-free film. The film build can reach 40–50 µm.

4.9
Outlook

Many attempts have been made in the past to bring electrocoat move into the automotive coating process. High film build cathodic coatings in which powder has been dispersed for outside application with a following application with an another electrocoat designed for throwing power have been tested in the production line at the end of the 1980s. Also a dry-on-dry double electrocoating process to replace the primer surfacer was tested on a production scale. This technology was based on a sufficient level of high-surface-carbon black in the electrocoat paint which generated a rather conductive film after baking and shrinking [39]. All attempts failed due to insufficient process reliability and costs. Most research and development work concentrates today on toxicology, emission, and cost. Due to continuous worldwide legal enforcement this will also be the task of the future when new toxicological data of the different components are available [40].

Nevertheless since the mid 1980s a number of scientists and technicians working for the paint suppliers are trying to find the next generation of paints in the automotive industry (see Chapter 14). Together with the steel industry's strategy to increase its added value to the steel the coil coating approach to replace pretreatment, the cathodic deposition process in paint shops is being researched mostly in North America and Europe. This idea is also welcomed by the automotive manufacturers because in this scenario the fix costs of producing cars are significantly reduced. On the other hand the situation would not affect the paint manufacturer too much because he will still supply paint, now to the steel industry.

A lot of projects on a bilateral basis or in research programs are still dealing with this approach. The first step is to eliminate any additional steps for corrosion protection like waxing operations for rocker panels or the comprehensive sealing of the car body. For this application there is some practical experience available on the basis of the so-called *presealed* coil coating material [41, 42]. These coil coatings having film thickness of 2–4 µm, are weldable and easily integrated into the standard body shops. The next step is considered as 'preprimed' which should be able to replace the pretreatment and the elctrocoat line in the paint shops. The progress is slow because of technically unsolved problems. One of the basic problems is the edge protection of the cut coil. The next is the conflict between corrosion performance and weldability. Good corrosion protection can be achieved only by film thicknesses of more than 12 µm which under today's technique is not weldable. A pilot project for car manufacturing has been cancelled recently.

On the other hand it is very clear that from a performance standpoint the electrocoat technology has in its life cycle past the peak point. This generates pressure on cost and reduces the technology to a commodity. The cost pressure is targeting material as well as processing costs.

Many car manufacturers stress this by using a new business structure. They make logistics, quality, and cost targets the responsibility of the paint supplier which is then added to the cost of the coated cars (see Chapter 13).

References

1 Burnside, G.L., Brewer, G.E.F. (1966) *Journal of Paint Technology*, **38**, 96.
2 LeBras, L.R., Christensen, R.M. (1972) *Journal of Paint Technology*, **44**, 63.
3 Beck, F. (1976) *Progress in Organic Coatings*, **4**, 1.
4 Hammann, H., Vielstich, W. (1981) *Elektrochemie II*, Taschentext 42, Verlag Chemie, Weinheim.
5 Beck, F. (1968) *Chemie Ingenieur Technik*, **40**, 575.
6 Beck, F., Guder, H. (1985) *Journal of Applied Electrochemistry*, **15**, 825.
7 Pierce, P.E. (1981) *Journal of Coatings Technology*, **3**, 52.
8 Streitberger, H.-J., Heinrich, F., Brücken, T., Arlt, K. (1990) *Journal of the Oil & Colour Chemists Association*, **73**, 454.
9 Guder, H., Beck, F. (1985) *Farbe+Lack*, **91**, 388.
10 Goldschmidt, A., Streitberger, H.-J. (2007) *BASF Handbook on Coatings Technology* (2nd ed.), Vincentz, Hannover, p. 481.
11 Suzuki, Y.-I., Fukui, H., Tsuchiya, K., Arita, S., Ogata, Y.H. (2001) *Progress in Organic Coatings*, **42**, 209.
12 Krylova, I.A., Zubov, P.I. (1984) *Progress in Organic Coatings*, **12**, 129.
13 Heimann, U., Dirking, T., Streitberger, H.-J. (1985) Proceedings Surtec, Berlin.
14 Krylova, I., (2001) *Progress in Organic Coatings*, **42**, 119.
15 Schwarzentruber, P. (2002) In *Macromolecular Symposia* (eds H.-J., Adler, K., Potje-Kamloth), Wiley-VCH, Weinheim, Vol. **187**, p. 543.
16 Goldschmidt, A., Streitberger, H.-J. (2007) *BASF Handbook on Coatings Technology* (2nd ed.), Vincentz, Hannover, p. 209.
17 Fieberg, A., Ehrmann, E. (2001) *Farbe+Lack*, **107**(6), 66.
18 US-Patent 3947339, PPG (1975)
19 Berger, V., Huesmann, T. (2004) *Journal für Oberflächentechnik*, **44**(3), 26.
20 Amirudin, A., Thierry, D. (1996) *Progress in Organic Coatings*, **8**, 59.
21 Sangaj, N.S., Malshe, V.C. (2004) *Progress in Organic Coatings*, **50**, 28.
22 Zhang, J.-T., Hu, J.-M., Zhang, J.-Q., Cao, C.-N. (2004) *Progress in Organic Coatings*, **51**, 145.
23 vanOoij, W.J., Sabata, A. (1991) *NACE Annual Corrosion Conference*, Cincinnati, Paper No.417.
24 Corrigan, V.G. (1993) *Metal Finishing*, **92**(10), 59.
25 Pietsch, S., Kaiser, W.-D. (2002) *Farbe+Lack*, **108**(8), 18.
26 Kleinegesse, R., Schauer, T., Eisenbach, C.D. (2003) *European Coatings Journal*, 34.
27 Haack, L.P., Holubka, J.W. (2000) *Journal of Coatings Technology*, **72**(903), 63.
28 Goldschmidt, A., Streitberger, H.-J. (2007) *BASF Handbook on Coatings Technology* (2nd ed.), Vincentz, Hannover, p. 356.
29 Biskup, U., Petzoldt, J. (2002) *Farbe+Lack*, **108**(5), 110.
30 Anonymous, (2006) *Journal für Oberflächentechnik*, **46**(3), 22.
31 Klocke, C. (2002) *Journal für Oberflächentechnik*, **42**(9), 32.
32 Boyd, D.W., Zwack, R.R. (1996) *Progress in Organic Coatings*, **27**, 25.
33 Nöles, H. (1998) *Journal für Oberflächentechnik*, **38**(9), 46.

34 Wolff, U., Thomas, H., Osterhold, M. (**2004**) *Progress in Organic Coatings*, **51**, 163.
35 Doroszkowski, A., Toynton, M.A. (**1989**) *Progress in Organic Coatings*, **17**, 191.
36 Schwarzentruber, P. (**2002**) *Farbe+Lack*, **108**(12), 24.
37 Shaw, A. (**2003**) *Modern Paint and Coatings*, **93**, 28.
38 Hall, W.S. (**1980**) *Journal of Coatings Technology*, **52**(663), 72.
39 Guder, H., Beck, F. (**1987**) *Farbe+Lack*, **93**, 539.
40 Drexler, H.J., Snell, J. (**2002**) *European Coatings Journal*, 24.
41 Lewandowski, J. (**2003**) *Journal für Oberflächentechnik*, **43**(9), 52.
42 Anonymous, (**2006**) *Journal für Oberflächentechnologie*, **46**(1), 46.

5
Primer Surfacer

Heinrich Wonnemann

5.1
Introduction

In modern automotive OEM (Original Equipment Manufacturer) coating, the 'primer surfacer' – often termed *primer* or *surfacer* – provides the bond between the primer coat and top coat. With almost 60 million [1] vehicles produced per year, the primer surfacer, despite low profit margins, has a high marketing potential of approximately 130 000 t per year. Three technologies have now been established:

- solventborne primer surfacers
- waterborne primer surfacers
- powder primer surfacers.

About 20 years ago, only solventborne primer surfacers were available. As shown in Figure 5.1, this product continues to dominate the world market. The substitution of the solventborne primer surfacer has been progressing in accordance with legal requirements to reduce solvent emissions. The European and NAFTA (North American Free Trade Agreement) economic regions have taken different routes in this respect (see Chapter 1). Whereas in Europe waterborne primer surfacers have become the standard technology, in the United States the powder primer surfacer has been introduced with a market share of about 26% in 2003.

In the 1950s, the emphasis was on functional characteristics so that both good corrosion resistance and good build were obtained: good build was necessary to mask the rough surface texture of the metal substrate. Various primers were used during this period. For example, both conventional dip coatings and spray primers were used with a subsequent primer surfacer. With the introduction of electrocoating for the primer coat, significant progress was made in ensuring durable corrosion resistance. Electrocoats fully took over from conventional dip coats. Around 1980, with the start of widespread introduction of the second-generation electrocoats, that is cathodic electrocoats (CED), the primer was increasingly the focus of rationalization considerations. With high-build electrocoats and improved

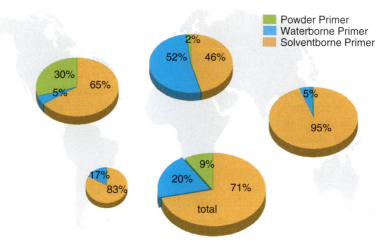

Fig. 5.1 Market share of different primer surfacers in 2003.

surface quality, a trend toward a 'primerless' coating system for passenger cars took hold in the United States. When two-coat finish systems were introduced in the mid-1980s, the saving at the time was deemed conclusive. This led to a painful experience in the United States. Not all base coat colors had enough hiding power to provide protection against the effects of ultraviolet (UV) radiation on the CED. Moreover, the film thickness of a base coat is limited by the capabilities of the application technology used. Since the electrocoat as primer is not weather-resistant, that is UV-stable, some chalking occurs with resulting delamination of the subsequent coat. Red and mica blue were affected in particular.

Volatile organic compound (VOC) standards in the United States, requiring that the attained emission reduction be maintained in the body shops, have forced the introduction of a primer surfacer with zero emission as UV barrier. The technology that was able to meet this requirement was the powder primer. Powder primers combined good technological characteristics with ecological advantages, which convinced Chrysler and GM. However, even before the introduction of this system, Chrysler was able to draw on experience with antichip powders.

To manufacture powder coatings, the raw materials – almost exclusively nonvolatile matter – are compounded by means of an extruder. By grinding, classifying, and straining, a powder with appropriate particle size – the powder primer – is obtained; VOCs are no longer used in this preparation. The powder primer is deposited electrostatically, and the film is formed in the stoving process [2]. Development has since shown that a high technological and optical quality can be obtained. Increased material consumption because of higher film thicknesses of 60–80 µm, when compared with liquid primer surfacers, is partly compensated by the fact that the overspray is recycled, so that the material transfer efficiency (TE) is almost 100% higher than that of liquid primers which are difficult to recycle [3]. In powder application, film thicknesses of up to 250 µm can be obtained, so that the underbody protection materials can be replaced in the area of the doorsills, which are particularly exposed to the risk of stone chipping.

Against this background, around 1980, the introduction of cathodic electrocoating and imposition of new environmental standards by legislation, combined with a change in thinking on quality and costs, shifted the emphasis to the visual quality of the coating, or in other words, to the vehicle appearance in Europe, particularly among German automobile manufacturers. At the same time, enormous quantities of solvent emissions from solventborne base coats were recorded. With the increasing introduction of waterborne base coats, the use of waterborne primer surfacers was also raised to a standard level. The pace of introduction of waterborne primers was set, in particular, by GM, VW, and DaimlerChrysler. Together with waterborne base coats, waterborne primer surfacers have also contributed significantly to the reduction of solvent emissions in automotive OEM coating.

Although automobile production has increased by 14% in Germany since 1980, emissions have been reduced to less than 50% since 1990 (see Figure 5.2). This achievement could be made possible by working in partnership with the coatings industry [4].

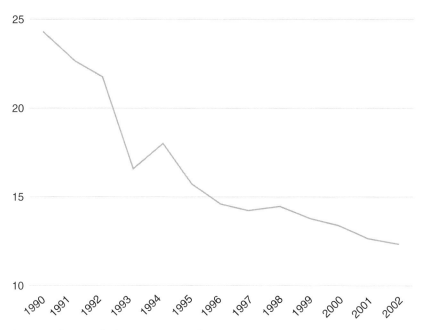

Fig. 5.2 Reduction of solvent emission in the German automotive coating process (source: German Automotive Association, VDA).

In Europe, the European Union (EU) is responsible for coordinating such processes with the help of state-of-art technology. The EU has entrusted this task to the 'European Integrated Pollution and Prevention Control Bureau (EIPPCB)' at the Institute of Prospective Technological Studies. This Bureau is also known as the *Seville Bureau* because of the location of its headquarters. At this bureau, committees consisting of representatives from national technical authorities and industry experts define the processes and methods that are considered to constitute the best available technology in Europe for achieving the required high-level of environmental protection. At present, the nonvolatile content of waterborne primers is in the range of 40–55%. To set the required stability and application properties, a solvent content of about 4–12% has, so far, been essential. The solid content of solventborne primers ranges from 60–65%.

However, legislative requirements do not necessarily lead to the use of waterborne or powder primer surfacers worldwide, solventborne primer surfacers continue to have the largest share of overall demand. This is particularly true for Asia where 97% of the demand is for solventborne primer surfacers. Since the nonvolatile content of 60% is relatively high in application consistency, the need for action is not as urgent for waterborne base coats as it is for solventborne ones. Primer surfacer application lines are more compact and are able to run at higher line speeds. They can be efficiently fitted with waste air cleaners to comply with the legal limits.

5.2
Requirement Profile

Primer surfacers have to satisfy an extremely wide range of requirements. The selection of suitable product formulas is based on the technical requirements, environmental compatibility, the framework conditions imposed by legislation, manufacturing processes, and finally, the application methods and their associated costs. Often, in the case of new projects, it is the cooperation between automobile manufacturers, plant suppliers, and coating manufacturers that results in the creation of a new product.

5.2.1
Legislative Requirement

To ensure full compliance in industrial-scale production, the automotive and coatings industries have to deal with a number of stringent legislative requirements. It covers all steps and processes of the business and primer surfacers are part of it. As already described, major impact in terms of reduced toxic risks has been seen in the last 20 years in solvent emission and chemical components of the formula (see Chapter 1). The development has not stopped yet [5].

5.2.2
Technological Requirements

5.2.2.1 Film Properties

The film properties of the primer surfacer must always be considered in interaction with the overall coating system, that is the CED (Cathodic electro disposition coating) and the base coat/clear coat. The customer normally requires the supplier to test and provide samples of the materials in accordance with the customer-defined specifications. The test categories include mechanical properties, climatic-technological properties, the characteristics of the liquid coatings, and information on the delivery packaging to ensure safe handling and use. If the test criteria are successfully satisfied, a basic approval is generally issued.

5.2.2.1.1 Mechanical Properties In the overall coating system, the primer is particularly important for stone chip resistance. Firstly, any shot-through to the bare metal must be prevented because perforations inevitably lead to corrosion. At the same time, good adhesion of the top coat is required to ensure minimal detraction from the visual appearance in the event of any chipping. The test methods for simulating these requirements on test-scale with sheet metal test panels are based on the experience of car manufacturers in the field. In principle, there are two groups of tests, the multistone impact method and the single-stone impact method. A few of the tests are listed below:
- stone chip test, DIN 55996-1, ASTM D3170,
- gravelometer, SAE (Society of American Engineers) J400,
- single-stone impact, DIN 55966-2, gunshot.

Because it is very difficult to resist repeated stone impacts throughout the life span of a vehicle, it is common for a suitable primer color to be used to mask damage to the top coat by stone chipping.

The surface hardness of the primer surfacer plays an important role in its compatibility with defect correction measures, that is sanding. A large number of test methods are available for describing the hardness in various ways, for example Buchholz penetration hardness conforming to DIN EN ISO 2815 and pendulum damping conforming to DIN EN ISO 1522.

Elasticity, by contrast, provides an indication of the potential stone chip protection capacity of the primer surfacer coat alone. It is generally measured using cupping test according to DIN EN ISO 1520 or mandrel bending test (cylindrical) according to DIN EN ISO 1519.

Adhesion to the electrocoat (CED) primer and to the base coat in the overall coating system is essential and is determined by the crosscut test according to DIN EN ISO 2409, ASTM 3359.

5.2.2.1.2 Climate-Related Technological Properties The exposure of the total automotive coating to constant or varying humidity and temperature conditions, as

Fig. 5.3 Delamination of top coats in a two-layer coating system in North America.

encountered during the life cycle of an automobile, is associated with osmotic effects. Surface defects such as blister formation and even loss of adhesion may occur in the overall coating system (see Figure 5.3) depending on the performance of the primer surfacer and its interaction with the total coating system. Depending on the specifications, there are constant and fluctuating stresses:

- VDA (= German Automotive Association) humidity test with constant or fluctuating climate conforming to VDA 614-1;
- water immersion test at increased temperature conforming to DIN EN ISO 2812-2, ASTM 870;
- autoclave test for stresses above 100°C with simultaneous humidity testing;
- cold check test (temperature change test);
- resistance to extreme climates.

Top coats that are transparent to UV cause the primer to be directly exposed to weathering. In case of disintegration of the film-forming agent because of UV light – a phenomenon frequently referred to as *chalking* – the result could be a drastic loss of adhesion and delamination of the top coat. The following tests are conducted to verify the overall coating system:

- Florida exposure test with subsequent adhesion test,
- accelerated weathering tests, for example SAE J1960, ASTM D4587.

Primer surfacers can also, occasionally, perform top coat functions in the vehicle interior at seams or when used as contrast colors. Consequently, the primer must also satisfy the requirements of top coats.

5.2.2.1.3 **Smoothness** Surface performance, concerned with the appearance of an automotive coating, is a demanding subject (see Section 12.1). Verification of visually evaluated aspect is difficult because individuals have different points of view [6]. To describe the quality on a physical basis, wave scan (Byk-Gardner) is a common method that generates numbers in short- and long-wave scale [7].

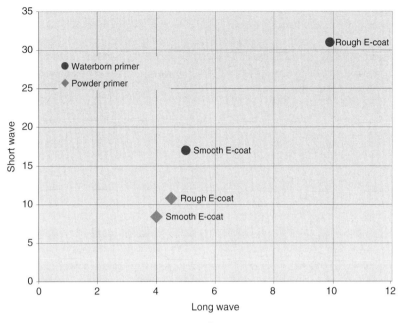

Fig. 5.4 Top coat wave scan measurements of automotive coatings using waterborne and powder primer surfacers on smooth and rough electrocoatings (E-coats).

Primer surfacers are designed to play an important role in the overall appearance of automotive coatings. In this respect, the evaluation of the primer surfacer performance has to be seen in combination with the total coating system and its interaction with the substrate. Depending on substrate roughness, the electrocoat surface appears to be either rough or smooth. In Figure 5.4, a powder primer with 60 µm and a waterborne primer with 35 µm of film build are applied on smooth and rough electrocoats. The roughness R_a – dominantly influenced by the substrate (steel) roughness – is described by perthometer measurements according to DIN EN ISO 4287: 2000-11:

Steel: $R_a = 0.6–0.8\,\mu m$
Smooth electrocoat: $R_a = 0.15\,\mu m$
Rough electrocoat: $R_a = 0.40\,\mu m$

The same waterborne base coat and two-component clear coat is applied on all primers.

The coatings are applied and baked horizontal. With the wave scan measurement, the long and short wave structure of the primer surfacers has been measured in this example and shown in a matrix diagram.

If both long and short wave can be reduced to small numbers, the performance would be improved. In optimization processes, long and short wave relation should be taken into account. If only short wave is reduced in the visual aspect, the long wave is boosted in the visual aspect. A relation of 1 : 2 for long wave to short wave appears to be best. In this example, the powder primer surfacer shows its advantage over liquid primers. Even rough electrocoat substrate is better covered compared with the waterborne primer and the smooth electrocoat. The reason for better build is film thickness of 60 µm and 100% solid content of the powder primer surfacer.

5.2.2.2 Product Specifications

The most important variable from the technical and commercial point of view is the nonvolatile content or nonvolatile matter as defined in DIN-EN-971-1. The methods for determining this variable – that is the specified drying or cross-linking temperature and time can vary from customer to customer according to DIN EN ISO 3251. In addition, for liquid primers, the viscosity at the time of supply and application are specified. For solventborne coatings, flow cups conforming to DIN EN ISO 2431 are suitable. For waterborne systems, which generally have marked non-Newtonian characteristics, measurement of the viscosity in Pascal second with a defined shear gradient and a rheometer conforming to DIN EN ISO 3219 is indispensable.

In powder primer surfacers, the particle size and the particle size distribution play an important role in defining the application properties. They are measured by a variety of methods (see Section 5.5).

The fineness of grind and grain size distribution of the pigments in the primer are generally determined by a brush-on method conforming to DIN EN ISO 1524. The density of the products to be applied is specified in accordance with DIN EN ISO 2811 in the case of liquid coatings and DIN 55990-3 for powder coatings.

For the packaging, storage, logistics, and transport, the stability features of the primer surfacer are decisive. Solventborne primer surfacers can be stored in steel or babbitt metal drums. For waterborne systems, plastic hobbocks are generally more suitable. Otherwise, stainless steel packaging, for example, in the form of containers, is necessary. Powder primers can also be stored and transported in low-cost 'big bags'. The specified temperature limits for storage are generally 5–30 °C.

Within these limits, the physical and chemical stability must be guaranteed for at least 6 months. In individual cases, for waterborne formulas in particular, stability of 3 months is specified.

5.2.2.3 Application

The requirements for application are highly complex. The testing and definition of application characteristics are therefore often not standardized. Application begins with

- delivery,

and involves several different steps and characteristics, including

- adjustment and handling in the paint mixing room,
- circulation stability,
- application,
- behavior – that is flow on the vertical and horizontal panel,
- cleaning of the assemblies,
- drying in the oven,

- corrections on the inspection line,
- further processing with the top coat.

Since the mid-1990s, many automobile manufacturers have handed over responsibility for the correct definition of these requirements to the coating supplier. At the level of coating supplier, application engineers are asked to examine the process and develop a method for simulating the process at a scale, which is lower than that of the test lab and pilot plant. With this methodology, the primer is then adjusted in its rheological characteristics during storage and application as well as in its evaporation characteristics and drying. Often, robots are used to readjust the coating processes. Test circulation systems simulate the stresses in the circulation system.

The application of liquid primer surfacers makes use not only of conventional pneumatic atomization but also of high-rotation bells [8]. The materials must be rheologically optimized to ensure their sprayability and to prevent any problems occurring due to solvent boils and the tendency to sagging. The same applies to powder coatings, where the rheological adjustment refers to their melting and coalescing at temperatures above $80°C$.

The primer surfacer must ensure good sandability of its surface after stoving to rectify defects in the body-in-white and the electrocoat. For this purpose, hardness and elasticity must be optimized. Any impurities because of surfactants in the process chemicals and materials must, if possible, be prevented from causing surface defects such as craters. Although this requirement is difficult to simulate in the lab, the primer surfacer must have a positive effect on the robustness of the coating process. At the body shop and in the refinish department, the repair concept must be validated. The minimum requirements of adhesion, climatic resistance, and stone chip protection must also be satisfied with the specified materials.

Often, bright top coats have a hiding power, which is more than the film of specified thickness, that is they are transparent. The color of the primer is one way of attempting to compensate for this problem. Consequently, a range of several colors in a primer surfacer line is common. Because the primer surfacer is also seen as a correction layer, high-gloss is often a customer requirement to ensure optimal defect identification.

The substrate structure that is generally determined by the panel quality and priming must be adequately hidden. The primer surfacer must have smoothness and sufficient 'build'.

Inadequate hiding of sanding marks, because of sanding of CED primer by mistake, often possesses a problem in wetting. Wetting is better on the sanded surface than on the unsanded surface. Consequently, the sanding scratches leave more visible marks.

At the same time, the inherent surface structure and the appearance, primarily the smoothness of the layer, must be optimal. Here, the absolute best appearance is not the most important feature. In fact, a better overall appearance of the vehicle can be produced with a suitable orange peel structure on vertical and horizontal surfaces. Suitable test methods (see above and Chapter 12) include the following:

- wavescan (Byk-Gardner);

- perthometer measurements according to DIN EN ISO 4287: 2000-11;
- gloss (DIN 67530) and haze measurements.

Once all the criteria have been satisfied in lab tests, a line test is conducted. The coating supplier produces a pilot batch that the automobile manufacturer uses in the running production line. The delivery consignment is approved only if the result of this line test is satisfactory. Despite the best preparatory work of development and application engineers, verification in the production process is always essential to prevent the risk of production failures.

More than the topcoats and possible primers, the primer surfacers are in the focus of cost analysis. Here too, coating material consumption alone is less important than the overall examination of the coating process. While in the past, automation and environmental compatibility were important; today, because of overcapacity in western countries and emergence of new markets with different framework conditions, for example in Asian countries, the focus of attention has switched to costs.

5.3
Raw Materials

For the primer, unlike the top coat, the functions like body or build, stone chip resistance, and elasticity are more important than the decorative effect. In this capacity, it supplements the properties of the overall coating system to provide the automotive industry customer with a high quality. For making liquid primer surfacers, the following four elements of the coating formula, namely,

- binders and resins,
- pigments and extenders,
- solvents (VOCs or water), and
- additives

are mixed together to form a heterogeneous mixture. The know-how of a paint manufacturer includes the formula and production process and also the dispersion properties.

The exact numbers of the product specifications may vary from supplier to supplier (see Table 5.1). It is striking that the formula for waterborne primer surfacers is more complex than solvent-based systems. Like the solvent, water also has a special role to play. Binders are available both as dissolved in solution and as dispersions. Other boundary layers also have to be stabilized. The powder primer is the exception here, in that it is characterized by its simplicity and the absence of any solvent. The formulation skill is to generate the complex performance profile of a powder primer surfacer while using fewer components compared to the liquid ones (see Section 5.5).

Table 5.1 Technical data for the three primer surfacer technologies

Properties	Units	Solvent-borne primer	Waterborne primer	Powder primer
Solid content	%	50–65	35–45	100
VOC	g l^{-1}	390–420	170–230	0
Density	g cm^{-3}	1.1–1.3	1.1–1.3	0.5 – 0.7[a]
Bake conditions	min °C^{-1}	20/130–165	20/130–165	20/160–190
Film build	μm	35–50	25–35	55–100
Viscosity (20 °C)	mPa at 1000 s^{-1}	60–100	60–100	Solid
Storage temperature	°C	5–35	5–35	5–30
Shelf life	Mo	6	6	12

a Bulk Density.

5.3.1
Resin Components

The basic properties of the paint like viscosity and reactivity and of the applied film like hardness and elasticity of a primer surfacer are set by the choice of the binder or resin (see Chapter 7).

Polyester is the basic component of almost all types of primer surfacers [9]. Polyesters are generally delivered in a solution with solvents. For powder systems, solid and solvent-free resins are required. Polyester synthesis is therefore carried out in the fused mass (melt), and the nonvolatile solid matter is then obtained by cooling and pastelizing.

For waterborne primer surfacers, polyesters must be supplied in the aqueous phase. This is done by neutralizing the carboxyl function with amine.

The esterizing process is an equilibrium reaction, which means that hydrolysis of the ester groups can occur in the presence of water. The oligomer is then broken down, and new carboxy functions are generated. The performance profile is modified, and the system becomes more acidic, which in the end can lead to the resin becoming insoluble in water. Waterborne polyesters are therefore, in principle, stable only to a limited extent.

The amino resins [10] include, in particular, the urea, melamine, and benzoguanamine (or glycoluril) resins. In the coating formula, both the methylol functions and ether functions of these products can react with hydroxyl and the carboxyl groups of the polyester resins. Amino resins are therefore used as cross-linking agents in many primer surfacer formulas.

Hexmethoxymethylamine (HMMM) is a very common basis of the cross-linker components in primer surfacers. The degree of etherization does not necessarily have to be complete. The melamine resin can condense with itself in the presence of free methylol groups. For solventborne coating systems, this process may be deliberate to set a higher hardness. The tendency to self-condense at low degrees of etherization can cause stability problems in waterborne systems. Consequently, in waterborne systems, mostly low-molecular melamine resins with a high degree of etherization are used. Simultaneously, reactivity is reduced; which can be compensated by the catalytic effect of the polyester's carboxyl function. If this is insufficient, the reactivity must, subsequently, be adjusted with acid catalysts.

Benzoguanamin resin and glycoluril resin improve appearance, gloss, and adhesion in solventborne primers. The drawback is loss of weathering stability with these raw materials, but in cases of absolutely nontransparent base coats, this may be irrelevant. Amino resins are important only for solventborne primer surfacers.

When using isocyanates to cross-link to polyurethane resins [11], components with at least two functional groups are required to start the cross-linking process with carbonic acids and polyalcohols. The high reactivity on one hand and the relatively high vapor pressure of low-molecular isocyanate monomers on the other do not allow their use as per plant health and safety requirements. Consequently, in the coating formula, we find modifications in which the functionality is blocked or the vapor pressure is reduced by increasing the molecular weight, for example, by trimerization. Isocyanates also form the basis for the production of polyurethanes, which are adducts of OH-functional binders with isocyanates as starting substance.

On the basis of hexamethylene diisocyanate (HDI), isophorondiisocyanate (IPDI), and toluenediisocyanate (TDI), the trimerized isocyanates are important for two-component primers in the field of low-temperature cross-linking, for example, plastic coating.

The isocyanurate ring of the three diisocyanates is thermally stable under technical coating conditions, whereas uretdiones made of two diisocyanates split at temperatures above 140 °C and react with OH groups. They are therefore ideal for use in powder systems. Cross-linking with OH polyesters then occurs without emitting products. Aromatic isocyanates are more reactive than aliphatic isocyanates. As already mentioned, for polyesters, cross-linking by the aromatic compounds is stiffer, whereas the aliphatic components confer plasticity and elasticity. Weathering stability of aliphatic isocyanates is generally out of the question. When using aromatic isocyanates, action must be taken to prevent the risk of delamination by UV breakdown reactions. IPDI has better film-forming properties compared with HDI. When wet-on-wet processes require good physical drying, they are to be preferred to HDI.

The primer surfacer is generally a single-component system – that is, it is applied as delivered. To increase the storage stability, the isocyanate groups must be blocked with suitable capping groups. The deblocking temperature, and therefore the minimum stoving requirements, can be adjusted by appropriate selection of the capping components (see Chapter 7).

In the waterborne system, the urethane group has a positive effect on dispersibility because of the interaction of the NH group with water. No additional carboxyl groups are required to make the system stable in water. It has been shown that exceptional film properties can be obtained with polyurethanes. Prepolymers for the production of polyurethane dispersions (PUR) can be formed not only with suitable diols but also with polyesters. For waterborne systems, anionic, cationic, and nonionic stabilized PUR resins are available. They can be given OH functions for further cross-linking reactions, for example with suitable hardeners such as free or blocked isocyanates, and can also be combined with amino resins.

With epoxy resins it is possible to create coatings with outstanding mechanical and technological properties and also efficient corrosion protection owing to their good adhesion to metal surfaces. For the formation of primer surfacers, the aromatic epoxies with bisphenol A as base unit are of particular significance. Low-molecular epoxy resins are still liquid in 100% 'as delivered' form. With two or more bisphenol A units, the epoxies are solid and therefore very important for the manufacture of powder coatings. The reaction partners in the various coating formulas are carboxyl and amine functions. The advantage of the positive technological profile of epoxies has to be set against the risk of delamination because the inadequate UV stability of the aromatic epoxies can lead to chalking. A thin-build top coat is not always sufficient to ensure good protection against UV radiation (see Section 5.1). For powder primers, alternatives such as cycloaliphatic resins and glycidyl methacrylate (GMA) acrylates come into consideration. Moreover, the existence of epoxy-functional light-resistant hardeners based on the isocyanurate ring should be taken into account. The best-known example, triglycidyl diisocyanurate (TGIC), was previously the standard hardener for powder coatings but is associated with physiological problems at the workplace and is therefore being phased out. Alternatives to TGIC as well as approaches to lower bake conditions are available [12] [13].

5.3.2
Pigments and Extenders

Unlike the pigments in top coats and base coats, pigments in primer surfacers are used not only to provide color but also to support the mechanical performance of the film. This aspect is reinforced by the use of extenders that contribute neither to the hiding power nor to the color, but affect the properties and performance profile beginning with gloss and stone chip resistance and continuing up to rheology. Because the demand for a variety of colors is of only secondary importance, we shall mainly examine white and black pigments and the most important extenders at this point.

The type, quantity, and quality of the pigments are factors affecting the paint and film performances. In addition to climatic resistance, for example humidity tests, the focus is on the mechanical properties of the coating for the primer. A good stone chip resistance will be reached as soon as cohesion and adhesion are balanced out. In order to avoid adhesion loss, the soft talc can help to get a preferable cohesion brake in the coatings layer. On the other hand, hard pigments like TiO_2 improve the impact of gravel by energy absorption as can be seen in respective tests of those coatings. The hardness scale [14] provides information for the appropriate selection of extenders (Figure 5.5). A pigment differs from the extender in its refraction index. The higher the ratio between the refractive index of the pigment and that of the binder matrix, the more significant is the pigment property in terms of hiding

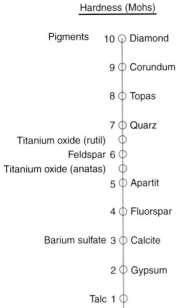

Fig. 5.5 Hardness scale according to Mohs.

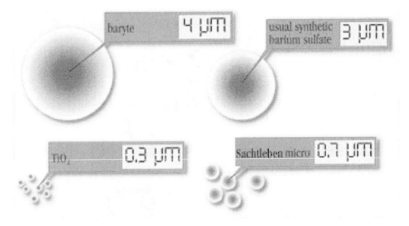

Fig. 5.6 Particle size of different extenders (baryte = natural barium sulfate).

power and color. Light is refracted and reflected in the coating. If the ratio is close to one, the pigment – the extender – is more transparent. In addition to the hardness and refractive index, particle size is also a major quality feature (Figure 5.6). This is particularly important with regard to the dispersion and to the binder demand for

Table 5.2 Refractive indices of selected pigments and extenders and their ratio to resins

Pigment/Extender	Refractive index	Ratio to polymer	Appearance
Polymer	1.48	–	–
Blanc fixe (synthetic BaSO$_4$)	1.64	1.11	Nearly transparent
Calcium carbonate	1.68	1.14	Nearly transparent
Zinc sulfide	2.37	1.60	White
TiO$_2$ – anatase	2.55	1.72	White
TiO$_2$ rutile	2.75	1.86	White

wetting [15]. The smaller the pigment, the higher the specific surface; and more binder, resin, and additives are needed for proper wetting.

5.3.2.1 Titanium Dioxide

Titanium dioxide with a density of 4.1–4.2 g cm^{-3} is the most important of the white pigments. Because of its high refractive index (see Table 5.2), TiO$_2$ provides good hiding power and owing to its hardness it provides good stone chip resistance. It is extracted from black ilmenite and is available in two crystal forms, anatase and rutile (Figure 5.7).

Rutile has higher refractive index and also higher chemical and UV resistance. The variety of types is obtained from the particle size and from the surface treatment, which may be inorganic or organic. The properties can also be modified

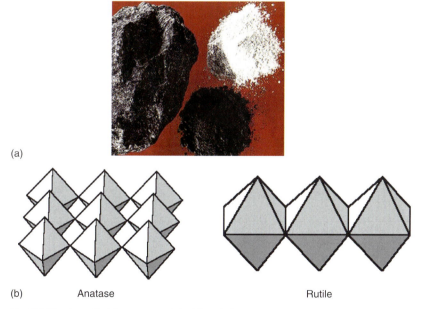

(a)

(b)   Anatase   Rutile

Fig. 5.7 Titanium dioxide: raw materials (a) and the two forms of crystals (b) anatase and rutile.

by insertions in the crystal matrix. The optimum hiding power is obtained with particle sizes conforming to the formula

$$D = \lambda/2, 1(n_1 - n_2)$$

where D is the optimal diameter, λ the wavelength and $(n_1 - n_2)$ the difference between the refractive indices of the resin matrix and pigment [6]. With an additional extender, the hiding power can be positively influenced.

5.3.2.2 Barium Sulfate

Barium sulfate, with a density of 4.1 g cm^{-3}, occurs as natural barite or precipitates as highly pure barium sulfate.

The precipitation process can adjust the particle sizes within a narrow size distribution. Otherwise, there are practically no impurities. Naturally sourced barite can have up to 10% impurities. Common barite types have 2–3% of strontium and calcium sulfate impurities bound into the crystal lattice. Both types are used in primer surfacer formulations dominantly based on cost considerations [16].

Owing to the low refractive index of 1.64, barium sulfate has practically no coloring effect in the coating film. It is, therefore, the ideal extender. It is absolutely inert in the presence of alkali and acid, and also has outstanding weathering stability and fastness to light. One of its most striking properties is the cost factor compared with titanium dioxide. It acts as a spacer improving the hiding power in dispersion with titanium dioxide.

5.3.2.3 Talc

Like barium sulfate, talc, with a density of 2.8 g cm^{-3} is also an extender. One of its characteristics is its relatively low hardness (see Figure 5.5). Talc has the form of platelets and is water repellent. Its lipophilic character promises good dispersion capacity. Talc is particularly used to adjust the fracture properties of the primer surfacer. If stone chipping in the boundary layers of the coating system causes flaking of the substrate, the use of talc additives can weaken the structure and thereby convert the adhesion fracture into a cohesion fracture. The low hardness and plate-type structure are the reasons behind this property. This is explained by the crystal morphology of magnesium silicate. Layers that can easily shift in relation to each other are formed.

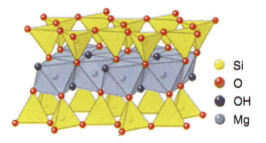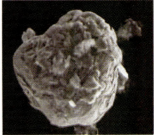

Fig. 5.8 Crystalline structure and particle of talc.

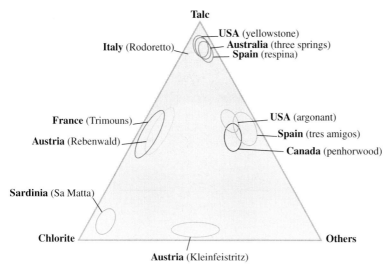

Fig. 5.9 The most important talc mining places in the world and their related type of impurities.

Although talc is chemically very inert, it has a good affinity for organic compounds. It is a naturally occurring mineral, which is ground to form talcum. Because it is formed by the weathering of other types of mineral such as dolomite and chlorite, it contains varying impurities as a function of the breakdown zone.

For the coatings industry, the preferred types come from Italy, the United States, Australia, and Spain. The particle size of the standard qualities ranges from 10 to 40 µm (Figures 5.8 and 5.9).

5.3.2.4 Silicon Dioxide

Silicon dioxide has a density of $2.2\,g\,cm^{-3}$. In 1942, it became possible to manufacture a particularly fine SiO_2 (also known as *pyrogenic silicic acid*) by high-temperature hydrolysis using a method developed by H. Klöpfer. Particle sizes were in the nano range, that is 0.007–0.020 µm. Because of its high specific surface area of approximately $200\,m^2\,g^{-1}$, SiO_2 forms agglomerates that require a considerable amount of space (see Figure 5.10).

As an extender and because of its tendency to interact with other substances, SiO_2 affects the rheology of the coating material. It is also used as a matting agent for obtaining matt surfaces. Pyrogenic silicic acid has one special application when used in powder coatings. During or after the grinding process, the extender is added to improve the fluidity and fluidization of the mix. The particles, which are several powers of 10 smaller than that of the powder, are deposited on the surface of the powder particles and act as spacers.

Pyrogenic silicic acid is available in both hydrophilic and hydrophobic form. The hydrophobicity is achieved by treating with silanes or siloxanes.

Fig. 5.10 Agglomerate of pyrogenic silicon dioxide.

5.3.2.5 Feldspar

Feldspar (china clay), with a density of 2.6 g cm^{-3} is, chemically, an aluminosilicate and is the most commonly occurring mineral.

The mineral is broken down and ground, and particle sizes starting from 10 μm are available. It is highly appreciated as extender because of its relatively high hardness (see Figure 5.5). It is also suitable for adjusting the sanding properties of primers.

5.3.2.6 Carbon Blacks

Depending on the manufacturing method and starting material, carbon blacks can be made with specific characteristics. The specific density is 1.86 g cm^{-3}. There are three manufacturing methods: the gas, furnace, and flame soot processes [17]. Gas carbon blacks have numerous acid chemical groups on the particle surface, whereas furnace carbon blacks are slightly basic. The main attention is on the particle size, which ranges from 5 to 150 nm.

The finer the particles of carbon black pigment, the greater is the light absorption, or lower is the light reflection, and the greater is the color depth. In other words, the primary particle size determines the carbon black concentration and the color depth in the coating. Because they are dissolved and considering that fine-particle pigments can have a determining influence on rheology, it is important to choose the right type of carbon black for the right color. In addition, dispersion capacity plays a key role. For better workability and improved handling, granulated types are available (Figure 5.11).

5.3.3
Additives

Additives are used to control and correct interactions inside the coating and in the boundary layers. As already mentioned, the use of additives in waterborne

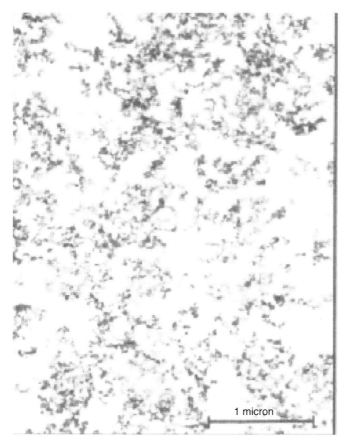

Fig. 5.11 Transmission electron microscopy (TEM) photograph of carbon black FW2 with an average primary particle size of about 13 nm.

systems is extremely complex because in addition to the resin and the pigment, water, in particular, because of its high surface activity and special physical characteristics, is responsible for an extra dimension of boundary layer effects. Thus waterborne primer surfacers have a more complex additive mixture than solvent-based ones:

- pigment wetting and dispersion agents;
- defoaming and deaerating additives;
- surfactants and substrate wetting additives; and
- rheology additives.

This list is supplemented by the additives, which are necessary for the resins or resins' mixturelike cross-linkers and for efficient application of all types of primer surfacers:

- catalysts
- electrical conductivity regulators.

These additives essentially act at boundary layers, partly the internal interfaces between pigment and binder matrix and partly the external boundaries – that is wetting of the substrate and surface to modify the surface tension. Additives are not always necessary as aids. In each case, effort should be made to minimize the use and quantity of additives because mutual interaction and overdosage can neutralize the desired effects. Silicones should be used with extreme caution, as there is a danger of poor intermediate coat adhesion.

5.3.3.1 Pigment Wetting and Dispersion Additives

These additives assist in the dispersion of the pigments in the binder and solvent. There is no defined or preferred chemical class of products. However, one characteristic is common to all additives of this type: they have one polar and one nonpolar end. Typical products include block copolymers with deliberately built-in polarity and molecular weight within a narrow range. These two features are in constant interplay with each other, so that the properties of both solvent-based and waterborne systems can be controlled. Typical examples include polyglycol ether, ethoxylated fatty acids, and polyurethanes.

5.3.3.2 Defoaming and Deaerating Agents

The problem of defoaming applies more to waterborne than to solventborne systems. Unwanted foam occurs when the air bubbles, like the pigments, are as well-wetted and stabilized as is expected of pigments.

With nonpolar substances, this effect can be remedied, and the air bubble opens. The nonpolar compound (blue in Figure 5.12 penetrates into the tensed double layer, which then leads to the bursting of the bubble.

Often, nonpolar aliphatic additives can assist this process. Depending on how compatible they are with the coating system, they can be added in the completion stage or else they must already be dispersed with the product during grinding. The defoaming or degassing of powder coatings is an entirely separate topic in itself. At this point, what is important is the included air that cannot escape fast enough during the melting and coalescing process, for which one standard agent used is benzoin. Waxes can also be successfully used.

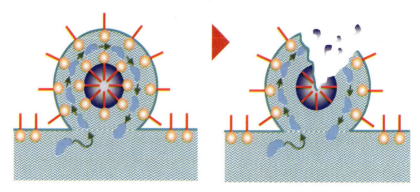

Fig. 5.12 Mechanism of defoaming additives on air bubbles with nonpolar additives.

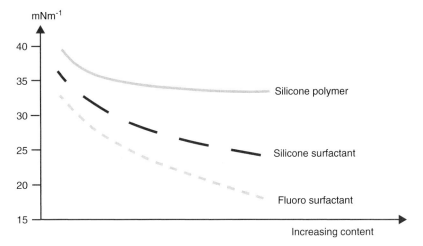

Fig. 5.13 Influence of surfactants on surface tensions of paints.

5.3.3.3 Surfactants and Additives for Substrate Wetting

Depending on the application and evaporation, different surface tensions occur at the boundary layers, with the result that forces are generated, disturbing the smooth flow (Figure 5.13). Here, the aim of additives is to compensate for these differences. Polyacrylates, silicone polymers, and fluorine-based surfactants assist the reduction of surface tension.

For flow control, in general, acrylates are used. To form a robust coating process, it is often necessary for the primer to compensate for inadequacies with regard to contamination. In this case, fluorine-based surfactants are useful, especially when a contaminated surface must be wetted.

5.3.3.4 Rheology Additives

Rheology is important for many characteristic features of primer surfacers. First, a high viscosity level prevents sediment formation during storage; and second, application requires as low a viscosity as possible to enable optimal atomization of the coatings. Then, when the coating is applied, it must not have any tendency to run on vertical surfaces. This property is controlled by means of structural viscosity, that is, pseudoplasticity and thixotropy. Thickeners are able to adjust these effects. With the aid of appropriate rheometers, viscosity can be measured as a function of shear rate. Specific modifications of urethanes, urea resins, and acrylates are suitable polymers that are able to obtain significant effects with the addition of only small percentages to the coating formula. With pseudoplastic polyurethane thickeners, it is possible to specifically affect the low-shear or high-shear range. In addition, the inorganic additives must be mentioned at this point including pyrogenic silicic acid, bentonite, or laponite.

Figure 5.14 shows waterborne systems with different thickeners. For each variant, the 'up' curve, that is the increase in shear speed, is shown together with the return ('down') curve. The measurement starts at low-shear stress at which the paint

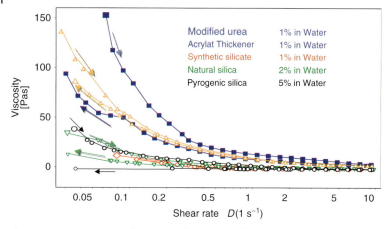

Fig. 5.14 Viscosity curves of primer surfacer with different rheology additives and increasing and decreasing shear rate.

system shows high viscosity. With increasing shear stress, the viscosity decreases. On removal of the stress the paints show a hysteresis. The viscosity level does not return to its starting point. One important aspect for paint developers to know is that the primer surfacer in the circulation system is permanently sheared by pumps and valves – that is the 'down' curve is relevant for an appropriate performance on vertical surfaces. As soon as the paint is applied, it is exposed to a very low level of shear stress and must provide sufficient viscosity in order to avoid sagging.

5.3.4
Solvents

Although the majority of surfacer compositions are based on blends of organic solvents and diluents, water is now playing a more significant role because of more and more stringent environmental controls.

In conventional solventborne surfacers, the term solvent is, probably, too general. Normally, there is a balance between solvents and diluents (aromatic hydrocarbons), which are used as thinners and also to lower the cost of the formulation. This balance is necessary to achieve satisfactory film appearance under particular application conditions within a specified solids content range.

Factors that affect the choice of solvent are numerous, that is solvency, viscosity, boiling point or range, evaporation rate, flash point, chemical nature, toxicity, and cost. In particular, odor, toxicity, and cost have become increasingly important.

In general, a spread of fast and slow evaporating solvent and diluents are incorporated to achieve the necessary balance. Slow evaporating solvents reduce popping and fast evaporating solvents are used to improve sag resistance. 'Solvent popping', sometimes designated as boil, is caused by the retention of excessive solvent/occluded air in the wet film, which on stoving escapes by erupting through the surface.

Table 5.3 Solvents and their important physical data for primer surfacer

Solvent	Evaporation rate (butyl acetate = 100)	Boiling point (°C)	Flash point (°C)
Aromatic hydrocarbons			
Xylene	75	138–141	23
Solvesso 100	21	165–179	38
Solvesso 150	7	190–210	66
Solvesso 200	3	230–280	99
Tetralin	<0.1	200–209	77
Alcohols and cellosolves			
Propylenglykolmethylether	76	120	36
Butanol	46	116–119	33
Butyl glycol	6.5	167–173	66
Pine oil	<0.1	195–230	77
Esters			
Butyl glycol		171	65
Butyl glycolate (hydroxyacetic acid butyl ester)		178–190	68
Butyl cellosolve acetate	3	184–196	88
Butyl diglycol acetate		246.8	108

In waterborne systems, organic solvents need to be water-soluble, which is provided by a high polarity. A certain amount of water-insoluble solvents can also be present.

Solvents used in primer surfacer compositions may be classified as follows (Table 5.3):

5.3.4.1 Aromatic Hydrocarbons: Diluents/Thinners

These are cheap, and in sufficient quantity aid spray atomization. They can be classed as diluents and, in excess, may cause destabilization of resin system. Normally, they are used for thinning and, as such, they are incorporated in controlled amounts in surfacer package. In waterborne systems, small portions of such water-insoluble solvents can have degassing effects.

5.3.4.2 Alcohols, Cellosolves, and Esters: Solvents

These are all oxygen-containing solvents and are true solvents for the resin system. The selection of alcohol, cellosolve, or ester will depend on a number of factors, particularly on application and stability. For example, butanol is particularly effective in stabilizing melamine and formaldehyde resins. Butyl glycol is one of the most important cosolvents in waterborne systems. It provides stability and improves interaction of resin and water phase and has an impact on rheology. Others like butyl diglycol and butyl diglycol acetate with higher boiling points are equally important for waterborne systems to improve spray dust melt in.

5.3.4.3 Tetralin or Pine Oil: Very High Boiling Additive Diluents

In most formulations for conventional primer surfacers, small amounts of such very slow diluents are usually required to keep the paint film sufficiently wet and open to allow lower boiling solvents to evaporate and prevent solvent popping.

5.4 Liquid Primers

5.4.1 Formula Principles

The product properties are characterized by two elementary requirements. First, stability and safe application of the product under industrial conditions must be guaranteed, and second, the specified and agreed film properties must be attained. The manufacturer must take into account both these aspects from the outset in the formula of the primer surfacer.

5.4.1.1 Application

For the coatings' users, when coating complex components such as automobiles, the factors of interest, in particular, are the cross-linking limits within which the product attains the specified film properties and the tolerances of the user-specific plant. The operating window defines the factors like oven temperature and time (Figure 5.15). Within these limits, the primer surfacer must meet the specifications of the customer. By means of temperature recorders, the oven curve is plotted to establish the specified/actual comparison.

The relevant temperature is the 'panel temperature' or 'metal temperature' – that is the temperature measured on the vehicle body, and not the ambient air temperature. Another important criterion is the heating rate – that is the rise in temperature after placing in the oven. In the formula, reactivity is adjusted by the

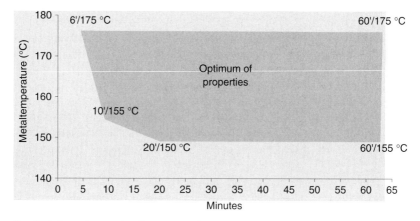

Fig. 5.15 Operating window for oven conditions.

choice of melamine resins or the use of acid catalysts. The reactivity of the blocked isocyanates is defined by the nature of the deblocking group (see Section 5.3.1). The reaction can be catalyzed by using suitable Lewis acids such as dibutyl tin laurate (DBTL).

5.4.1.2 Rheology

Viscosity is influenced by binders, pigments, extenders, and solvents. In pure solventborne primers with roughly the same nonvolatile content, the basic rule is that the higher the molecular weight of the binder, the higher the viscosity. The viscosity η can then be adjusted by means of the solvent concentration [18].

$\eta = \tau * dv/dx$ (Newtonian liquids)

η = viscosity (Pa s)

τ = shear stress $(N/m^2 = Pa)$

The adjustment has to be more finely differentiated when the main focus is on interfacial effects between boundary layers. In solventborne primer surfaces, these interfacial effects concern pigments and extenders. Depending on the shear gradient, interactions occur between the pigments. These interactions are effective at low- shear, whereas at high-shear the forces between the particles are increased. The system becomes viscous. In other words, viscosity is dependent on the rate of shear. The higher the shear, the lower is the viscosity. Often, the system requires a minimum shear stress to start to flow. The 'flow limit' τ_0 must be attained in order to trigger the interactions. In this case, the material is said to have plastic viscosity. These plastic viscosity properties are adjustable and deliberate. On delivery, the product must have no sediment. In other words, at low-shear rates, the viscosity must be high. On application, the viscosity must be low to ensure that the droplets formed in the atomization process are small (Table 5.4).

After deposition on the metal, an optimal viscosity level must, if possible, be set to ensure good flow without generating runs. Besides the pigment–binder ratio, the key characteristics here are the pigment–volume concentration and the pigment–surface concentration. Extenders with high specific surface and low rheological effect are suitable, for example pyrogenic silicic acid and layered silicates.

In waterborne primer surfacers, water plays a special role. The resins can be present in form of colloids, that is PUR-dispersions. The interactive attractive forces

Table 5.4 Shear rates D of different life stages of liquid paints

	$D\ [s-1]$
Sedimentation	10^{-3}
Flow	10^{-2}
Spray application	10^4
Dispersion	10^4

Table 5.5 Comparison of basic material and application data of waterborne versus powder primer surfacer

	Film build (μm)	Transfer efficiency (%)	Film density (g cm^{-3})	Solid content (%)	Consumption (14 m^2 car exterior) (g)	Solvent (g)	Waste (g)
Liquid primer surfacer	35	75	1.5	50	1960	157	490
Powder primer surfacer	60	98	1.5	100	1286	0	26

that the colloid particles have depend on resin concentration. Under shear stress, these forces can be destroyed, which is the basic mechanism of pseudoplasticity. The lower the concentration, the lesser attractive are the forces that are generated. Normally, high molecular weight of resins provides high viscosity and low solid content. In case of dispersions, high molecular resins can be at higher concentration – that is solid content at moderate and low viscosity. Consequently, it is possible to formulate primer surfacers using polyurethane dispersions that are solely film forming without requiring cross-linking by stoving, while still providing respectable film properties. Rheology is adjusted by the extenders already mentioned and can also be controlled with thickeners (see Section 5.3.3.4). In solvent-based as well as in waterborne paints, extenders like bentone, talc, or silica are used to control rheology, sagging resistance, and atomization behavior. As already described, the use of adjusting additives is more complex for waterborne than for solventborne primers. Organic polyurethanes and polyacrylic carbonic acids as associative thickeners are the most common additives.

The product properties must be stable from the start of manufacturing up to application or the date of expiry. In general, a storage stability of 6 months is required. Product stability is affected by physical changes, for example in dispersions and chemical changes by the cross-linking agents. Storage stability is essentially expressed in the stability of the rheology. In solventborne primers, viscosity is increased by the reaction of amino resins with the polyester. The result is an increase in molecular weight, which in turn leads to a increase in viscosity. This process is controlled by adjusting the reactivity and by selecting appropriate solvents. Melamine resin can be stabilized by additions of butanol.

In waterborne primer surfacers, the ester groups are subject to hydrolysis with water. The molecular weight is decreased, and free carboxyl groups are generated. Both the viscosity level and the pH value are reduced. In extreme cases, the system can coagulate.

Waterborne primers based solely on polyester–amino resins often have a continuous molecular reduction of the polyester. The resulting fall in pH value and viscosity can, in principle, be counteracted by subsequent neutralization. At the same time, solvation is improved with polar solvents, for example butyl glycol. These corrections also cause the viscosity to rise again. The reduction in molecular weight remains and can have a negative effect on the mechanical and technological properties.

For electrostatic (ESTA) application, it is necessary to adjust the electrical resistance of solventborne primer surfacers. If the resistance is too high, the deposition characteristics are adversely affected. The result is that the application equipment becomes spoiled, and the TE falls. The resistance can be increased with polar substances (e.g. amines). Waterborne primers are naturally more conductive, and generally there is no need to adjust the resistance. In a coating line, despite the 'clean room' concept, either the coating itself or the surface to be coated may contain impurities, resulting in cratering. Surface tension can also be adjusted with a suitable additive concept. Impurities on the substrate and also in the primer surfacer itself are wetted, and therefore, they no longer detract from surface quality. The level and type of pigmenting can also have a positive effect by increasing viscosity during film forming.

The most important film properties of primer surfacers in automobile production include the flow and substrate hiding power. Flow is influenced by a whole range of factors:
- applied droplet spectrum
- wetting of the substrate and of the top coat
- film thickness
- flow characteristics in the stoving process
- substrate structure.

The influence of rheology on the droplet spectrum and therefore on flow has already been mentioned. The higher the film thickness, the better is the flow. On the other hand, higher film thickness has disadvantages in terms of costs and increased tendency to sagging on verticles.

The effects of surface tension have to be taken into account after coating while still in the wet state. The primer wets the substrate if its surface tension is less than that of the substrate.

In the stoving process, depending on the increase in temperature, viscosity is reduced until the cross-linking reaction starts. The lower the molecular weight of the binders, the lower are the minimum viscosity levels attained by the coating, and the same applies to the pigment concentration. The lower the nonvolatile content, the greater is the shrinkage during film forming. The inevitable result is that the substrate texture is repeated in the surface (telegraphing effect). In solventborne primer surfacers, the solvents have a decisive influence on the flow properties.

In waterborne primers, cosolvents – that is the addition of organic solvents, are necessary at this stage to control the flow phase.

The primer surfacer as a correcting layer requires sufficient hardness to ensure sandability. Hardness is generally adjusted by means of the ratios and reactivity of the resin and cross-linking agent components. For example, the amino resin content has a decisive influence on formulas containing this resin.

Adhesion is established by wetting and subsequent physical and chemical interactions with the substrate. The bond between boundary layers by means of chemical binding leads to better adhesion than by physical interactions alone, for example van der Waals forces and hydrogen bridging bonds. Nonpolar additives active in the boundary layer can have an adverse effect on wetting, and therefore on adhesion. In this respect, critical protection must be provided against the effects of fluorine-containing surfactants, silicones, and siloxanes.

Particularly good adhesion is obtained with aromatic groups containing PUR resins and also with epoxies. However, epoxies have the drawback of inadequate weathering stability.

The crosshatch test alone is not sufficient to determine adhesion: it must be combined with climatic tests such as humidity and weathering tests. Electrolytes in the film generate an osmotic pressure. In its extreme effect, this pressure can lead to blistering and loss of adhesion. Waterborne primers are more critical than solvent-borne primers in this respect because of their high proportion of polar groups. In a coating system with UV-transparent base coats, adhesion is evaluated after a suitable accelerated weathering test (2000 h CAM180) or after Florida exposure test.

Good stone chip resistance is obtained by optimizing the adhesion characteristics to the electrocoat as well as to the top coat thereby damping the impact stress. An outstandingly good performance profile is obtained with PUR components. TiO_2 also has a positive effect. The thinner the builds, the lower the damping effect. The automobile customer does not accept any shot-through to the metal, which in the end leads to corrosion. A calculated rupture point should therefore be installed in the primer. This is achieved by means of soft layered silicates such as talc. Stone chipping will then lead to a cohesion break.

5.4.2
Manufacturing Process

From the point of view of process technology, it is possible to produce primer surfacer materials based on organic solvents and waterborne systems using essentially the same method. However, different requirements are laid down for the resistance of the materials that come into contact with the product. Unlike solventborne primer surfacers, the material chosen for the production of waterborne primer surfacers must be chemically resistant to alkalis and acids. In general, the machines are therefore made of stainless steel.

Production is generally divided into two stages:
1. Establishment of a mill base formula consisting of binder components, additives, solvents, and solid components in a premixer, which is typically fitted with a dissolver disc and has a high installed power-to-weight ratio. Usually, the product is predispersed and ground in the thermostable and often special grinding binders such as polyester or PUR. For this purpose, the dispersion resin and the necessary quantity of solvents, together with any dispersion additives, are first prepared. The pigments and extenders are added and

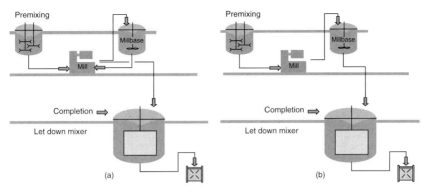

Fig. 5.16 Material flow of different grinding processes for mill bases (pigment dispersions) (a) reticulation mode, (b) passage mode.

separately predispersed, according to the particular type, with a dissolver. The correct consistency of the premix is decisive for the dispersion result. Depending on the dispersion process, the mill base components are then pumped through a grinding mill in passage mode, cascade mode (several mills in sequence), or recirculation mode and are dispersed in the mill with the aid of grinding media (Figure 5.16). The criterion for terminating this stage can be, for example the measurement of the fineness of grind according to Hegmann or Garmsen or a planned energy input. Suitable process parameters to be monitored and regulated include heat conduction, the volumetric flow of the feed pump, and the level of grinding medium.

2. Mixing of the dispersed mill base components with the other (including heat-sensitive) components of the formula, the let down. For this purpose, agitator vessels with appropriately designed agitators are used. Usually a large number of different agitator systems are used, such as blade agitators, propeller agitators, pipe discs, or inclined blade agitators. Depending on the system formula, either the mill base or the let-down components are fed first, the components that are still lacking are then pumped in, and the actual mixing process is performed.

The primer surfacer batch in the mixer is defined by extracting a representative sample to test the product properties.

The process of filling the bins with the final product is carried out using appropriate filling plants. The process always comprises a filtration stage. Typically, the primer surfacer batches are cleaned by filtration in passage mode. Cartridge-type filters or bag filter systems are used with the filter fineness appropriate to the products and filter areas appropriate to the batch size.

5.4.3
Application

Maximum reliability of application is required for primer surfacers. Often a single primer surfacer line supplies two top coat lines. Line speeds of up to 9 m min^{-1} are common. To increase the efficiency of the application equipment in terms of quantity and TE, a trend toward automation by overhead systems for robot coating can be discerned. These systems enable the spray pattern to be even better adapted to the body contour.

The common principle of all application processes for liquid primer surfacers is that they are atomized in the air and are transferred to the metal (Figure 5.17). The following spray application methods are used:
- single-substance nozzles—airless application
- dual substance nozzles—pneumatic application
- pneumatic atomization with electrostatic charging
- centrifugal atomization—electrostatic high rotation

The process normally selected is electrostatic high rotation. This process provides optimal atomization and TE. Processes for atomization by compressed air, airless method, and also the high-volume low-pressure method (HVLP) are only of secondary importance.

The success and quality of a spray coating depends primarily on the type of application. On the other hand, this processing stage, depending on the selected method, represents a major proportion of the waste material generated by a paint shop. 'Overspray' is the term for particles that are not deposited on the workpiece but fly past the metal during spraying. However, all coating particles must be deposited if they are not to be emitted into the atmosphere with the exhaust air. If overspray is successfully lessened, both the quantity of emitted waste and the quantity of new paints purchased will be reduced.

Both reductions will directly save costs. Overspray-free application methods based on other principles, such as rolling, casting, dipping, or flood coating are of no

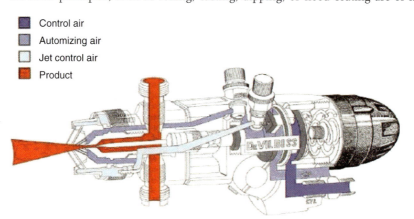

Fig. 5.17 Diagram of a pneumatic spray gun (source: DeVilbiss).

significance for the application of primer surfacers because either the geometry of the vehicle body does not permit them or else their effectiveness in terms of variable colors is insufficient. In all cases, application equipment that limits overspray is preferred. At the same time, high-quality atomization is demanded to obtain good flow. In automobile manufacture, electrostatic coating with the aid of high rotation has prevailed because it achieves the best reduction of overspray from all spray equipment. An electric field is built up between the vehicle body (earth potential) and the application device. The coating droplets, electrostatically charged with the aid of an electrode, move with their charge along the field line and are thereby applied. In conventional solventborne systems, the primer surfacer has such low resistance that the high voltage can be applied directly in the coating feed inside the spray unit to start atomization into droplets. By contrast, waterborne systems are much more conductive. In order to prevent the high voltages of up to 80 kV breaking through to the coating supply and to the supply lines, either a potential separation must be structurally established before applying the high voltage or else the primer surfacer must first be atomized into coating droplets before they are charged in the transfer phase. In the latter case, air ions are generated with the aid of externally applied electrodes. These ions then charge the coating droplets.

High-rotation atomization takes place using high-rotation bells with a diameter of 60–100 mm. The coating is accelerated on the bell plate and is atomized at the outer edge owing to the generated centrifugal forces. Rotation speeds of 20 000 rpm are standard. High-performance atomizers can rotate at speeds up to 60 000 rpm (Figure 5.18).

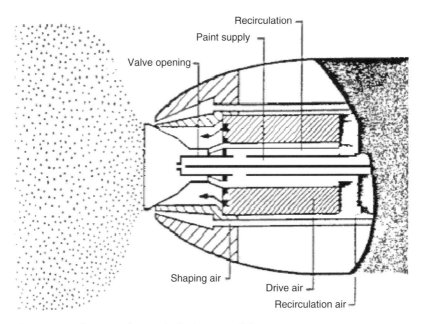

Fig. 5.18 Liquid paint application by high rotational discs.

The charge is collected on the vehicle body and flows off via the earth. TE of up to 90% is possible. TE of 85% for solvent-base primers and 75% for waterborne primers can be realistically expected. Normally, waterborne primers are exposed to high voltage with external charging at the bells or guns, which is less effective than the internal charging. But external charging is used more frequently with waterborne primers because of costly technical installations for avoiding electrical shortages in the paint supply system owing to the high conductivity of the paints. The layer smoothness and the flow quality depend on how finely the bell atomizes the coating droplets. Low viscosity is required for this purpose. As soon as the coating strikes the panel, sagging should be avoided. For this purpose, viscosity must be as high as possible. As the shear forces are much higher in the atomization process than after contact with the metal, both requirements can be met by intrinsic structural viscosity.

5.5
Powder Primer Surfacers

5.5.1
Formula Principles

The performance profile of a powder primer surfacer is marked by the melting characteristic of the binders and the density due to the type and degree of pigmentation. Additives are used to adjust surface tension, flow, and degassing. The application properties and the flow and gloss are decisively influenced by the manufacturing process. The problem of dispersing the pigment in the binder must be solved without additional solvents. In a manner comparable to the manufacture of plastics, the pigment is dispersed in the fused mass of the melt. A constant balancing act is necessary between reactivity and dispersion. The solid obtained after cooling is ground to approximately 22–25 µm.

For powder primers, two classic formulation options have been established: the carboxy–epoxy primer surfacers and the *PUR* primer surfacers.

In the carboxy–epoxy primer surfacers, acid polyesters (see Section 5.3 and Chapter 7) are combined with epoxy resins. Products of this type are termed *hybrid powders*. The carboxyl group reacts with the epoxy group during thermal cross-linking. Curing ranges from 160 to 190°C can thereby be obtained.

In principle, acid acrylates are also suitable. One typical example is provided by epoxy-functional acrylates with the aid of GMA, which are generally formulated with polycarbonic acids or their anhydrides. Good weathering stability and lower stoving temperatures have to be compared against higher costs and poor technological characteristics.

OH-functional polyesters, which cross-link with blocked isocyanates, are the classic components of PUR primer surfacers. The most widely distributed blocking agent is caprolactam (see Section 5.3.1). Alternatives with lower deblocking temperatures are also available. In the meantime, this class of cross-linking agent is increasingly being replaced by isocyanates blocked by uretdion. This cross-linking agent does not generate any decomposition substances that are released in the oven. Powder primers on a PUR base have better weathering stability than hybrid powders.

The main focus of development for powder primer surfacers, as for liquid types, is flow. The reason is that, by experience, powder primers already have the best performance in terms of their technological properties, such as stone chip resistance and elasticity. As a first priority, the flow is determined by the viscosity characteristics during the stoving phase. With a rise in temperature, the viscosity of the applied powder film falls to a minimum, the point at which the cross-linking reaction starts, leading in turn to a rise in viscosity.

In the examples shown in Figure 5.19, the viscosity minimum and the extension of the curve can be clearly seen. This test is normally performed at a heating rate of $5°C\ min^{-1}$. The purpose of the curve is to determine the position of the viscosity minimum and the rate at which viscosity increases again. It is easy to demonstrate that the lower the viscosity and the broader the melting range, the better the system flows. The most important factor is the molecular weight of the resins: the lower it is, the lower is the melt viscosity. This is limited in order to provide sufficient solid handling properties to a resin glass transition temperature, T_g, of not below 40–45°C. Unlike liquid primer surfacers, three important properties are adjusted by means of additives:

1. Degassing: Benzoin, like the addition of waxes, is a classic degassing agent. It reduces the surface tension of the fused mass to a gas bubble, thereby enabling air to escape during the melting process.

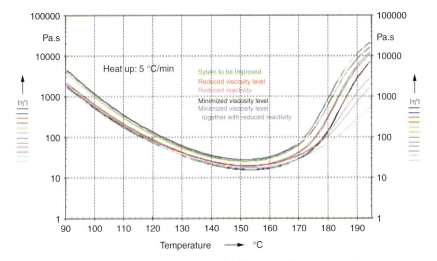

Fig. 5.19 Viscosity versus temperature curve of different powder primer surfacers.

2. Surface defects: Polyacrylates are used as flow-control and anticratering agents. Because liquid components are difficult to integrate, they are reduced to salts with silicic acid and are then available in solid form to an extent of approximately 40–50%. Resin manufacturers supply master batches of flow-control agents manufactured by adding the flow-control agent to the resin melt after synthesis of the resin.
3. Flowability: The additives traditionally used for this purpose are Al_2O_3 and also SiO_2, mostly in the form of pyrogenic silicic acid (see Section 5.3.2.4). Extremely fine modifications of pyrogenic silicic acid from 10 to 15 nm, and of a specific surface area of 100–400 $m^2\,g^{-1}$ are added in concentrations of 0.05–0.3%. The surface of the powder is coated with the additive, and the silica acts as spacer. The fluidity and fluidization of cohesive powders can be significantly improved by this method. The particular know-how required is the process of mixing the additive to the powder (Figure 5.20).

5.5.2
Manufacturing Process

The powder manufacturing process is characterized by essential differences compared to the manufacture of liquid primer surfacers. First, the homogenization procedure is continuous and not discontinuous, as in batch mode. Otherwise, with certain exceptions, solids must be handled in the same way as additives. The

Fig. 5.20 Pouring test of powder coatings.

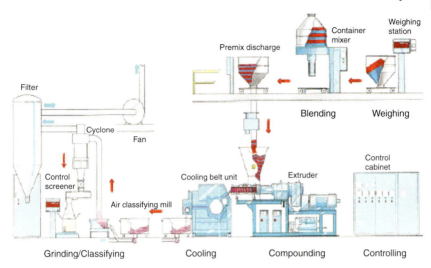

Fig. 5.21 Process steps of powder manufacturing.

process begins with weighing in and premixing the solid base materials (Figure 5.21). All components of the formula must be weighed in because subsequent completion, as with liquid coatings, is no longer possible. Dispersion, which then also leads to pigment wetting, takes place in the continuously working extruder. Like a meat mincer, the friction energy imparted by the worm-screw is so high that the binders are melted by the generated heat, and the pigments are wetted in the molten mix. After leaving the extruder, the molten mix is rolled out and cooled. The mix is subsequently broken into chips and finally ground to a powder [19]. Common extruders include the single screw, double screw, and planet screw extruders (Figure 5.22).

All three variants are common. The decisive factor for quality is process control of material throughput, temperature, and screw speed [20]. Normally, the chips are then ground in classifier mills (Figure 5.23). The product to be milled is fed either by means of gravity via a feeder or pneumatically via an injector. The feed material and internally recirculating product are presented to the internal air classifier wheel, where ground products and fines in the feed are classified and discharged from the mill.

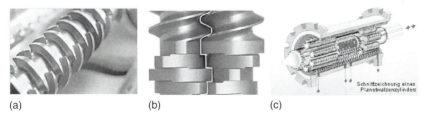

Fig. 5.22 Examples of extrusion screws (a) single screw, (b) double screw, (c) planet screw.

164 | *5 Primer Surfacer*

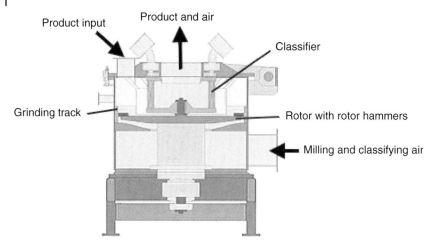

Fig. 5.23 Classifier mill for powder grinding process.

In the grinding process, rotor hammers accelerate the product against the grinding track. The geometry of the hammers and grinding track are designed such that the angle and velocity of impact create optimal particles. The milling and classifying gas is drawn into the housing and flows through the gap between the grinding rotor and track. Process gas conveys the ground particles through the guide ring directly to the variable speed air classifier.

The classification step of powder manufacturing creates coarse and fine product fractions in an air classifier by the drag force provided by an airstream and the mass force provided by a rotating classifier wheel. Dispersion takes place when gas enters the classifier at high velocity through a spiral housing and wing beater located at the bottom of the machine.

The separation of coarse and fine particles is accomplished by an adjustable classifier wheel. The fines exit the machine via the centrally located classifier wheel. Coarse particles are rejected by the classifier wheel and discharged through the bottom of the classifier housing. Coarse product and feed material are then returned to the grinding rotor. The product that meets the set requirements (classifier speed, air volume) is then removed from the mill by process gas [21]. After grinding, the product is sieved. Conventional vibrating screens operate reliably up to a mesh size of $\geq 150\,\mu m$. In vortex-type machines, particle sizes of $\leq 100\,\mu m$ are obtained. These values are unconvincing for the automotive industry. The choice is therefore the ultrasonic sieve, which can provide particle sizes of $\leq 60\,\mu m$ (Figure 5.24) [22].

Just as viscosity is important for liquid primers, particle size distribution is the key factor for powder primers (Figure 5.25). The higher the fines content, the greater is the van der Waals force of attraction between the particles. The result is that the material agglomerates and becomes impossible or extremely difficult to apply.

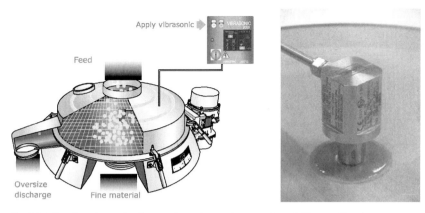

Fig. 5.24 Powder ultrasonic sieve and an ultrasonic generator sticking on the sieve screen.

Particle size is measured by sieve analysis and also by laser defraction spectroscopy [23]. In this method, the powder is dispersed in air and channeled through the laser beam. The particle size is calculated from the amount of light scattering, and is generally specified in the form of a cumulative particle size distribution (see Figure 5.26). Characterization of the particle size distribution cannot be given only by the average particle size. Minimum of two indicators need to be given to describe such a distribution as long as it is not bimodal – that is consists of a mixture of two different particle size distributions. With $d\ 0.1$ it is given the particle size of those 10% particles smaller. In this case, it represents all the particles smaller than 9.9 μm. $d\ 0.2$ is related to 10%, $d\ 0.5$ is related to 50%, and $d\ 0.9$ to 90%. The 50%

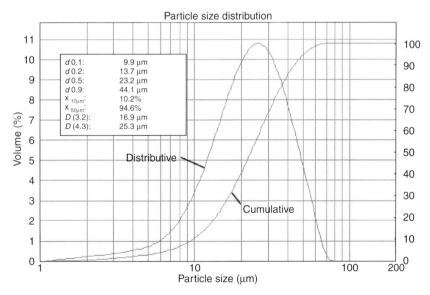

Fig. 5.25 Typical particle size distribution of a powder primer surfacer.

value $d\,0.5$ is not the average, but it is the so called 'median'. $x_{10\,\mu m}$ or, in this case, $x_{50\,\mu m}$, give the amount of particles smaller than 10 or 50 µm. The average particle size can be calculated by number, diameter, surface, or volume. In case of average by volume, it is

$$D\,(4.3) = \Sigma(V_i^* d_i)/\Sigma(V_i)$$
$$= \Sigma(d_i^4)/\Sigma(d_i^2)$$

It is remarkable that the average by volume $D\,(4.3)$ is much higher than the average by surface $D\,(3.2)$! In order to investigate agglomeration, which is normally due to the van der Waals attraction forces, the value $D\,(3.2)$ should be discussed. To study milling parameters volume (= mass*density), the value $D\,(4.3)$ is more important because it has the direct relation to energy and impulse.

The width of the distribution is calculated by the span in the following way:

$$\text{span} = (d\,0.9 - d\,0.1)/d\,0.5$$

In the given sample, it is 1.47. The higher the span, the broader is the particle size distribution. It simply describes the difference of coarse and small particle size fractions in relation to the median. The smaller it is, the lesser the coarse particles and in the same time less small particles are part of the distribution. In the compromise of good leveling and good workability, it can help to verify improvement steps.

5.5.3
Application

The relatively small powder particles of approximately 25–35 µm have a high specific surface area. Interactions occur between the particles, namely, the Coulomb forces of repulsion and van der Waals forces of attraction. These attraction forces increase in direct proportion to the specific surface area, thereby also increasing the tendency toward agglomeration. The standard method of applying powder coatings requires fluidization of the powder to make it pneumatically transportable. A reduction of 20% in the fines content of a powder distribution can lead to an immediate 80% reduction in its surface area.

Figure 5.26 shows the relationship between the average particle size, density, and cohesive properties.

The higher the average particle size, the lower the tendency to agglomerate. Powders of type C are cohesive and difficult to apply. By contrast, type A has optimal flowability. Groups B and D are no longer relevant when considering powder primers. However, the higher the density, the better is the fluidization. The graph clearly shows that finer powders are also possible with higher densities. Flow and TE have the opposite demands on powder size. In other words, the finer the particles, the better are the flow properties obtained. The higher the density, the lower is the TE on horizontal surfaces. There is, therefore, a classic

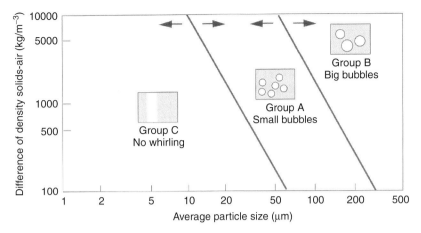

Fig. 5.26 Fluidization diagram (source: BASF Handbook).

problem of optimization between the parameters of density and particle size. High-density pigments and extenders include titanium dioxide and barium sulfate. For bright-colored pigments, iron oxides are used. If a black or anthracite color is desired, the density is often controlled by means of barium sulfate.

Powder coatings are not atomized as is the case with liquid coatings: at best, we can speak of dispersion in air. The question of how well the atomizer splits the product into droplets is no longer relevant. The powder must already be packaged with the optimal particle size on delivery. We are now dealing with a solid for which it is impossible to adjust the viscosity. The powder must be rendered transportable and so, as already mentioned, it is fluidized. Air is blown through the porous floor of a tank, generating a swirling vortex of powder termed *fluidized bed*. Powder and air behave like a fluid and are then transportable. The distinction is drawn between 'high dense conveying' and 'low dense conveying'. In low dense conveying, velocity rates of 20–30 m s^{-1} are achieved. By means of Venturi nozzles, the powder is drawn in, accelerated, and transported to the atomizer via hoses. The ratio of powder to air by mass is approximately 1:1, that is relatively little powder must be transported with a high volume of air. At high speeds, the energy dissipation that can occur due to friction is so high that agglomeration can occur. Low-radius bends in the powder supply line must be avoided to reduce friction. Recently, new transport systems and the digital high dense conveying technique (DDF) have been developed in the field of high dense conveying systems [24]. Less volume is transported, and so the conveying speed is reduced.

Conventional powder guns for applying powder coatings are single-substance nozzles because blasting with air is no longer necessary (Figure 5.27). Often, the gun is a simple pipe with a baffle plate or a slit nozzle at the outlet. Like the high-rotation process for liquid coatings, powder bells are also available. Here, the powder particles are distributed by the centrifugal forces of rotation. The speed of these bells is a power of 10 lower than that for liquid coatings, first to prevent agglomeration and secondly because atomization into droplets is no longer necessary.

5 Primer Surfacer

Fig. 5.27 Powder primer surfacer application with corona powder guns on a car body (source: BASF coatings).

Two electrostatic charging methods exist, tribocharging and corona charging (Figures 5.28 and 5.29).

Tribocharging exploits the fact that the solid powder is not conductive. By rubbing a nonpolar surface, the particles receive a positive electrostatic charge and are therefore attracted to the earthed metal panel. Consequently, no high-voltage supply is necessary. Tribocharging is particularly suitable for overcoming any problems of Faraday cages formed by the metal to be coated. Some trials have been made to coat doorframes because of the fact that no "backspray" occurs. But problems with respect to handling of tribo guns by robots have not been overcome. Today, tribocharging is not common in automotive OEM applications.

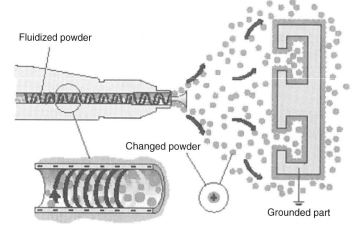

Fig. 5.28 Diagram of tribocharging and application of powder.

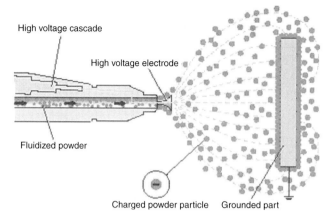

Fig. 5.29 Diagram of corona charging and application of powder paints.

In the common corona charging (see Figure 5.29), an electrode generates air ions at the outlet of the spraying device. These ions carry the charge to the powder particles. Naturally, there is an excess of air ions.

These ions are deposited with the powder particles on the vehicle body. At high voltages with low distances to the metal and high powder film thickness, this effect can get out of control.

The excess charge in the film currently being applied causes the already applied powder to be blasted off again.

This phenomenon is termed *back ionization* [25]. Once the powder is applied, it is essential that the charge is maintained. Adhesion to the substrate functions only because of the electrostatic charge. This explains why application becomes difficult if the relative humidity reaches approximately 75% or more because discharging occurs. There is, therefore, no reason for a minimum limit on air humidity. However, experience shows that back ionization also increases at <40% RH.

There is no clear-cut general answer to the question as to which charging method is the better one. However, in each case, the stability and reproducibility of the application process is better with corona charging; therefore this is the common method used in automobile manufacture.

For the large outside area of cars, the bell application with corona charging is the right choice for efficient powder application because of increased material output. In modern plants, mostly robots are in place, whereas in other plants reciprocators are used. For interior areas, single or double guns as well as robots are used.

Normally, the first time efficiency or material TE of the powder is low and generally less than 50% for complex parts. High-performance plants for the automotive industry are optimized in terms of flow rate, number and distances of powder guns from the body, and voltage to attain 70–80% TE.

In most cases, the powder overspray is recirculated and is therefore, not wasted. In order to control this process reliably, the relative humidity must not exceed 60% because otherwise the handling of the recycled powder will be more difficult as a result of the overspray. It should be noted that the content of fine particles is higher in the overspray than in the delivered product. The result is reduced fluidity and flowability. The particle size distribution on delivery must take into account the increased fines content in the process.

For the stoving process, it makes no difference for the powder primer surfacer whether the coating has a flask oft of 1 minute or 20 minutes. However, the plant engineer must take into account that any loose powder particles must be removed before they enter the oven.

Powder primers can go through the curing oven with extremely steep heating rates. The rate of heating up of 3–10 minutes is optimal for the flow properties. Temperature peaks should be avoided because otherwise reactions in the electrocoat can lead to gassing. Provided that reactivity is correctly set, the rheology and melting properties of the powder primers are the reasons for their extremely robust performance. There are no solvent popping or problems of sagging.

The stoving phase is defined on the basis of the curing window specified or preset for the primer. With the present technology, a metal temperature of ± 10 to 15°C can be attained at all points of the body. This represents a challenge, in particular for powder primers, because their appearance and flow pattern must not be damaged by any fluctuations in film thickness or differences in temperature. At temperatures of 80–100°C, the applied powder melts. The decisive flow phase begins at approximately 150°C and is completed after 1–2 minutes.

The automobile manufacturer expects the body paint work to have a uniform overall appearance. Polyester-epoxy primer surfacers provide the best basis. Differences ranging from 60 to 100 µm in film thickness (rocker panel with 150 µm to provide extra stone chip resistance), temperature differences between 160 and 190°C, and the difference between horizontal and vertical surfaces have only a minor influence on the flow of these primer surfacers.

5.6
Process Sequence

The standard process of automotive OEM coatings (see Chapter 1) begins with degreasing and cleaning the body-in-white followed by pretreatment. The first coating is the electrocoat. Then, in many cases, sealing compounds and seam sealing, often still based on polyvinyl chloride (PVC), are applied, and bitumen mats are fitted for insulation and damping. As a rule, after cleaning the surface, the body is placed in the gelling oven to gel the PVC at approximately 120°C. Gelling is necessary if automatic cleaning is part of the coating process. The machines therefore consist of an overhead gantry carrying rotating emu feathers placed in ionized air. If the sealing compound has not been gelled, it would stick to the feathers.

After applying the primer surfacer, depending on the system, it is cured at 140–170°C.

In the past, and again today, this standard process has been repeatedly called into question, and optimizations and cost savings expected from new sequences have been tested in pilot plants.

Considerable development work has been devoted to the wet-on-wet application of electrocoat and primer surfacer. The aim of this method is to achieve savings on the primer surfacer stoving process. In principle, this method is possible in the laboratory. However, in the overall process, other methods also have to be found for applying the sealing compounds and fitting bitumen mats. If the electrocoat is not cured, not only is this layer initially moist but also residual water remains in cavities and gaps, and is liable to run. Consequently, the use of water dryer cannot be avoided. The uncertainty in evaluating the process reliability is the main obstacle to this process.

Chrysler has long applied powder primers dry-on-wet as antichip coats in the doorsill area.

The 'wet-on-wet' process of primer surfacer and top coat [26, 27] is fundamentally more attractive than the wet-on-wet process with electrocoat. The application of PVC sealants and bitumen mats are more easily solved because of the fact that they are applied before the primer surfacer coating process and cured with it. To ensure that the process is successful, the primer surfacer and top coat materials must be sufficiently compatible. However, this occurs only when the materials are supplied from a single source, as it is a question of the overall system. Daimler has implemented this process in partnership with suppliers for painting the A-class car in Rastatt [28]. The base coat–clear coat application is already a wet-on-wet process. A low-build waterborne primer surfacer in the base coat color is applied with approximately 20–25 µm, and is predried in an intermediate dryer. The waterborne base coat is then applied and also dried in intermediate dryer to apply the clear coat. In this special case, a powder slurry is used – that is a clear coat with almost zero emission.

The latest developments for top coat application aim to provide a coating that already has the necessary primer surfacer properties during the first base coat application. The second application with a different coating is then used to confer color. In this case, we are faced with the difficulty that the first base coat layer cannot be termed the *primer surfacer*, while the second coat cannot be called the *base coat*.

All variations of this process sequence share the requirement that the quality of the body-in-white and of the electrocoat must be optimal in terms of defects and flow. Otherwise, correction is possible only after the top coat has been applied.

5.7
Summary and Future Outlook

On the global scale, the actual dominance of the solventborne primer surfacer with a figure of 98 000 t is significant. Nevertheless, in the long term, for ecological reasons, it will be replaced by waterborne or powder primer surfacers. The speed at which these processes will take place is hard to estimate because it will doubtlessly be dependent on legislation and the overall economic situation. It is, therefore, not surprising that the largest proportion of solvent-borne primer surfacers is found in Asian countries. Moreover, we must remember that the ecological framework conditions are laid down at national level. However, most automobile manufacturing groups operate at the global level. The quality standard and quality philosophy defined in the source country are exported abroad. So it is also not surprising to find a substantial proportion of waterborne primers in South America because European groups, in particular, produce in that part of the world. It is also striking that powder technology is established in the United States with 25% of the worldwide application in the automobile manufacturing industry, while hardly any waterborne primer surfacers are found there. However, these figures do not reflect the fact that Peugeot recently introduced colored powder primers in two of its factories [28].The decision-making matrix for new processes includes

- costs
- environmental aspects
- quality and
- process reliability.

The fact is that investment is being made in new coating lines or replacement equipment must also be taken into account.

Costs and Environmental Aspects

The greater complexity of the formula for a waterborne primer surfacer together with the significantly greater use of additives results in a higher nonvolatile matter price. By experience, commercial powder polyester resins are cheaper than waterborne alternatives.

Taking the example of a $14\,m^2$ exterior surface of a car body, the consumption is much higher with waterborne primer surfacers. However, the price could be approximately one-third lower because the nonvolatile content is lower. The overall costs must also include the costs of waste treatment due to the low TE. These costs also include the disposal of wastewater from paint shop washing.

The process costs should also be lower for the powder process because the spray booth can be operated in recirculating air mode. The energy costs for air conditioning are much lower and are one-tenth of the comparable costs for waterborne primers. Powder primers are remarkable for their good build. It has not yet been verified whether steel qualities with deeper roughness can be used. This would lead to decisive possibilities for optimization.

With waterborne primers, integrated processes in combination with the base coat can be implemented. This integration is virtually impossible with powder primers. There is also the possibility of implementing dry-on-wet processes with the electrocoat.

Quality and Process Reliability

A body coated with powder primer has a 'wet gloss' top coat appearance. However, all powder coatings have a typical long-wave character. The question of whether the appearance 'pleases' is unfortunately a question of taste and cannot be answered rationally. By contrast, the short-wave texture of a waterborne primer surfacer is not noticeable for the automobile customer.

The technological quality of a powder system, in terms of elasticity and resistance, is undisputed. By experience, resistance to multistone impact is better than for waterborne primers. The same does not apply to single impact or gun shot test. Here, a thinner build has quality advantages.

The compatibility with other connected coatings is undisputed for both waterborne and powder primer surfacers. The same applies to seam sealing and PVC coatings. Plastic hang-on parts based on sheet molding compounds (SMC) or thermoplastic polyamide/polyporpyleneoxide blends (PA/PPO) like Noryl are critical in combination with powder primers (see Section 19.1). Owing to the melting characteristic of the powder, outgassing leads to problems. Waterborne primer surfacers are produced in the batch process, while powder production is continuous. Quality assurance is therefore less of a problem for the waterborne primer surfacer than it is for the powder primer surfacer. Samples taken from the process must be generally representative. A powder sample is always only a random sample taken from the process. Assurance can only be provided by statistical evaluations.

In the end, the user opts for a particular system. Here, it is essential to determine how the systems will interact in future with the production process and to identify the future aspects to be considered. The waterborne primer surfacer fits into the process without problem when replacing the solventborne primer. Users who opt for a powder system will have to adopt new working procedures for the handling of powders.

Assuming that all solvent-borne primer surfacers can be replaced, the above considerations of the global trend would indicate a forecast in favor of the waterborne primer surfacer boosted by the trend in the integrated paint process [29].

References

1 VDA publication and BASF unpublished study
2 de Lange, P.G. (**2004**) *Powder Coatings – Chemistry and Technology*, Vincentz, Hannover.
3 Thomer, K.W., Vesper, H. (**1997**) *I-Lack*, **65**(4), 212.
4 Autojahresbericht (**2004**), Verband der Automobilindustrie e.V.(VDA), 150.
5 May, T. (**2004**) *Metalloberflache*, **58**(4), 44.
6 Wicks, Z.W. Jr., Jones, F.N., Pappas, S.P. (**2001**) *Journal of Coatings Technology*, **73**(917), 49.
7 Biskup, U., Petzoldt, J. (**2002**) *Farbe+Lack*, **108**(5), 110.
8 (a) Hoffmann, A. (**2002**) *Journal für Oberflächentechnik*, **42**(3), 54. (b) Biallas, B., Stieber, F. (**2002**) *Farbe+Lack*, **108**(12), 103.
9 Poth, U. (**2005**) *Polyester und Alkydharze*, Vincentz, Hannover.
10 van Dijk, H. (**1998**) *The Chemistry and Application of Amino Crosslinking Agents or Monoplasts*, Wiley Surface Coatings Technology, Volume V Part II.
11 Vandevoorde, P., Van Gaans, A. (**2005**) *European Coatings Journal*, 22, 09.
12 Grenda, W., Weiß, J.-V. (**2000**) *Farbe+Lack*, **106**(6), 97.
13 Rijkse, K. (**2001**) *Modern Paint and Coatings*, **91**(4), 36.
14 Goldschmidt, A., Streitberger, H.-J. (**2007**) *BASF Handbook on Basics of Coating Technology* (2nd ed.), Vincentz, Hannover, p. 396.
15 Goldschmidt, A., Streitberger, H.J. (**2007**) *BASF Handbook on Basics of Coating Technology* (2nd ed.), Vincentz, Hannover, p. 203.
16 Lückert, O. (**1994**) *Pigmente und Füllstoffe*, **5**. Auflage, Vincentz, Hannover.

17 (a) Ferch, H. (1995) *Pigmentruße*, Vincentz, Hannover. (b) Degussa, Technische Information TI 1227, *Pigmentrusse in Wasserlacken*.
18 Goldschmidt, A., Streitberger, H.-J. (2007) *BASF Handbook on Basics of Coating Technology* (2nd ed.), Vincentz, Hannover, p. 125.
19 Pietschmann, J. (2003) *Industrielle Pulverbeschichtung*, 2nd edn, Vieweg, Wiesbaden.
20 de Lange, P.G. (2004) *Powder Coatings – Chemistry and Technology*, Vincentz, Hannover, p. 262.
21 de Lange, P.G. (2004) *Powder Coatings – Chemistry and Technology*, Vincentz, Hannover, p. 275.
22 de Lange, P.G. (2004) *Powder Coatings – Chemistry and Technology*, Vincentz, Hannover, p. 309.
23 Moser, J. (2003) *Journal für Oberflächentechnik*, **43**(10), 54.
24 Franiau, R.P. (1999) *Polymers Paint Colour Journal*, **189**(4423), 18.
25 Wegner, E. (2004) *Coatings World*, **9**(10), 44.
26 Dössel, K.-F. (2004) *Journal für Oberflächentechnik*, **44**(9), 52.
27 (a) *Journal für Oberflächentechnik* (1999), **39**(9), 22. (b) Klasing, J. (2000) *Fahrz+Kaross*, **53**, 10.
28 Anonymous, *Journal für Oberflächentechnik* (2003), **43**(9), 48.
29 Dössel, K.-F. (2006) *Journal für Oberflächentechnik*, **46**(9), 40.

6
Top Coats

Karl-Friedrich Dössel

6.1
Introduction

While pretreatment and electrocoat primer are required for corrosion protection, and primer surfacer for leveling of structure and stone chip protection, it is the function of the top coat layers to give color and durability to the coating system. 'You can have any color as long as it is black' is a statement attributed to Ford when advertising the model T, while today approximately 1000 new colors come to the market each year, and a color databank covering the last 30 years may hold 25 000–40 000 entries. With the widespread use of effect pigments – aluminium flakes, micas, and all other types of interference pigments – color design capabilities have become unlimited. To protect these pigments from the environment, a clear coat is applied over the colored base coat. These clear coats offer protection from extensive sunlight, scratches, and all types of chemical attack. Top coats have been and still are the major source of emissions from an OEM (Original Equipment Manufactures) paint shop. This led to the development of high solids (HS), waterborne and powder base coats, and clear coats. The emissions of volatile organic compounds (VOC) have thereby been reduced from $150\,\mathrm{g\,m^{-2}}$ coated car surface (1970) to $<35\,\mathrm{g\,m^{-2}}$ in modern paint shops.

6.2
Pigments and Color

Pigments are used to bring color to the car. While soluble dyes could also be used, their light fastness typically is not sufficient for automotive applications. A pigment is defined as 'any colored, black, white, or fluorescent particulate solid, which is insoluble in (and essentially unaffected by) the vehicle in which it is incorporated'. A pigment will retain its crystal or particulate structure throughout the coloration process. It will alter the appearance of an object by the selective absorption and/or scattering of light (Figure 6.1).

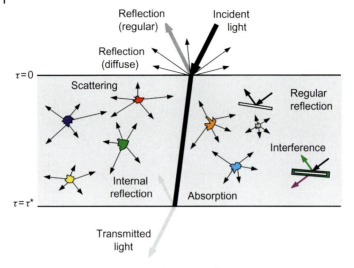

Fig. 6.1 Optical phenomena in colored paint films.

Pigments are classified as follows:
- organic pigments
- inorganic pigments
- aluminium pigments
- effect pigments (non-aluminium)
- TiO_2, carbon blacks
- functional pigments – corrosion, extender, nanoparticles, matting agents.

Out of a total global production of 250 000 t of organic pigments, 40 000 t are used in coatings and 5 000 t in the segment of automotive coatings, bringing color to approximately 60 million cars produced in the year 2006.

Attributes used to describe color are as follows (see Section 12.1):

Hue	red, yellow, green, blue, violet, and so on.
	Hue is the term used for the classification of colors.
Lightness	pale colors, dark colors.
	Lightness is the differentiating scale between white, gray, and black.
	Gray axis is the term used to specify lightness.
Saturation or chroma	vivid colors, dull colors. Saturation is the attribute of color defining how far away the color is from the gray axis

With aluminium effect pigments for metallic shades, the lightness becomes dependent on the observation angle, which is called *lightness flop*. With mica effect pigments for pearlescent shades or with interference pigments, the color (hue) also has a dependence on angular observation, in which case it is called *color flop*.

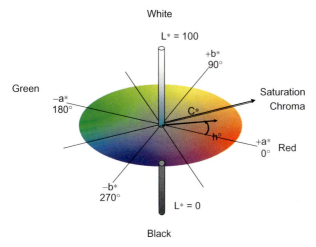

Fig. 6.2 Cie-l*a*b* description of color.

Besides coloristic attributes, pigments are described by their technical properties [1]:

In Liquid Paint
- dispersibility
- rheology
- flocculation, sedimentation
- storage behavior.

In Final Coatings Film
- light fastness
- weather fastness
- solvent resistance
- chemical resistance
- heat resistance
- bleeding resistance.

The properties of pigments in paint are described by their chemical composition in the form of a color index, and by the physical shape and size of the pigment particle, which is determined partly by the pigment manufacturing process and partly by the paint making process in dispersion or milling stages. Finally, dispersants or grinding resins are used to bind to the pigment surface and support the dispersion process and stabilize the final dispersion versus settling of the pigment. The triangle in Figure 6.3 shows some of the relationships.

Among the inorganic pigments, TiO_2 white is, by far, the most important one. The photo catalytic activity of TiO_2 leads to a rapid degradation of the organic binder matrix, so the surface of the TiO_2 particles has to be covered by an inorganic

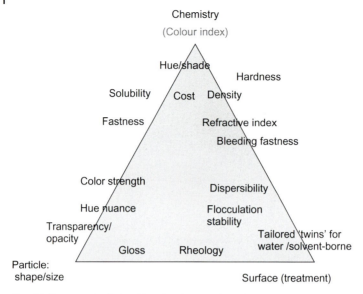

Fig. 6.3 Terms and relationships for pigments in paint.

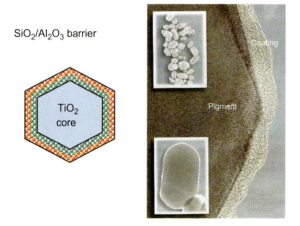

Fig. 6.4 Surface modification of TiO_2.

coating to prevent the matrix from being photo oxidized (Figure 6.4). This is done by applying layers of SiO_2 and Al_2O_3.

Within the group of effect pigments, aluminium flakes are, by far, the most important segment as silver color cars made up for approximately one-third of the global car production in 2006.

The metallic effect is caused by the reflection of light at the surface of the aluminium particle. Larger particles are better reflectors leading to higher flop and brightness, while smaller particles show less flop as the amount of light scattered at

Table 6.1 Main organic pigment classes based on color for automotive coatings (Source: DuPont)

Yellows/Golds
Bismuth vanadate (PY184)
Tetrachloroisoindolinone (PY110)
Isoindoline (PY139)
Opaque/Transparent Yellow iron oxide (PY42)
Benzimidazolone (PY154)
Azo opaque, hostaperm yellow H5G (PY 213)
Ni-azo Fanchon fast yellow Y-5688 (PY 150)

Oranges
Benzimidazolone (PO36)

Reds
Perylene (PR179)
Quinacridone (PV19 gamma, PR122, PR202)
Diketo-pyrrolo-pyrrole (PR254)
Opaque/Transparent Red iron oxide (PR101)

Violets
Perylene (PV29)
Quinacridone (PV19, beta crystal form)
Dioxazine (PV23)

Blues/greens
Indanthrone (PB60)
Copper phthalocyanine (PB15, PG7, PG36)

Blacks/white
Carbon black (PB7)
Titaniumdioxide (opaque and micro)

Effect pigments
Aluminum flake
Iron oxide coated aluminum
Pearlescent pigment
Color variable pigments
Graphitan pigments

edges increases as a nondirectional reflection. With coarser aluminium pigments, the individual particles become more visible, leading to graininess or texture.

As aluminium reacts with water to form aluminium hydroxide and hydrogen gas, the aluminium surface has to be passivated for use in waterborne base coats (Figure 6.5). The chrome treatment gives very good gassing stability, but chrome has become a substance of concern for some carmakers. Silica treatment results

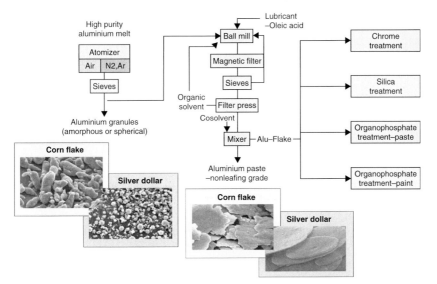

Fig. 6.5 Aluminium flake manufacturing steps and treatments for passivation.

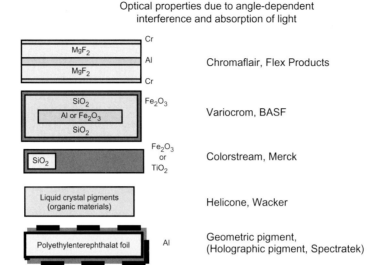

Fig. 6.6 Different types of color variable pigments.

in a stable protection of the aluminium particles from hydrolysis even under the conditions of shear stress in a circulation line. Organophosphates can be applied on the aluminium paste or can be added during paint manufacturing. Although this type of passivation is not as stable as chrome or silica treatment, it usually is sufficient for automotive applications and has a shelf life of 3 months.

The metallic appearance also depends on the orientation of the metal flakes in the coating film, the particle shape, the transparency of the binder matrix, and the presence of other colorants. Disoriented flakes will also lead to deterioration in surface appearance and increase in short wave structure or haziness of the coated surface [2].

New effect pigments are based on light interference with layers of materials having different indices of refraction, and layer thickness being in the order of the wavelength of light of approximately 500 nm (Figure 6.6) [3].

6.3
Single-Stage Top Coats (Monocoats)

Until 1970, most cars were painted with solid color paint as the only top coat layer in a 1-coat–1-bake system. While initially these coatings were based on alkyd resins and were not very durable, later they came to be based on thermoplastic acrylic enamels, which had slightly better outdoor durability. At the same time, aluminium pigments were used to give a metallic effect. The durability was not sufficient, which then led to the introduction of base coat–clear coat as 2-coat–1-bake systems.

In 2006, alkyd-based monocoats were still being used to a minor extent in Europe, and to a larger extent in most other areas globally for entry-level passenger cars, light commercial vehicles, vans, and trucks. The dominant color in this segment is white.

Table 6.2 Material and process data of single-stage top coat systems (Source: DuPont)

	Medium solids	Waterborne
Solid content (%)	45–55	45–60
VOC (g l^{-1})	380–450	100–150
Applied viscosity (s) DIN4/23 °C Pneumatic spray	30–35	35–40
Applied viscosity (s) DIN4/23 °C ESTA bell spray	26–30	60–65
Film build (μm)	35–50	35–50
Flash off (min)	5	5
Predrying in oven	–	10 min 80 °C
Baking	30 min 130 °C	20 min 130 °C

To meet more stringent environmental regulations, some truck makers like MAN and DaimlerAG started to introduce waterborne monocoats in their plants in the late 1990s.

While most truck makers would use a coating system comprising e-coat, primer surfacer, and a solid colored monocoat, in the segment of light commercial vehicles, which makes up for approximately 50% of the car production in the Asian developing countries, the primer surfacer is very often eliminated for cost reasons.

6.4
Base Coats

Base coats bring the color to the car. They are applied over the primer surfacer and covered by the clear coat layer to protect it from the environment.

There exist three main base coat systems in the paint shops of the automotive industry worldwide:
- medium solids (MS)
- high Solids (HS)
- waterborne.

North America predominantly uses HS, whereas waterborne base coat is the preferred technology in Europe (see subsequent chapters).

In a typical paintshop layout, the car body enters the top coat line and passes an emu station (feather dust off) followed by base coat application to the internal areas, either by manual or robot application, which is then followed by the external ESTA (Electrostatic) bell application. For solid colors, the full film build is applied

6 Top Coats

Table 6.3 Material and process data of base coat systems (Source: DuPont)

		Medium solids	High solids	Waterborne
Solid content (%)	solids	25–40	45–60	20–45
	Effect	15–25	40–50	15–25
VOC (g l^{-1})		450–600	250–400	100–150
Applied viscosity (s) DIN4/23 °C		20–30	15–20	35–60
(mPas) (1000 rpm)		40–50	30–40	60–120
Film build µm	solids	15–25	20–30	15–25
	Effect	10–15	15–20	10–15
Flash off before clear coat		2–3 min 23 °C	3–5 min 23 °C	3–8 min 50–80 °C

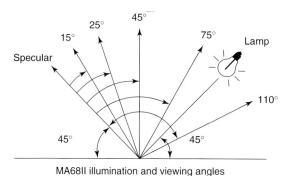

MA68II illumination and viewing angles

Fig. 6.7 Observation angles in a specific device for effect color measurement (source: X-Rite).

in this step, while for effect colors a second base coat layer is applied usually by pneumatic spray or a second bell application (see Chapter 8).

Base coat film thickness depends on the hiding power of the base coat, which is again dependent on the pigmentation. For silver shades, which exhibit good black–white hiding, typical film build is around 10 µm, for white shades around 20 µm, and for yellow/red shades up to 30 µm.

For effect color shades, the orientation of metallic or mica flakes parallel to the substrate controls the lightness difference, that is, flop between perpendicular and inclined observation angle (Figure 6.7) [4].

Flop index is the measurement of the change in reflectance of a metallic color as it is rotated through the range of viewing angles. A flop index of 0 indicates a solid color, while a very high flop metallic or pearlescent base coat–clear coat color may have a flop index of 15–17.

$$\text{Flop index} = \frac{2.69(L^*_{15°} - L^*_{110°})^{1.11}}{(L^*_{45°})^{0.86}} \tag{1}$$

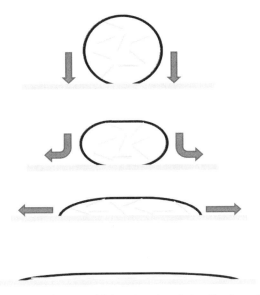

Fig. 6.8 Mechanism of flake orientation during film formation.

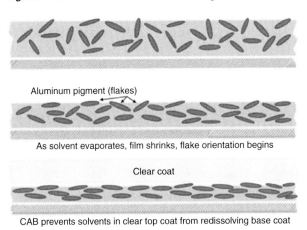

Fig. 6.9 Support of flake orientation by film shrinkage.

Two effects or models are used to explain the flake orientation (Figures 6.8 and 6.9). The first model looks at the dynamics of the paint droplet hitting the substrate surface.

As the droplet hits the surface, the momentum is redirected from the perpendicular to the lateral direction. Under the influence of this lateral force, the droplet spreads and under this lateral flow, the effect pigments become oriented [2].

The second model or additional mechanism is based on the theory that the shrinking film provides a uniform and flat flake orientation during evaporation of the solvent. Flake orientation is influenced by the spray process. Electrostatic bells and pneumatic guns create different effects, which have to be considered for color matching, for example in repairs [5]. The observed difference between these application processes can be related to the mechanisms above, as electrostatic bell application tends to be less wet and droplets have less momentum when reaching the substrate, so flake orientation is less good for electrostatic bell application.

6.4.1
Base Coat Rheology

Flake orientation is strongly influenced by rheology of the paint. Solventborne base coats formulated with soluble polymers and pigments show a Newtonian flow behavior (see Figure 6.10). Without the addition of rheology control agents, these base coats show poor flake orientation and insufficient film build on vertical areas without sagging. Microgels, wax dispersion, and urea-based SCA (sag control agent) are added to the formulation to introduce some level of shear-dependent rheology to overcome these problems (see high solid base coat curve in Figure 6.11). The most common type of SCAs are crystals of the urea-type addition product of benzylamine and hexamethylenediisocyanate. An appropriate level of pseudoplasticity or thixotropy is required to provide enough compatibility of the wet film and the spray dust during the sequence of spray processes.

During the application process, the base coat material experiences a sequence of different shear rates. While in the storage tank the paint is not agitated, which is represented by a shear rate close to zero, the high viscosity prevents the pigments from sedimentation. Then, as the material is slightly agitated and pumped into the circulation line, the viscosity drops rapidly to facilitate pumping.

At the spray nozzle or bell edge where the shear rate is highest, the viscosity has to be very low to allow for good atomization – that is, appropriate droplet formation. As the droplet hits the surface of the car and deforms and flattens out, the viscosity has to be low again to support the flake orientation process (see above). After this leveling is complete, the shear rate drops to zero again, and the viscosity rises back to high levels thereby preventing the paint from sagging.

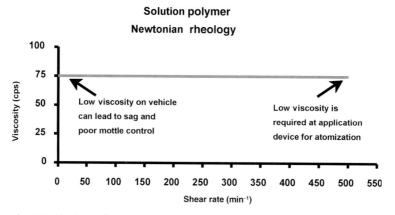

Fig. 6.10 Rheology of solventborne base coat without rheology control.

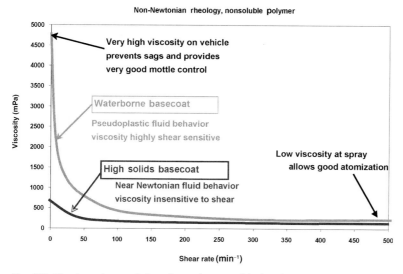

Fig. 6.11 Rheology characteristics of waterborne and high solids base coats.

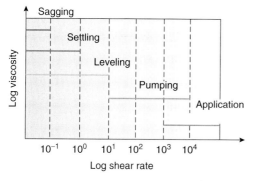

Fig. 6.12 Typical areas of shear rates (min^{-1}) during storage and application of a spray paint.

Most waterborne base coats exhibit some degree of thixotropy. In this case, the rebuilding of viscosity after applying shear stress does not follow the same curve, but shows a time delay for the recovery. This gives the paint some more time to level out and improve appearance.

The figure 6.12 shows the different levels of shear rate experienced by the paint during storage and application.

6.4.2
Low and Medium Solids Base Coat

Effect base coats were initially formulated at solid levels of 10–15% (low solids) and later at 15–20% (medium solids) having a spray viscosity of 100 mPa at 1000 rpm. Polyesters are used as a main binder in combination with CAB (cellulose acetate butyrate), SCA, and wax. In these formulations, the flop index is strongly dependent on the solid content, where lower solids yield higher flop. This indicates that the model explaining the flake orientation with film shrinkage is most applicable here.

Solid color shades in this technology typically have 25–45% solids.

6.4.3
High Solids (HS) Base Coats

High solids coatings formulations were developed from medium solids solvent-borne coatings formulations. In response to EPA (Environmental Protection Agency in USA), the total solids of coatings had to be increased and the amount of organic solvents reduced. To maintain good spraying properties, the molecular weight of the resins used was decreased. Lowering the molecular weight of the resin reduces the viscosity, and therefore, less solvent is required to reduce the coating to application viscosity. As a negative side effect of this, these coatings show less physical drying and are more sensitive to sagging.

To prevent sagging in high solids coatings, the coating must be either pseudoplastic showing shear thinning at higher shear rates or have a yield stress showing high viscosity behavior below a certain shear rate. The common method to induce shear thinning and yield stress in HS base coats is the addition of microgels formed by nonaqueous emulsion polymerization. Polymer microgels and organoclays are the common 'rheology control agents' in HS coatings because they do not increase the high shear viscosity of the coating significantly, which is critical for achieving good atomization during spray application [6].

6.4.4
Waterborne Base Coats

Waterborne base coats have become the main base coat technology for all new paint shops built after the year 2000. Since their first industrial use in 1987–1988, they have captured most of the market in Europe, United States, and Japan. This success is driven partly by the environmental benefits, and partly by their superior performance and robust application properties.

The commonly used amines are DMEA (N,N-dimethylethanol amine) in Europe and AMP (2-amino-2-methyl-1-propanol) in the United States. Waterborne base coats neutralized with DMEA cannot be used with US-type HS clear coats based on acid catalyzed cross-linking of acrylic polyols with hexamethoxymethylmelamine (HMMM) type melamine resins because amine interferes with the acid catalyst (Figure 6.13).

The main binder is typically a partially cross-linked core shell emulsion polymer based on acrylics and/or polyester. Melamine resins act as reactive diluents or cosolvents and are of the HMMM or butylated type. The melamine reacts with the binder during the clear coat baking step and introduces some cross-linking into the base coat film.

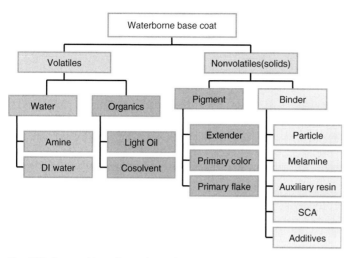

Fig. 6.13 Composition of waterborne base coat.

Auxiliary resins are either waterborne polyesters or polyurethane dispersions. These cobinders are introduced mainly to improve chipping performance, flow, and appearance.

The rheology of the waterborne base coat partially comes from the binders used, but is mostly created by either alkali swellable thickeners (polyacrylic acid type) or a slurry of layered silicate [7].

Several additives perform all kinds of functions such as defoaming, wetting of substrate or wetting of clear coat. Stability against hydrolysis of the aluminium flakes under conditions of shear is of great importance for metallic formulations. Insufficient stabilization will lead to hydrogen evolution and a degradation of the metallic effect.

6.4.5
Global Conversion to Waterborne Base Coat Technology

Figure 6.14 shows the contribution of emissions from the different paint layers and technologies. Solventborne base coats are, by far, the biggest contributors to the emissions from a paint shop.

Replacing medium solids solventborne base coat by waterborne base coat reduces the total emissions by 35 g m^{-2}. With the EU regulations requiring <35 g m^{-2} (25 g m^{-2} for some countries) emissions for new paint shops and <60 g m^{-2} for old ones, medium solids base coats can only be used in combination with costly incineration of spray booth air.

Because of its better rheology control (see above), waterborne base coats also allow for more brilliant colors and flop. This leads to a global trend toward waterborne base coat (Figure 6.15).

While all paint shops in Germany use waterborne base coat, some plants in Europe continue to use medium solids base coat with incineration or HS base coat.

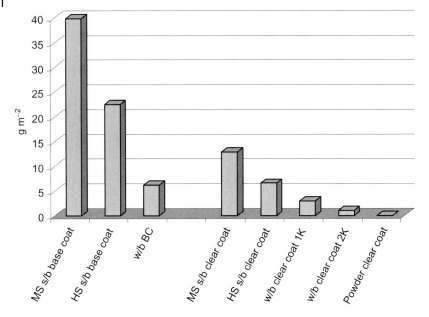

Fig. 6.14 Contribution (g m^{-2}) to solvent emissions from base coats and clear coats for an automotive coating (w/b = waterborne, s/b = solventborne).

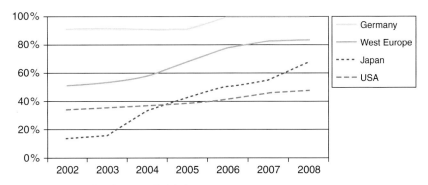

Fig. 6.15 Historical and expected global conversion to waterborne base coat technology for automotive painting processes.

In Japan, all carmakers have active programs to switch over to waterborne base coat. In the United States, all European and most Asian transplants, as well as most Chrysler plants, use waterborne base coat, while GM and Ford continue to remain mostly on HS base coat technology. Developing countries like China, Korea, and Brazil have approximately 20% of their plants on waterborne base coat currently and are also expected to convert to this technology.

6.4.6
Drying of Base Coats

In the coating process, base coat and clear coat are applied 'wet-on-wet' – that is, the clear coat is applied without fully drying the base coat layer (see Chapter 8). Full drying and curing is achieved only in the clear coat–baking step. The short (1–8 minutes) intermediate drying of the base coat is called *flash off*, and typically leads to a solid content >90% in the applied film. This dehydration process is not a simple diffusion–evaporation process as model calculations have demonstrated [8]. In reality, this often leads to several steps of material and process optimization to avoid solvent or water popping.

When waterborne base coats were introduced into the automotive industry in 1988, the flash off process between base coat and clear coat application was considered most critical. At that time, a sequence of 3-minute infrared (IR) radiation [9], 5-minute convection at 80 °C, and subsequent cooling to <40 °C was considered to be the optimum process. The advantage of the IR zone was its capability for faster energy transfer into the wet film and the heat being generated from the bottom to the top of the film supposedly leading to less popping. As the IR absorption varies by color, the radiance level was adjusted per color to achieve more uniform heat transfer. The convection zone was operated at 4–5 m s^{-1} air velocity (measured at the nozzle) to get the necessary solid content of the base coat film.

Today the IR zones have been eliminated from most OEM lines, and air velocities in the convection zone have been increased up to 15 m s^{-1}. Although the heat of evaporation is much higher for water than for the solvents used in solventborne base coats, the energy required to evaporate approximately 1 kg of water from a car versus the heat capacity of 1000 kg of steel and the good thermal conductivity of steel is negligible. In this case, the speed of the drying process is not limited by heat transfer, but kinetically by the removal of water vapor from the film–air interface. When drying waterborne base coats on plastic substrates with their low heat capacity and low thermal conductivity, large temperature drops are observed and heat transfer becomes more important.

6.5
Clear Coat

6.5.1
Market

While technologies for e-coat, primer surfacer, and base coat are rather homogeneous, the number of chemistries used, and the number of forms of supply, is biggest with clear coats:
- SBCC1 = 1K solventborne clear coat (all types)

- SBCC2 = 2K solventborne clear coat PUR
- WBCC = 1K waterborne clear coat
- PCC = Powder clear coat.

Figure 6.16 shows the distribution of clear coat technologies in Europe and globally for 2005.

6.5.2
Liquid Clear Coats

As can be seen from Figure 6.16, liquid clear coat is the dominant form of application on a worldwide basis. This is because of the fact that application techniques as well as the chemistry are well understood and optimized, and guarantee excellent performance for a complete automotive coating. The chemistries vary somewhat according to different market needs or customer specifications (see Table 6.4). Nearly all systems are based on acrylic resin backbones, which are functionalized either with OH groups for cross-linking with melamine resins or reactive polyurethanes, with epoxy groups for cross-linking with carboxyl groups, or with carbamate groups cross-linking with melamine resins (see also Chapter 7). They are mostly formulated as one-component systems that need to be sufficiently stable for storing and handling at room temperature.

6.5.2.1 One-Component (1K) Acrylic Melamine Clear Coat
1K liquid clear coats is the most common technology used in the automotive industry. It is typically based on combinations of acrylic polyols (Ac) and amino

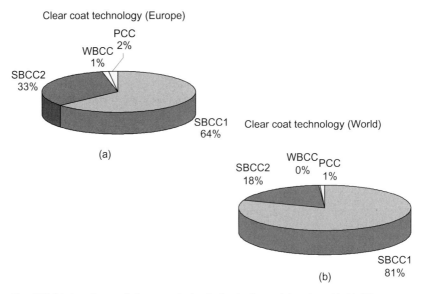

Fig. 6.16 Market shares of clear coat technologies in Europe (a) and worldwide (b).

Table 6.4 Mateial and process data of actual clear coat systems (Source: DuPont)

	1K PUR MS	Carbamate HS	Epoxy acid HS	Silane HS
Solid content weight %	42–48	50–58	50–55	54–64
VOC g l^{-1}	380–450	250–400	300–400	100–200
Applied viscosity sec DIN4 23 °C	30–35	40–60	40–60	35–40
Film build μm	35–45	45–50		45–50
Flash off	3–8 min 23 °C	3–8 min 23 °C	3–8 min 23 °C	3–8 min 23 °C
Baking	20 min 145 °C	20 min 150 °C	20 min 140 °C	20 min 150 °C

cross-linking agents (MF, melamine resins). Such systems are cured for 12–20 minutes at temperatures ranging from 130–150 °C. When acid catalyzed for lower bake repair the temperatures can be reduced to 90–120 °C. The principal chemical reaction of curing involves condensation of the alkoxylated melamine resin with the hydroxyl group of the polyol as illustrated in Figure 6.17. As shown, an ether cross-link is formed. This type of bonding is sensitive to hydrolysis at pH conditions

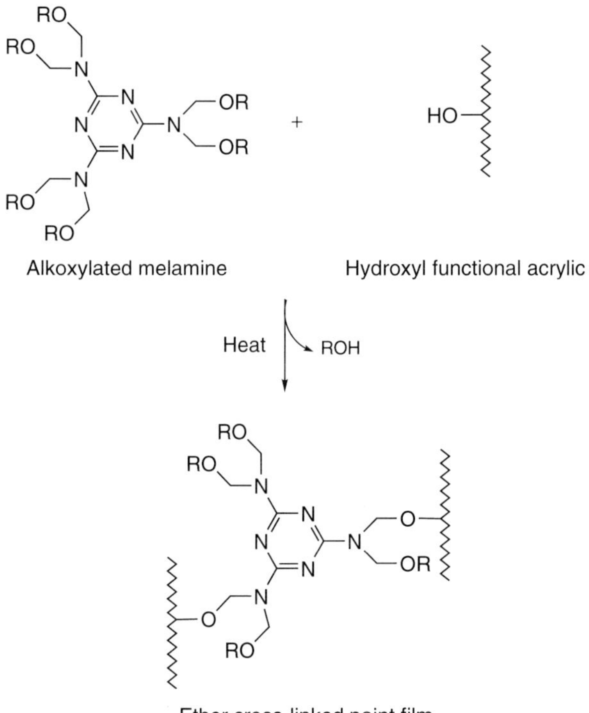

Fig. 6.17 Cross-linking reaction of 1K (AcMF) clear coat.

below six, which then leads to a chemical degradation of the clear coat surface. Subject to the environment, this material will be eroded away from the surface, leading to reduction in gloss and permanent visible damage. Because of their good cost–performance balance, acrylic melamine clear coats are still the dominant global clear coat technology.

6.5.2.2 Acrylic Melamine Silane

Starting in the mid-1990s, 1K clear coats with improved acid etch resistance were introduced to the market. This version of the 1K acrylic melamine clear coat has silane groups on the polyol chain [10]. These silane groups will hydrolyze during the clear coat baking step to form silanole groups, which then condensate to build an additional cross-linking via siloxane structures. The increased cross-linking density, as well as the chemical stability of the Si–O–Si network, leads to an improvement in etch and scratch resistance (see Section 12.4). Silane containing acrylic melamine clear coats have a relatively high market share in Europe and USA.

6.5.2.3 Carbamate-Melamine-Based 1K Clear coat

Another technology used to improve etch resistance is based on primary carbamate functional polymers [11] cross-linked with melamine resins to provide thermosetting urethane coatings (see Figure 6.18). Till today, carbamate clear coats are rarely used outside the United States.

6.5.2.4 One-Component Polyurethane (PUR) Clear Coat

As the 2K polyurethane (PUR) clear coat offers advantages over 1K AcMF clear coat with respect to acid etch resistance, clear coats were developed using blocked isocyanates. Malonate esters and dimethyl pyrazole are the preferred blocking agents for oligomeric polyisocyanates based on HDI (hexamethylenediisocyanate) and IPDI (isophoronediisocyanate) for appropriate bake conditions and low yellowing tendencies. In most cases, these blocked polyisocyanates are used in combination with melamine resins. Baking conditions are 130–150 °C and catalysis for lower bake cure is not very efficient (see Chapter 7).

1K PUR clear coats offer a good balance of etch and scratch resistance and are mostly used by European car manufacturers.

6.5.2.5 One- (and Two-) Component Epoxy Acid Clear Coat

The cure reaction of a gycidyl-functional acrylic polymer with an aliphatic polycarboxylic acid proceeds as shown in Figure 6.19 to give a β-hydroxy polyester.

The acid etch resistance of epoxy acid clear coats is among the best commercial clear coats and certain OEM-specific acid etch tests, for example, the Toyota zero etch test, almost mandate this chemistry. Recently, mar resistance has been improved by increasing the cross-linking density by higher number of functional groups on the polymer chains. When formulated as a 1K system, the shelf life is rather limited. In some cases epoxy- and acid-components are supplied separately (2K) to the automotive paint shop. Epoxy acid clear coat chemistry is mostly used by Japanese carmakers.

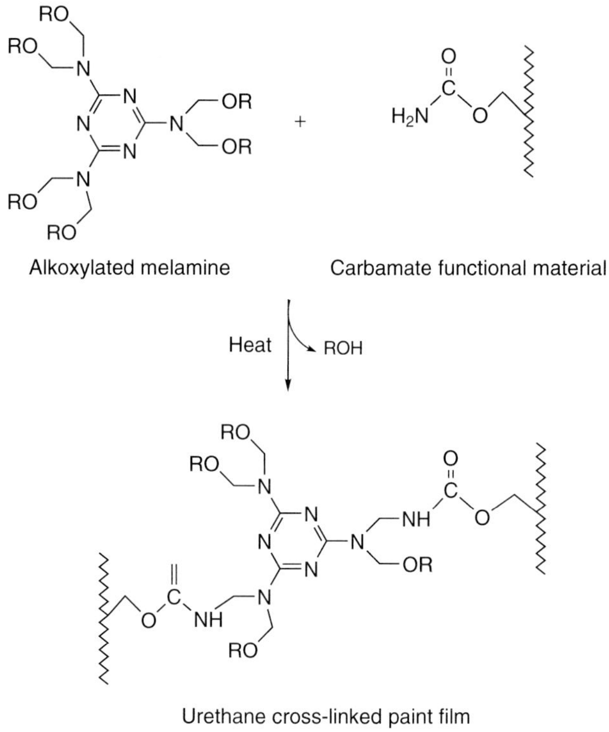

Fig. 6.18 Cross-linking of melamine carbamate clear coat.

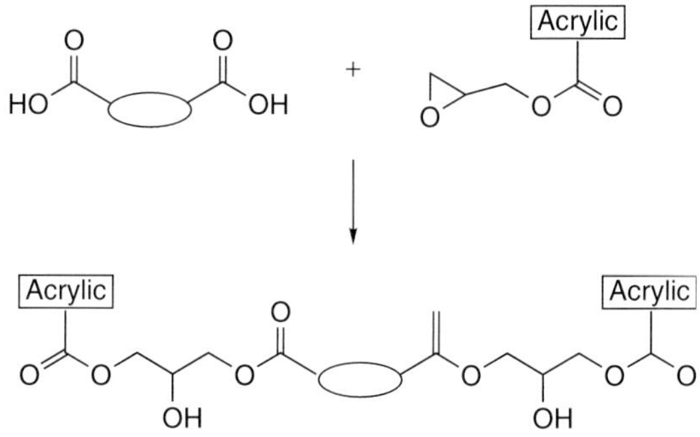

Fig. 6.19 Cross-linking reaction of an epoxy acid clear coat.

6.5.2.6 Two-Component (2K) Polyurethane Clear Coat

2K polyurethane clear coats are widely used in the European car industry, but are used to a lesser extent globally (see Figure 6.16). The chemical reaction by which the coating cures involves reaction of the polyisocyanate with the hydroxyl group of a polyol to form a urethane cross-link as illustrated in Figure 6.20. The reactivity of the hydroxyl isocyanate reaction leads to a pot life of typically 4–8 hours, which prohibits one-component packaging. However, this reactivity does enable cure over a broader

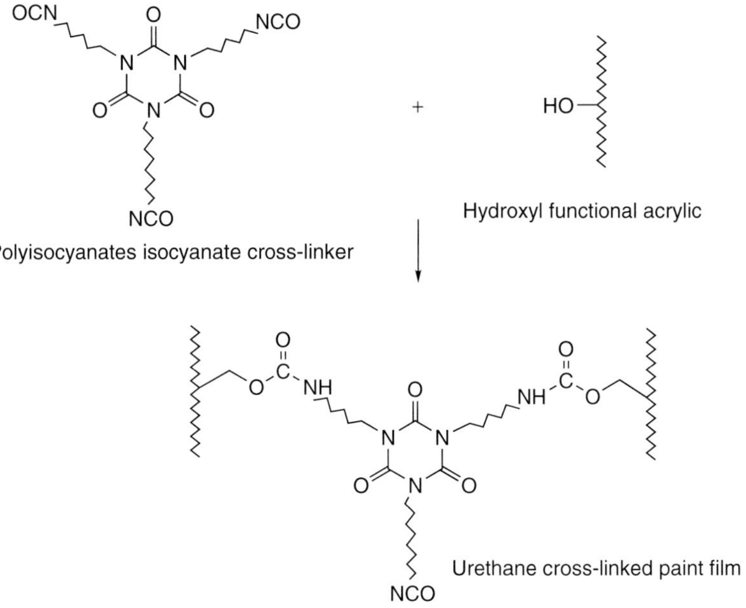

Fig. 6.20 Cross-linking reaction of 2K-PUR clear coat.

range of curing conditions. Such systems are cured at 120–150 °C, but are also used in lower bake repairs at 80–100 °C. In contrast to hydroxyl-melamine-based coatings, hydrolysis plays a minor role in degradation of a urethane coating. These coatings show improved appearance and acid rain resistance.

The low molecular weight polyisocyanates have a high tendency to redissolve the base coat, which leads to a misorientation of the effect pigments, also described as *strike-in* or *mottling*. In combination with waterborne base coats, the effect is small as these are made of higher molecular weight latex binders, which are more stable toward strike-in. The acrylic and polyester polyols are in development for further VOC reduction [12].

In most paint shops, the mixing of the polyol component and the polyisocyanate cross-linker is done in a static mixer (Kenics mixer) with both components being pumped in a 3 : 1 ratio [13].

6.5.2.7 Waterborne Clear Coat

Waterborne clear coat was first introduced in the automotive industry in 1990 at the Opel plant in Eisenach, Germany. Today, the chemistry is based on a waterborne polyester-acrylate cross-linked with a blocked isocyanate and melamine resins.

In 1997, DaimlerChrysler started using a waterborne powder slurry in its Rastatt plant, Germany. The initial product was based on a GMA(glycidyl methacrylate) acrylic powder clear coat, which was dispersed in water [14]. As of now, this product

Table 6.5 Material and process data for waterborne clear coats (Source: DuPont)

	1K (low VOC)	1K (zero VOC)
Solid content weight %	40–41	36–37
VOC g l^{-1}	130–140	Approximately 0 (zero)
Applied viscosity sec DIN4 (23 °C)	30–32	60
Film build (μm)	35–45	35–45
Flash off	5 min 23 °C	2 min 22 °C
Baking	2 min 50 °C/7 min 80 °C/24 min 150 °C	5 min 50 °C/7 min 80 °C/24 min 155 °C

has been replaced by a solvent-free emulsion based on acrylic polyol, melamine, and blocked isocyanate.

6.5.3
Powder Clear Coat

Powder clear coat is an environmental friendly technology, as it does not emit any organic solvent during its application [15]. Besides this, powder clear coat has the following other advantages:
- direct recycling: the collected overspray powder can be directly used for the original coating process.
- no waste, waste water, or paint sludge from the clear coat application.
- no use of solvents for cleaning of application equipment or spray booth: just vacuuming.
- reduction of total energy: air supply to spray booth can be reduced by higher recycling rate, no VOC, very low toxic aspects.
- same film thickness and similar appearance on horizontals and verticals.

No other technology in OEM coatings can offer direct recycling and 'zero waste' operation, although optimized e-coat operations come close (see Chapter 4). One-hundred percent solids of the paint require lower flow rates at the applicators and help to improve transfer efficiency to >90%. While the average usage of clear coat is 3 kg per car for medium solids and 2 kg per car for HS, it is 1.5 kg per car for powder clear coat. It remains a future target to reduce the film build of powder clear coat from the current level of 65 μm in 2007 to 40–50 μm commonly used with liquid clear coat. UV-powder clear coat (see Section 6.5.5) may be a way to achieve this target.

Table 6.6 Material and process data for powder clear coat (Source: DuPont)

Properties	Data
Output per bell (g min^{-1}) external[a]	200–250
internal[a]	120–150
Voltage (kV)	80
Film build (μm) typical	55–65
best appearance	80–85
Flash off waterborne base coat	7 min 70 °C
Baking	20 min 145 °C
Repair	Panel or spot repair, 2K or powder

a areas

In the 1970s, GM Framington and Ford Edison plants started to coat cars with powder top coat. Several thousand cars were made using polyester hybrid powder, but appearance and durability of this chemistry were not competitive versus the acrylic enamels, and powder coatings thus disappeared from the OEM top coat lines, but moved into the primer surfacer area (GM Shreveport, 1980). Around 1990, GMA- based acrylic powder clear coats were developed. DDDA (dodecane-diacid) was mostly used as the cross-linker, leading to stable ester bonds in a polyaddition reaction (see Section 6.5.2.5 and Chapter 7). The initial commercial application of acrylic powder clear coat was at Harley Davidson, USA. In 1993, the US automotive industry formed the Low Emission Paint Consortium (LEPC), which later built a pilot line at the Ford Wixom plant in Detroit and studied powder clear coat from 1996–2000. In 1996, BMW built a powder clear coat line at its Dingolfing plant, Germany; and currently it is running five lines, two in Dingolfing, two in Regensburg, and one in Leipzig, all of which are in Germany [16].

The level of appearance, cleanliness, and reproducibility required in the manufacturing of powder clear coats for automotive coatings differ somewhat from products being commercialized in the general industry (Figure 6.21).

In the first manufacturing step, granules of resin, cross-linker, additives like ultraviolet absorbers (UVA), hindered amine light stabilizer (HALS), and additives for degassing, flow and leveling are loaded according to the formulation into a mixer and are dispersed with a rotating blade. This mixture is continuously fed into an extruder where the cross-linker is finely dispersed in the resin under the influence of temperature and shear. The molten mixture is extruded onto a cooling belt, where it solidifies and is broken into chips. These chips are loaded to a mill and ground to a well-defined particle size distribution. A cyclone classifier and sieve are used to eliminate oversized particles, which would stick out from the clear coat film, and fines, which lead to poor application properties such as clogging and spits.

As in a standard application process using waterborne base coat, the powder clear coat is applied after a heated flash off of the base coat at 6 minutes and 60 °C. To achieve good flow and appearance, clear coat film builds must be 60–70 μm.

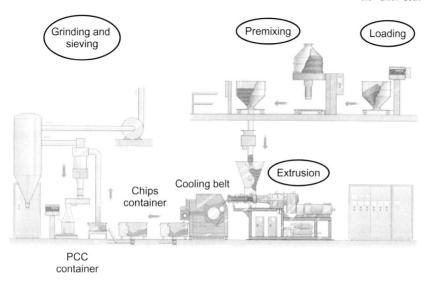

Fig. 6.21 Manufacturing sequence of powder clear coat.

Fig. 6.22 Bell application of powder clear coat.(source: BMW)

The melting and cross-linking of the powder clear coat is done at 20 minutes and 145 °C.

While BMW initially used spray guns to apply the powder clear coat, these were replaced in 2002 by powder bells, which allowed higher flow rates (less applicators) and led to a lower defect rate. Also in 2002, BMW introduced DDF (Digitale Dichtstrom Förderung, Ramseier Technologies AG) pumps for powder dosing and supply, which significantly reduced the consumption of air in the fluidization system improving the dosing accuracy and transfer efficiency.

A unique design of an OEM coating process is implemented at MCC Smart in Hambach, France. After normal cathodic electrodeposition coating, the metal frame is coated with powder primer and baked. Then a powder base coat (three colors including silver) and a powder clear coat as the final layer for gloss and durability are applied. Other exterior panels such as door panels or fenders are produced at a subsupplier with colored plastics and liquid clear coat [17].

6.5.4
Top Coat Performance

Automotive coatings are specially designed to fulfill the demanding requirements of the automotive industry (see Chapter 12). Besides gloss and appearance, the most important targets are as follows:
- environmental etch resistance
- durability
- scratch resistance.

It is mainly the clear coat that has to deliver these film properties.

6.5.4.1 Enviromental Etch
During the late 1980s, it became apparent that the environment was capable of severely damaging the automotive top coats. This was especially evident at the automotive import storing areas on Blount Island in Jacksonville, Florida. Many cars were severely damaged that they could not be sold 'as is' owing to extreme etching phenomena of the paint surface. It has since been well established that the local precipitation in this region and other polluted areas worldwide contain high concentrations of acid and other airborne pollutants, both organic and inorganic. In fact it is common for acid rain to have a pH of 4.5 [18]. The combination of acid rain and the high temperatures in Florida is sufficient to cause deep acid etching of automotive top coats in a short span of only a few months in the field. For this reason, many companies have adopted Jacksonville as a 'worst case' exposure site for automotive top coats.

There are two basic reasons why this problem became obvious during the late 1980s. The first relates to the introduction of base coat–clear coat systems, which have high gloss and DOI (distinctness of image). The new levels of gloss and DOI enable the eye to pick up small defects that are not yet readily visible in lower gloss–DOI coatings. The second reason for the increasing problem is related to the strict VOC emission regulations. Melamine–polyol clear coats dominated the automotive industry, and to meet VOC regulations, lower molecular weight polymers were formulated with higher concentrations of melamine cross-linkers, rendering the coatings more susceptible to acid rain damage.

Environmental etch is a clear coat appearance issue with the formation of permanent water spots or nonremovable marks from bird droppings, tree resin, or other chemicals getting into contact with the car surface. The physical damage resulting from etch is associated with the localized loss of material or deformation of the clear coat surface resulting in visible pitting of the clear coat surface. The acid etch phenomenon is believed to be primarily based on cross-link hydrolysis as a result of acid rain exposure, while marks from bird droppings are often explained as a pure deformation without any chemical change [18].

Table 6.7 Typical data of etch testing results for different clear coats in Jacksonville, Florida

Clear coat technology	Jacksonville etch ratings
Acrylic melamine	8.0–10.0
Acrylic melamine silane	5.0–7.0
Acrylic melamine carbamate	5.0–7.0
Epoxy acid	4.5–6.0
Acrylic urethane	4.5–6.0

Today the acid etch resistance is commonly determined by field exposure of test panels outdoors, commonly in the described industrial harbor area at Jacksonville, Florida. Following exposure, the panels are evaluated and rated by visual inspection and assigned a number from 0 (best, no etch) to 10 (worst), which categorizes the severity of the etch into three groups defined as: (i) 0–3, imperceptible to the customer; (ii) 4–6, perceptible to the critical customer but repairable by polishing; and (iii) 7–10, severe damage, customers will complain, repairable only by repaint. The exposure always is done in a 16–week period from May to September. From year to year, exposure conditions vary with the weather and are neither controllable nor reproducible. Obtaining relevant data then requires internal standards and exposure data over several years. Still this test has the highest acceptance across all carmakers and data exist over more than 15 years, while acid etch lab tests, for example, the gradient bar test, are not standardized across the industry and vary considerably between the different car makers. Rankings obtained with one test method may not correlate with rankings obtained by another lab method.

While Table 6.7 points to the weakness of the acrylic–melamine ether bonds toward an acid catalyzed hydrolysis, other factors influencing this test result are as follows:

- the glass transition temperature (Tg) of the cross-linked film: higher Tg leads to better ratings in all systems;
- the hydrophobicity of the film: lower water permeability is better;
- the cross-link density: higher cross-link density is better; [19]
- UV durability: better UV durability of the resin matrix is better.

6.5.4.2 UV Durability of Clear Coats

HALSs are used together with UVA in automotive clear coats to prevent UV-induced degradation of the polymer network. The UVA are typical of the 2-hydroxy phenyl benztriazole or the 2-hydroxy-s-phenyl triazine types (Figure 6.23).

UVAs are added to absorb the ultraviolet radiation in the wavelength range of 290–400 nm to protect the coating from UV-induced degradation. Usually the absorbers have little absorption >380 nm to avoid imparting a yellow color to the products. The presence of a strong intramolecular hydrogen bond between an O–H or N–H group and other oxygen- or nitrogen-containing groups allows

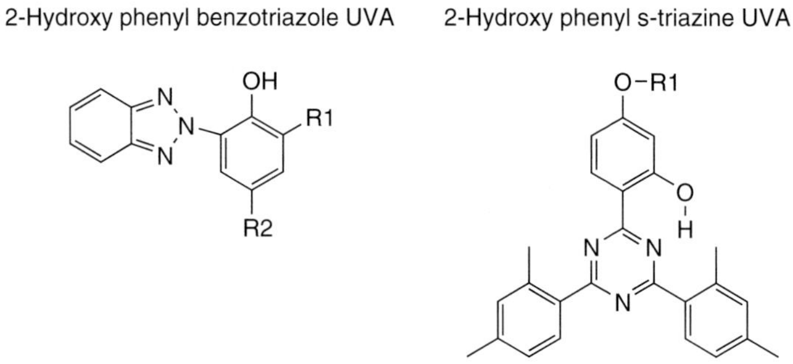

2-Hydroxy phenyl benzotriazole UVA **2-Hydroxy phenyl s-triazine UVA**

Fig. 6.23 Chemical structures of commonly used UV absorbers.

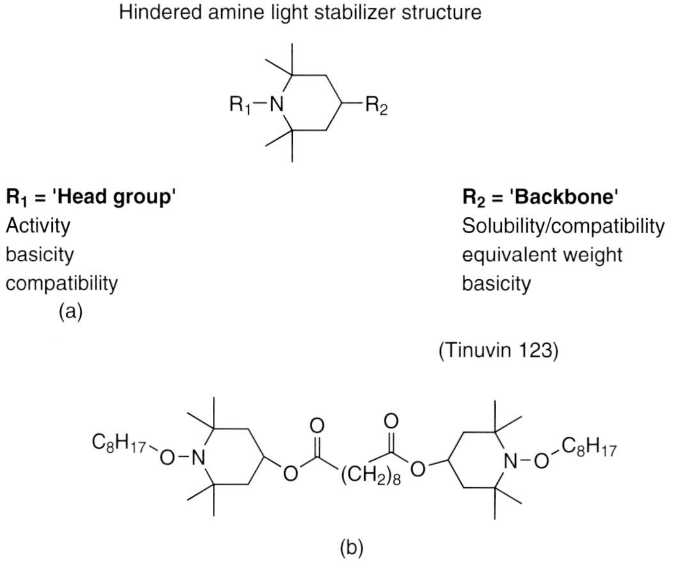

Hindered amine light stabilizer structure

R_1 = 'Head group'
Activity
basicity
compatibility
(a)

R_2 = 'Backbone'
Solubility/compatibility
equivalent weight
basicity

(Tinuvin 123)

(b)

Fig. 6.24 Basic chemical structure of hindered amine light stabilizer (a) and a typical example of commercial product (b) (source: Ciba).

the absorbed energy to be dissipated harmlessly as heat. UVAs have been shown to be slowly getting depleted from the clear coat by about 50% in 4 years by photochemical degradation and diffusion from the coating. UVAs are synergistically complemented by hindered amine stabilizers. All HALS currently used for coatings are based on the chemistry of the hindered piperidine ring [20]. The substituents on both the nitrogen ('head group') and position-4 ('backbone') of the ring affect both light stabilizing effectiveness and secondary properties such as basicity, solubility, and compatibility (Figure 6.24).

The HALS complement the activity of the UVA by scavenging free radicals that are formed either on the surface of the coating, where the UVA cannot effectively shield the polymer due to Beer's Law, or within the coating due to formation of

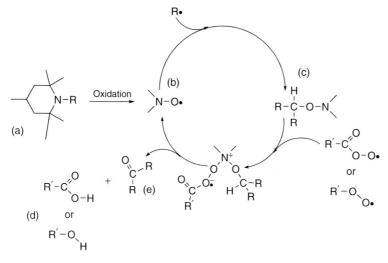

Fig. 6.25 Denisov Reaction cycle for clear coat stabilization by hindered amine light stabilizers

radicals from photolysis at wavelengths the UVA may not absorb efficiently. As result, they slow down the proliferation and associated destruction caused by alkyl and peroxy radicals.

The HALS effectively scavenge free radicals through a series of chemical reactions, in which they both react with and are transformed by various reactive radical species. However, rather than simply being consumed, as is the case with phenolic antioxidant radical scavengers, they have been shown to be transformed through a cycle of reactions that allows the regeneration of the various reactive scavenging species. Figure 6.25 shows the mechanistic cycle of the HALS chemicals.

6.5.4.3 Scratch Resistant Clear Coats

During recent years, scratch resistance has become a major area of clear coat R&D and test method development in the laboratories of the automotive coating suppliers, the automotive industry itself, and many research facilities [21–25]. Especially on dark color shades, smaller or bigger scratches become very visible. Marring is the English term used to describe the fine scratches introduced, for example, by car wash machines. These fine scratches are only 1–2 µm wide and a few hundred nanometers deep. With these dimensions being in the order of the wavelength of light, the observer will not see the scratches, but will only see the light scattering at the scratches. In many cases, these scratches are just surface deformations and level out over time and temperature (reflow).

Under the impact of keys or shopping carts, the clear coat surface starts to break, material is ablated, and a recovery in the form of reflow effects is no longer possible. Figure 6.27 shows the mechanical behavior of polymers under increasing load.

These film properties are best tested by the nano indentor method (see Section 12.4).

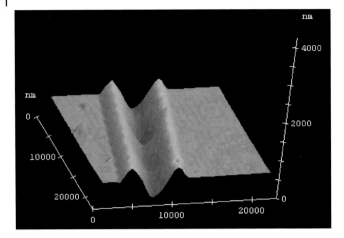

Fig. 6.26 Car wash scratches by scanning tunneling microscope (source: Dupont).

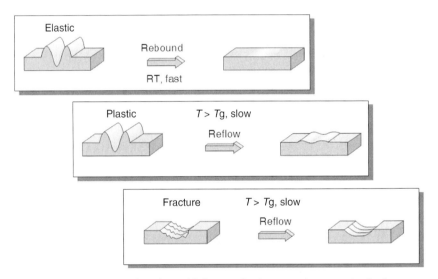

Fig. 6.27 Schematic view of mechanical behavior of polymers under increasing load.

From the nano scratch experiment, and in an earlier stage, also from microscratch experiments, [26] the resistance to plastic flow and the fracture limit (load at which fracture starts) can be determined. It is observed that 1K clear coats typically show a higher resistance to plastic flow (they are 'harder'), while 2K clear coats commonly show a higher fracture resistance as they are more resistant to hard impact scratches.

Scratch resistance can be related to cross-linking density and elasticity of the polymer network (Figure 6.28).

Type A represents a highly etch resistant clear coat having a poor scratch resistance because of low cross-link density. Type B represents a highly flexible clear coat, as used for plastic coatings, having improved scratch resistance, but poor

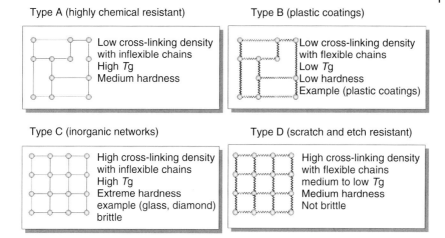

Fig. 6.28 Types of polymer network structures and their application in automotive clear coats.

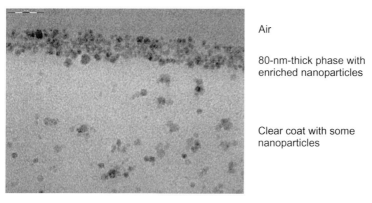

Fig. 6.29 Transmission electron microscopy (TEM) picture of clear coat with self-stratifying nano particles (source: Dupont).

etch resistance. Type C represents a glasslike inorganic network, which has good scratch resistance, but is too brittle to be used on automobiles. Type D represents a new type of scratch and etch resistant clear coat.

Another concept for scratch resistant clear coats is based on the incorporation of hard nano particles into a flexible polymer matrix [27]. Fumed silica has a primary particle size of 10–50 nm, which is the target range, but the large polar particle surface leads to unwanted rheology and agglomerates are causing haze. Haze can be related to Rayleigh scattering, which is proportional to the particle size and the difference in refractive index. More recently, nano particles have been made by a sol process. This leads to a more uniform particle size distribution and less haze. This concept can be combined with a self-stratification, which places most of the nano particles at the clear coat surface (Figure 6.29) [28].

6.5.4.4 Application Properties

Top coat appearance and color are strongly dependent on the application conditions. The application of coatings in the automotive industry is not standardized and takes place in approximately 1000 different paint shops worldwide using different application equipment, application processes, and conditions. Coating materials and application process have to be perfectly matched to obtain high-quality coating results and good productivity. One method to predict the process properties of a coating system at an early stage of the development is the DuPont Fingerprint Analysis System. This method is based on the experience that the film thickness of a coating represents a critical factor in the application process on which many paint properties depend (see Figure 6.30).

The method will analyze this dependency in the form of a correlation between statistically measured surface properties and film thickness.

The precise and reproducible coating of test panels is an important prerequisite. This is done with a laboratory-scale application robot using pneumatic gun and/or electrostatic bell atomization techniques for application. The system has to be very flexible to be able to simulate all variations of application conditions at all car plants. Still it is not always possible to set up exactly the same equipment and process found in the industrial line within the lab. In this case, panels are sprayed in the industrial automotive coating line, the same paint material is then taken to the lab and sprayed on the laboratory-scale application robot, and a correlation function is set up between the evaluation data measured from the two sprayout experiments.

When evaluating the optical properties of the paint film, the sample size of measurement points has to be sufficiently high to yield statistically relevant results. Automatic data acquisition using 500 measurement points per panel and collecting up to 2500 individual data on one panel was proved necessary.

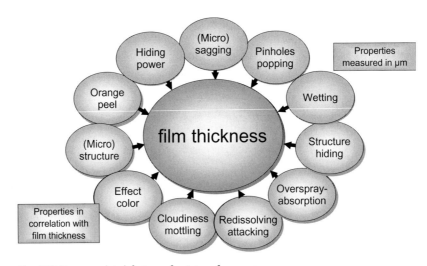

Fig. 6.30 Process related factors of paint performance

Improving the application robustness of coating materials versus variations in film build and process conditions is a key driver to reduce paint defects in an industrial coating line. The 'fingerprint' analysis is used in product development and product and process optimization. The following are the statistically recorded measurements:
- film thickness
- angular dependent lightness – flop
- long wave, short wave structure, image clarity
- color, gloss, haze.

From these data the following application properties are evaluated:
- process hiding
- sagging tendency
- pinholes, popping
- wetting
- overspray absorption
- mottling
- effect and color
- appearance.

Figure 6.31 shows two examples of base coat fingerprints. The graph on the top left shows significant variations in the observed brightness over film build. The brightness increases with film thickness until process hiding is achieved at 6 μm. While film thickness over 10 μm is achieved by two spray passes, poor overspray absorption can be seen in the transitional phase from the first to the second spray pass. Brightness variations in the form of mottling are a consequence of the inferior

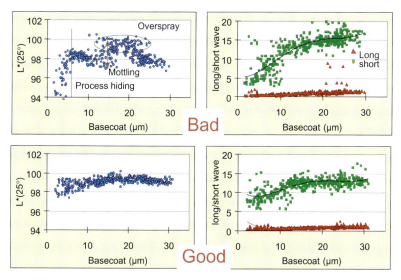

Fig. 6.31 Test results of 'fingerprint' studies of different base coats.

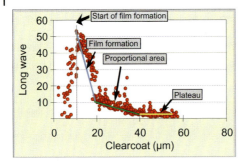

Fig. 6.32 Typical film forming characteristics of a clear coat by the 'fingerprint' method.

absorption of overspray. The graph below shows a more robust paint performance, where process hiding is achieved at 2 µm and brightness is virtually independent of film thickness. An improved absorption of overspray and superior stability during application can be expected in this case.

The two examples on the right hand side of Figure 6.31 compare the long and short wave structure of a base coat as a function of film thickness. The base coat of the top right graph shows unacceptable short wave structure at higher film builds, whereas the material below shows more stable performance.

Figure 6.32 shows an example of a clear coat tested by the 'fingerprint' method. Wetting of the substrate and film formation occurs at film builds between 10–20 µm. Between 20–40 µm, appearance of the panel is improved proportional to the clear coat film thickness.

Clear coat appearance then stays constant up to 60 µm indicating a stable paint performance. Popping or sagging may still occur at even higher filmbuilds and would be indicated by an increase and strong variability of the long wave measurement data.

6.5.5
Future Developments: UV Curing

As discussed in the previous chapter, scratch resistance is related to the cross-linking density of the polymer matrix.

Standard 1 and 2K clear coats have a cross-link density of 8–12 MPa. With UV curing, much higher cross-link densities are obtainable. Cross-link densities above 30 MPa no longer lead to better scratch resistance. (See Fig. 6.33).

The high cross-link density of UV-curing clear coats offers advantages in scratch and etch resistance [29]. The cross-linking is based on the UV-light- initiated radical polymerization of C=C double bonds. While all OEM clear coat lines use heat to cure the paint at typically 20 min at 140 °C, UV-curing paints complete curing in seconds. This allows coating lines with UV curing to build on a much smaller footprint.

Limited or no curing in shadow areas is a disadvantage of UV technology. Complex three-dimensional shapes like that of a car will always have areas not

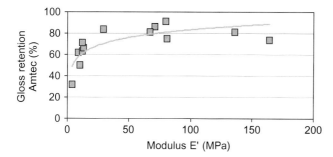

Fig. 6.33 Scratch resistance versus cross-link density expressed by loss modulus E'.

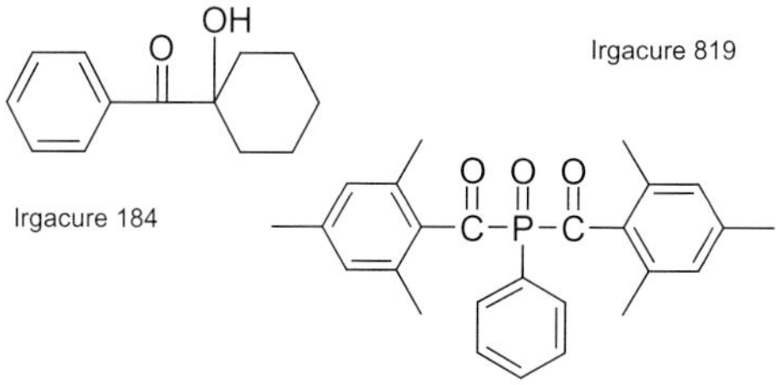

Fig. 6.34 Photoinitiatores used in UV clear coats (source: Ciba).

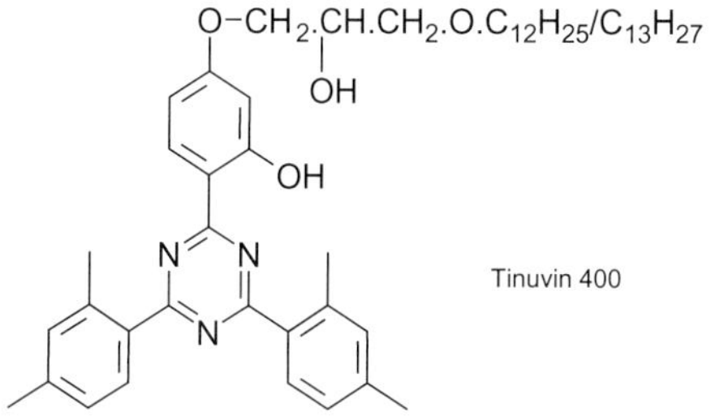

Fig. 6.35 UV absorber used in UV clear coats (source: Ciba).

directly exposed to the UV light. This consideration led to the development of 'dual-cure' clear coats using an isocyanate NCO- polyol cross-linking mechanism in addition to the radical UV-initiated polymerization reaction [30]. While the exterior visible areas of the car achieve full cross-linking by both mechanisms, interior shadow areas still meet basic film performance requirements through the formation of PUR structures from NCO and polyol.

Combinations of these photoinitiators offer the right balance of surface cure and bulk curing.

Within a project consortium involving Audi–Volkswagen, BMW, Daimler-Chrysler, DuPont, Duerr, Fusion UV systems, and Volvo; an UV pilot line was built in 2003, which allowed car bodies to be made complete [31]. The work done in the pilot line demonstrated a reduction of the length of the clear coat oven by >50%.

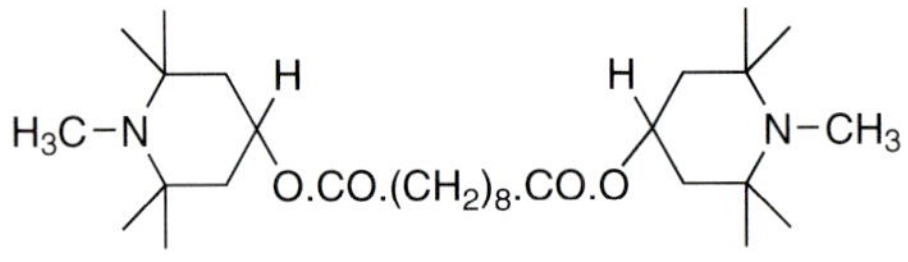

208 | 6 Top Coats

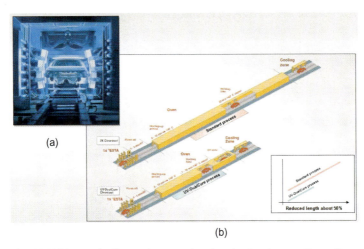

Tinuvin 292

Fig. 6.36 HALs used in UV clear coat (source: Ciba).

(a) (b)

Fig. 6.37 UV cure plot line and process length reduction (source: Dupont).

Although UV cure offers 'cold' curing capability, the waterborne base coat still requires temperatures >100 °C to evaporate all water and solvents and to achieve some degree of cross-linking for best properties of the coating system. A short baking of the base coat can probably not be eliminated.

6.6
Integrated Paint Processes (IPP) for Top Coat Application

6.6.1
Wet-On-Wet-On-Wet Application (3 Coat 1 Bake) of Primer Surfacer–Base Coat–Clear Coat

The first plant to use a waterborne paint system in a process of waterborne primer surfacer, waterborne base coat, and waterborne clear coat without intermediate baking was Mercedes Rastatt, Germany [32]. By eliminating the primer surfacer baking step, which typically would require 20 minutes at 160 °C, the total line length as well as the energy consumption is reduced. Furthermore, the sanding of primer surfacer is eliminated, reducing the size and headcount in the paint shop. In 2007, this type of process is being used in Mercedes, Rastatt, and BMW mini, Oxford, based on an all waterborne paint system.

The same process, but based on medium solids base coat and clear coat, was developed in Japan (Mazda) and a HS base coat and clear coat analog was developed in the United States. A commercial application is expected in 2007.

6.6.2
Primerless Coating Process

In the base coat spray booth, the base coat is typically applied in two stages. For solid colors, only the first ESTA bell application is used, while for metallic base coats, 60% of the total film build is applied using ESTA bells and the rest using pneumatic atomizers. In some cases, bell–bell application is also used.

For the primerless system, an isocyanate is added via a 2K Kennics mixer in the first step of base coat application. This activator modifies the characteristics of the base coat layer so that it adopts the function of a primer surfacer, which is dominantly chip resistance and the UV barrier for the electrocoat primer. In this case, it is a 'primerless' or 'wet-on-wet-on-wet' finish [33].

In the primer–top coat application, one of three spray booths and the whole primer surfacer drying–oven section are eliminated. The space required and the capital expenditure are reduced by about 30% in comparison to the present standard, as also the entire primer surfacer logistics – mixing room, paint circulation systems and primer surfacer–base coat color sequence control – are no longer necessary. The complexity of the painting process is clearly reduced. Another advantage is the reduction of material consumption. Compared to the 'wet-on-wet-on-wet' process, the primer spray booth and primer flash off zone are no longer needed.

The process of seam sealing, underbody protection, and sound deadening, which are normally integrated after the electrocoating step and baked in the primer oven have to be adapted. In most cases the current PVC ovens continue to be used.

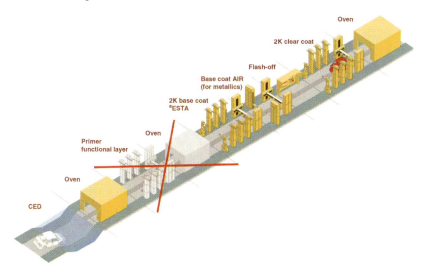

Fig. 6.38 Primerless coating concept using 2K application in the first basecoat stage.

The primerless coating process using isocyanate (2K) is mostly named *EcoConcept* and running at VW in Puebla, Mexico [34], and at VW in Pamplona, Spain, while a 1K version is called *IPP* II (integrated paint process) and running at BMW Mini in Oxford, UK [35].

References

1. Gee, P., Gilligan, S. (**2006**) *European Coatings Journal*, 5, 44.
2. Sing, L.P., Nadal, M.E., Knight, M.E., Marx, E., Laurenti, B. (**2002**) *Journal of Coatings Technology*, **74**(932), 55.
3. Liu, W., Caroll, J.B. (**2006**) *Journal of Coatings Technology*, **3**(10), 82.
4. Cramer, W.R., Gabel, P.W. (**2001**) *European Coatings Journal*, 34, 7–8.
5. Biallas, B., Stieber, F. (**2002**) *Farbe+Lack*, **108**(12), 103.
6. Boggs, L.J., Rivers, M., Bike, S.G. (**1996**) *Journal of Coatings Technology*, **68**(855), 63.
7. Kalenda, P. (**2002**) *Pigment and Resin Technology*, **31**(5), 284.
8. (a) Henshaw, P., Prendi, L., Mancina, T. (**2006**) *JCT Research*, **3**(4), 285. (b) Bosch, W., Cuddemi, A. (**2002**) *Progress in Organic Coatings*, 44, 249.
9. Bopp, M.L. (**2004**) *Journal für Oberflächentechnik*, **43**(3), 38.
10. Yaneff, P.V., Adamsons, K., Ryntz, R.A., Britz, D. (**2002**) *Journal of Coatings Technology*, **74**(933), 135.
11. Green, M.L. (**2001**) *Journal of Coatings Technology*, **73**(918), 55.
12. Vandevoorde, P., vanGaans, A. (**2005**) *European Coatings Journal*, 09/, 22.
13. Goldschmidt, A., Streitberger, H.-J. (**2003**) *BASF Handbook on Coatings Technology*, Vincentz, Hannover, p. 558.
14. (a) Woltering, J., Kreis, W., Streitberger, H.-J. (**1997**) *Proceeding Congress Surcar 97*, Cannes, (b) Clark, P.D. (**1995**) *Powder Coating*, **6**(3), 35.
15. Könnecke, E. (**2002**) *Journal für Oberflächentechnik*, **42**(3), 64.
16. Türk, T. (**2001**) *Journal für Oberflächentechnik*, **41**(9), 16.
17. *I-Lack* (**1998**), **66**(1), 65.
18. Henderson, K., Hunt, R., Spitler, K., Boisseau, J. (**2005**) *Journal of Coatings Technology*, **2**(18), 38.
19. Flosbach, C., Schubert, W. (**2001**) *Progress in Organic Coatings*, 43, 123.
20. Gerlock, J.L., Kucherow, A.V., Nichols, M.E. (**2001**) *Journal of Coatings Technology*, **73**(918), 45.
21. Schmid, K.H. (**2001**) *Chemie in Unserer Zeit*, **35**(3), 176.
22. Ryntz, R.A. (**2005**) *Journal of Coatings Technology*, **2**(18), 30.
23. Shen, W. (**2006**) *Journal of Coatings Technology*, **3**(3), 35.
24. Köhn, F. (**2007**) *Metalloberflache*, **61**(1-2), 12.
25. Meier-Westhues, U., Klimmasch, T., Tillack, J. (**2002**) *European Coatings Journal*, 9, 258.
26. Osterhold, M., Wagner, G. (**2002**) *Progress in Organic Coatings*, 45, 365.
27. (a) Fernando, F. (**2004**) *Journal of Coatings Technology*, **1**(5), 32. (b) Vu, C., Laferte, O. (**2006**) *European Coatings Journal*, 6, 34.
28. Patent EP 1204701, PPG (**2005**)
29. Schwalm, R., Beck, E., Pfau, A. (**2003**) *European Coatings Journal*, 39, 1–2.
30. Beck, E. (**2006**) *European Coatings Journal*, 4, 32.
31. Siever, L. (**2003**) *Journal für Oberflächentechnik*, **43**(9), 28.
32. Klasing, J. (**2000**) *Fahrzeug+Karosserie*, 53, 10.
33. Wegner, E. (**2004**) *Coatings World*, **9**(10), 44.
34. Dössel, K.F. (**2006**) *Journal für Oberflächentechnik*, **46**(9), 40.
35. Svejda, P. (**2006**) *Metalloberflache*, **60**(11), 12.

7
Polymeric Engineering for Automotive Coating Applications

Heinz-Peter Rink

7.1
General Introduction

Coating materials consist of a large number of individual substances: polymers, cross-linking agents, rheology, flow and structuring additives, reaction accelerators or inhibitors, light protection agents, pigments, and effect agents. The additives mainly support the application and film-forming processes, while the role of the pigments and effect substances is to meet the visual requirements of the coatings.

However, the specific technological requirements in automotive coatings made from paint, such as mechanical properties, acid or scratch resistance, and flexibility or resistance to yellowing, must be achieved by means of the polymers or polymer systems and cross-linking systems used in the coating materials.

This is because polymers and cross-linking agents contain complementary functional groups that react with each other under the relevant specified reaction conditions to form huge three-dimensional networks that can be compared to the skeleton of the cured coating. It has to be mentioned for the sake of completeness that a considerable level of self-curing also occurs with most common cross-linking agents (see Figure 7.1). The three-dimensional networks have a very significant impact on the property profile of the coating. This makes for a very vivid network image. It becomes clear that not only parameters such as the closeness of the mesh and the number of points of linkage, but also the type of 'mesh material' used must, of necessity, have a decisive effect on the property profile. The very high scratch resistance currently targeted in automotive paints, for example, is a function of both the glass transition temperature and the cross-link density [1]. The properties of such networks can even be measured now using the latest physical methods such as dynamic mechanical analysis [2, 3] and nano-indentation [4].

The polymer system and cross-linking agent are, therefore, the key components of the paint. Consequently, the apparently almost limitless options for variations in terms of the polymer systems open up for the paint chemist to manufacture

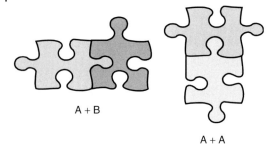

A + B

A + A

Fig. 7.1 Schematic complementary reaction and self-reaction.

customized polymers as optimally adapted as possible to the relevant specification profile, which has resulted in the coining of the term *polymer engineering*.

The aim of this introduction is to attempt to provide a guide enabling the reader to find a route into the relevant polymer chemistry via the concrete requirements of the desired automotive coating system. This will then be examined in greater detail in the subsequent sections in order that the strategic ideas underpinning polymer engineering may be understood.

The first issue here is undoubtedly the question of the conditions under which the coating is to or can be formed from the generally liquid paint, that is, the issue of the drying conditions. This question is, of course, also closely connected with the question of the type of cross-linking and, thus, primarily of the cross-linking agent.

A distinction is drawn in paint chemistry between two cases; the two-component systems in which the polymer and cross-linking agent are only mixed at the moment of use and the one-component systems in which the polymer and cross-linking agent are already premixed in the finished paint.

For drying conditions below 80 °C, two-component systems based on polymers and cross-linkers that can react very quickly with each other, that is, that have very reactive functional groups, are used almost exclusively. This is the only means of guaranteeing satisfactory curing at low temperatures and achieving both the desired cost-effectiveness and performance in the coating. The clearest examples of this are the isocyanate-based cross-linking agents, which are described in greater detail in Section 7.7.2.3. These systems are used for in-plant repair of automotive coatings.

Higher stoving temperatures reduces dwell time of the coating process and enable one-component systems to be used. In this case, the polymer cross-linking agent system has to be designed such that no reaction occurs at low temperatures, yet the cross-linking reaction can be initiated at higher temperatures. One-component systems permit easier inventory control and this means that mixing errors at the user's end can be avoided. However, more energy, and possibly a longer curing time, is required for the curing process.

There is a range of cross-linking agents available for formulating one-component systems. This includes melamine and benzoguanamine resins as well as cross-linking systems based on blocked isocyanates. Section 7.7 presents the common cross-linking agents in their respective application fields.

The remaining specification profile of the coatings is largely determined by the type of polymer used. This is where polymeric engineering begins from the paint chemist's point of view.

If the first question, therefore, concerned the film-forming conditions, the second question to be answered for automotive applications is, without doubt, the issue of the desired paint film in the paint system. This is because the respective paint film embodies very special combinations of properties that have to be covered by the polymer system.

Certain classes of binder and resins embody such combinations of properties better than others, as a result of which particular polymer types from the polymeric engineering toolbox are selected in practice, in preference to others for particular films in the paint system.

For example, polyacrylics (see Section 7.2) are used in the automotive industry especially in topcoats, which is the result of the outstanding gloss retention and excellent UV and chemical resistance of these materials. At the same time, the components used to manufacture this polymer class permit excellent management of the property profiles of the coatings to meet the very rigorous demands in visual terms (gloss, haze), in combination with mechanical properties such as resistance to stone chipping and scratching. In addition, however, polyacrylics are also used in functional layers such as primer surfacers or as additives.

However, the dominant polymer class in the functional layers, for example, the primer surfacer, is made up of polyurethane-based resins (see Section 7.4). They produce very good physical drying properties in the coating materials, have excellent pigment wetting properties [5, 6], exhibit better resistance to chemicals than polyesters, and provide coatings with a very broad portfolio of mechanical properties since they can generate substructures by means of phase formation [7]. Polyurethane microgels are also of interest from the viewpoint of controlling the rheological properties of coating materials [8–10].

Epoxy resins (see Section 7.6) are inextricably linked with the corrosion-protection coats in automotive paint systems. As coating material, they are used in the form of dispersions in electrochemical dip coating processes. This involves applying a wet film to pretreated metal bodies in an electrical dip coating process, with this film being stoved at temperatures of between 165 and 185 °C. The process used nowadays is the fully automated cathodic deposition of the polymer on the pretreated metal substrate. This process is therefore known as *cathodic electrodeposition* (CED).

The last polymer class is that of the polyesters. It is not possible to assign their roles as clearly as the other examples mentioned above, to a particular film coat. However, this class is a good example of a special tool in the paint chemist's toolbox. This binder class lends striking properties to coatings such as excellent gloss and flow, while also offering a good mechanical profile. The application window extends from high-gloss coatings on dashboards to use in topcoats, solvent-borne base coats, primer surfacers, and powder coatings. A further important application is the production of polyurethanes in coating materials for the automotive industry, which is dealt with in greater detail in Section 7.4. Polyesters are, furthermore, the common basis for powder coatings, which have gained importance in automotive coatings processes (see Chapter 5).

Another important question that has to be answered to achieve targeted control of the polymer design concerns the desired solvent system for the paint.

Various methods, the more common of which are mentioned briefly here, have been developed for achieving water solubility for the different binder classes during the development of waterborne materials.

The standard process for manufacturing acrylics in water, for example, is emulsion polymerization (see Section 7.2.2.2). Polyesters are generally transformed into a water-dispersible form by modifying the polymer molecules with polar functional groups (see Section 7.3.2), whereas with polyurethanes, self-emulsifying systems are of particular importance (see Section 7.4.2) in addition to external emulsifying systems. The polymer molecule has to be specially designed depending on the particular method. The paragraphs cited illustrate the polymer design routes more precisely.

If the decision is taken in favor of a solvent-based system, steps must be taken to clarify how, for example, the wish for high-solids systems influences the polymer design. An even greater challenge is undoubtedly the formulation of completely solvent-free systems, that is, powder coatings (see Section 7.7.3), which require both special polymers and cross-linking agents.

The decision-making matrix used in the past provides the basic outlines within which polymeric engineering then has to fine-tune the product. The choice of monomers for building the desired polymer, the molecular weight of the polymer, the molecular weight distribution of the polymers, the number and configuration of the network-forming reactive groups, and the branching of the polymers are all parameters that help to achieve targeted management of the property profile.

The above-mentioned variable parameters are discussed below using the various binder classes and are compared with the corresponding specification profile.

7.2
Polyacrylic Resins for Coating Materials in the Automotive Industry

The polyacrylic resins (polyacrylics) in question are copolymers that are formed by radical polymerization in solution (solution polymerization) or water (emulsion polymerization) of acrylic monomers, methacrylic monomers, and other radically polymerizable compounds like styrene or vinyl ethers (Figure 7.2).

7.2.1
Managing the Property Profile of the Polyacrylic Resins

As already stated in the introduction, acrylics are often chosen as the polymer class if outstanding resistance properties like UV resistance, chemical resistance, and gloss retention are to be achieved. This can be attributed to the particular chemical

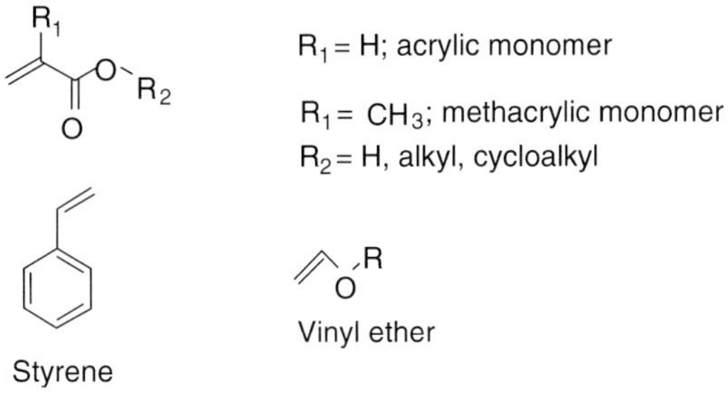

7.2 Polyacrylic Resins for Coating Materials in the Automotive Industry

R_1 = H; acrylic monomer
R_1 = CH_3; methacrylic monomer
R_2 = H, alkyl, cycloalkyl

Styrene

Vinyl ether

Fig. 7.2 Monomers for radical polymerization.

structure of the polyacrylics as the polymer backbone is based in this case on chemically stable, and therefore also hydrolysis-stable, carbon–carbon bonds.

Another feature of the acrylics is that the property profile of the coatings generated by them can be particularly effectively managed depending on the choice of monomers used in their structure.

For this reason, the most commonly used structure/effect relations used in polymer engineering for acrylic polymer design will first be presented since these have a general significance and are valid for both waterborne and solvent-based paint formulations.

Probably the most important variable for characterizing acrylics is the glass transition temperature T_g, which is the temperature that represents the turning point in a coating's curing curve [11]. High glass transition temperatures in the copolymers are associated with high surface hardness and thus also with increased scratch resistance, while the adhesion properties, tendency toward cracking, and impact resilience of the coating are influenced in exactly the other direction by the glass transition temperature. The glass transition temperature of polyacrylics can be approximately calculated after *Fox* [12] with the aid of the glass transition temperatures of the homopolymers of the corresponding monomers.

$$1/T_g = (\acute{\omega}_1 \times T_{g,1}) + (\acute{\omega}_2 \times T_{g,2}) + \ldots + (\acute{\omega}_n \times T_{g,n}) \qquad (1)$$

where

T_g = glass transition temperature of the copolymer consisting of 1 to n monomers in Kelvin;
$\acute{\omega}_1$ to $\acute{\omega}_n$ = weight of the respective monomer as a proportion of the polymer weight; and
$T_{g,1}$ to $T_{g,n}$ = glass transition temperatures of the homopolymer of the monomer 1 to n in Kelvin.

Divergences occur here in particular in the event of low molecular weights and deviations from the linearity of the polymer.

A few qualitative rules illustrate the above formula. Thus the copolymerization of styrene results in a higher glass transition temperature in the coating, which brings advantages in terms of surface sensitivity. However, excessively high styrene content results in drawbacks in terms of the coatings' UV resistance [13]

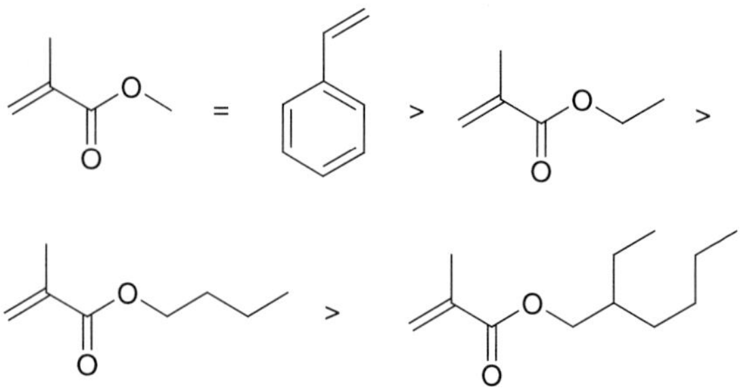

Fig. 7.3 Monomers decreasing glass transition temperature of the homopolymers.

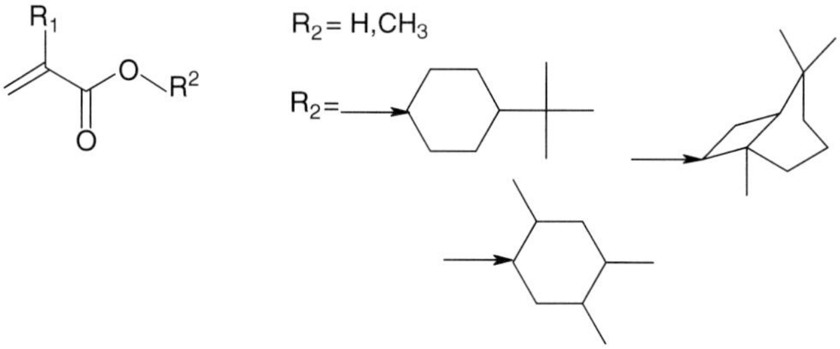

Fig. 7.4 Special monomers for high T_g polyacrylics with cycloaliphatic ester groups.

and flexibility. Hence it is preferred to produce a higher glass transition temperature via methacrylates such as methyl methacrylate, which does not exhibit this disadvantage of styrene (Figure 7.3).

However, the glass transition temperature of the polymers is overwhelmingly influenced by the type of ester group in the acrylic monomers used. Thus short alkyl groups result in polymers with higher glass transition temperatures, while those with longer chains result in polymers with lower temperatures. Monomers with cycloaliphatic groups have a somewhat special position. Such monomers result in polyacrylics with relatively high glass transition temperatures that can, however, be processed in organic solvents to make low-viscosity initial solutions [14] (Figure 7.4).

This has a certain significance when formulating high-solids paints. The glass transition temperature is also important in the manufacture of polyacrylic dispersions by means of emulsion polymerization [15]. The features involved here are the minimum film-forming temperature, the physical drying, and the usability properties such as block resistance. A further important characteristic for acrylic polymers is the acrylate/methacrylate ratio and the proportion of further copolymerized compounds such as styrene or vinyl compounds. The copolymerization of styrene with (meth)acrylates improves, for example, the gloss and chemical resistance of coatings.

It is, of course, not the equipping of the monomers with functional groups located in the lateral chains of the polymer backbone after polymerization that has a very significant effect on, for example, the conceivable cross-linking chemistry and the Cross-link density and thus on the mesh size of the three-dimensional

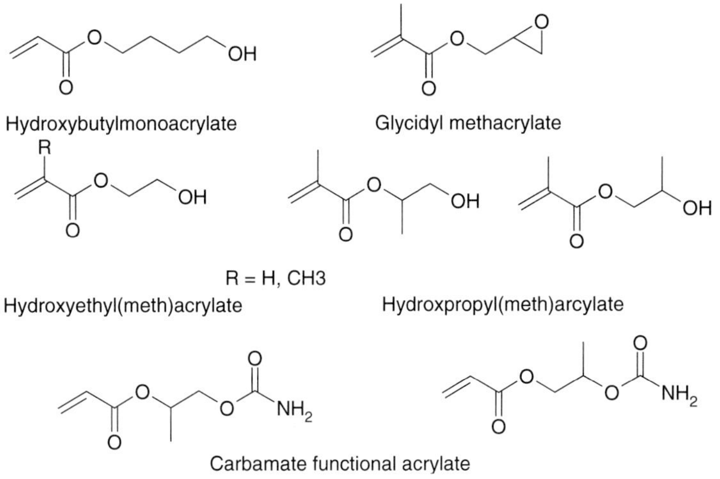

Fig. 7.5 Monomers carrying cross-linkable groups.

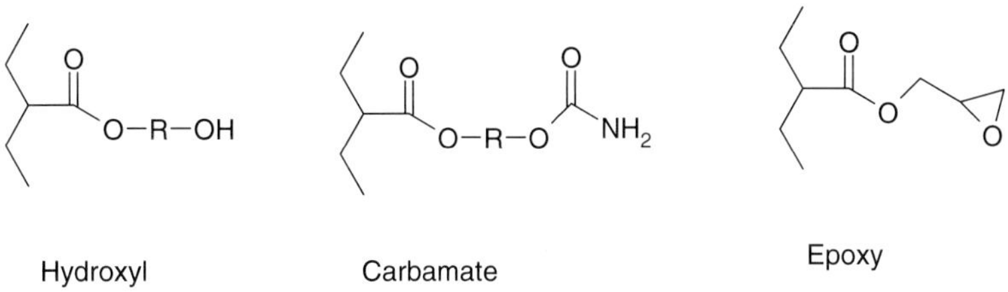

Fig. 7.6 Reactive side groups in polyacrylic backbones.

coating network mentioned initially. Polyacrylics with, for instance, hydroxyl- or epoxy-functional groups are particularly widely used (Figure 7.5).

This involves cross-linking hydroxy-functional polyacrylics with polyisocyanates, blocked polyisocyanates, melamine resins, and tris(aminocarbonyl)triazines (see Section 7.7.2), while di- or polycarboxylic acids or their blocked derivates [16] are used to cross-link epoxy-functional polymers. In order to manufacture waterborne polyacrylics, ionizable or other hydrophilic groups must be present in these lateral chains, with the result that in this case (meth)acrylic acid or (meth)acrylates with relatively long ether groups are proportionately copolymerized.

Apart from the monomer composition of the polymers, there are other important variables that affect the property profile of acrylic-based coatings (Figure 7.6).

For example, the molecular weights, their distribution, and the linearity or branched structure of the polymers play a major role in the coatings' performance.

The physical properties of the coating materials are also fundamentally affected. The molecular weight, for example, significantly influences not the physical drying but also the viscosity of solvent-based paints. Branched polymers result in a higher viscosity in the corresponding initial solution compared with linear polymers of the same molecular weight. This is extremely important especially with reference to the formulation of high-solids paints [17].

The molecular weight distribution of polyacrylics also has an impact on the viscosity of initial polymer solutions. Molecular weight distributions that are as tight as possible with mean molecular weights in the range from 1000 to 3000 are preferred in order to provide high-solids coating materials for the automotive industry.

7.2.2
Manufacturing Polyacrylic Resins

Various processes for manufacturing acrylics are described below. The manufacturing process should also be regarded as a polymer engineering tool to enable particular properties to be targeted and achieved in the final coating. The first property is the solvent system for the paint that is to be applied later. But the achievement of certain molecular weights, molecular weight distributions, and certain structures in the polymer are also generally characteristically associated with the choice of manufacturing process.

7.2.2.1 Manufacturing Polyacrylic Resins by Means of Solution Polymerization

The standard method for manufacturing polyacrylics in solvents uses semibatch processes [18]. However, continuous solvent-free processes [19, 20] are also described (Figure 7.7).

The process solvent used results in effective dissipation of the process heat and guarantees low viscosity in the reaction mixture, as a result of which thorough mixing is guaranteed. The usual method employed in the process is to feed the process solvent into the reactor and heat it to the reaction temperature. The monomer mixture and the initiator solution are then metered and fed in, which moderates the heat tonality. The maximum reaction temperature can be limited

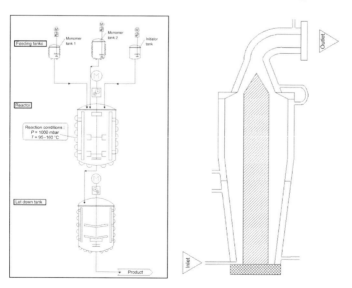

Fig. 7.7 Radical manufacture of polyacrylic resins in semibatch or continuous processes.

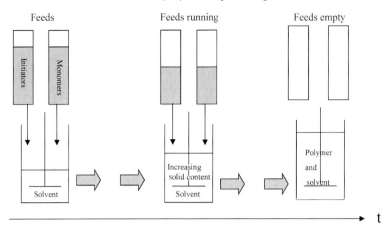

Fig. 7.8 Manufacturing procedure in semibatch processes of polymerization.

both by the boiling point of the process solvent – hot cooling – and by an appropriate technical control system (Figure 7.8).

The process solvent usually remains in the product, so it must also be suitable for the manufacture and later application of the coating materials.

Polyacrylics manufactured by means of polymerization in solution have mean molecular weights M_n of 1000–50 000. The molecular weight distribution varies between 2.0 and 20 depending on the design.

It should be mentioned at this juncture that work has been in progress over the last 10 to 15 years on manufacturing polyacrylics with as tight a molecular weight distribution as possible in order to produce low-viscosity polymer solutions and products that melt extremely well. The processes involved are group transfer polymerization [21], on the one hand, and controlled/live radical polymerization, on the other. The term 'live' in this context means that no termination and transfer reactions occur. Mechanistically speaking, the reactions involved are polymerization moderated with 2,2,6,6-tetramethylpiperidin-N-oxide (TEMPO) [22], metal-catalyzed atom transfer radical polymerization (ATRP) [23], reversible addition fragmentation chain transfer (RAFT) [24], and the addition fragmentation method with diphenylethene (DPE) [25]. To date, however, no polyacrylic manufactured using any of these methods has reached the stage of being used widely in coating materials for the automotive industry.

Radical polymerization in solution uses a range of organic initiators that decompose to form radicals under heat action. The initiators are chosen depending on the planned use of the polyacrylic and the technical reaction management of the polymerization process. There is therefore a wide choice out of azoinitiators, alkyl peroxyesters, dialkyl peroxides, and hydroperoxides (Figure 7.9).

The radicals produced by the initiator decomposition start the polymerization process, in the course of which they are partially bound in the polymer, remain in the solution as decay products, or evaporate.

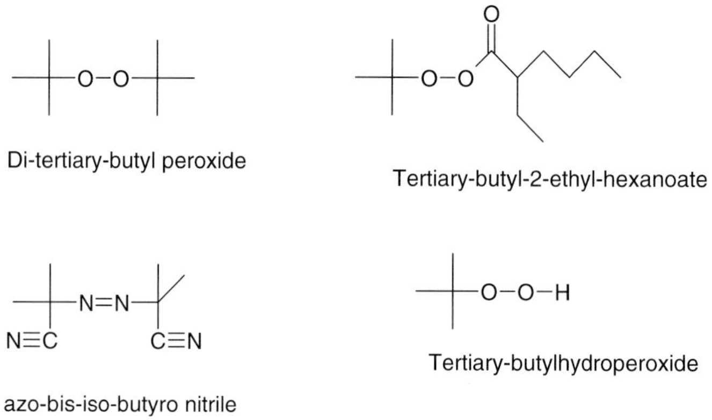

Di-tertiary-butyl peroxide

Tertiary-butyl-2-ethyl-hexanoate

azo-bis-iso-butyro nitrile

Tertiary-butylhydroperoxide

Fig. 7.9 Variety of initiators for radical polymerization.

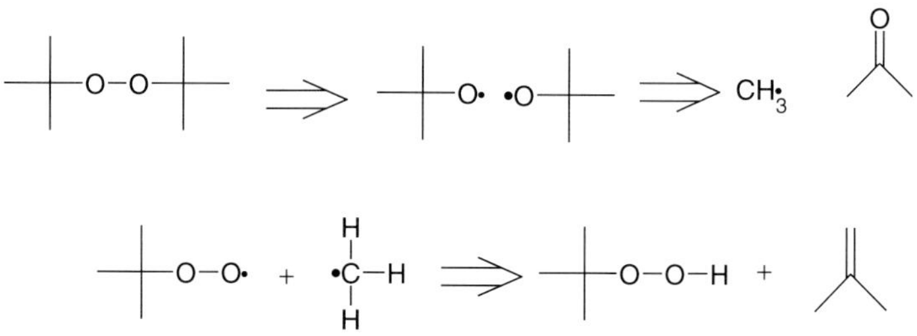

Fig. 7.10 Decomposition of di-*tert*-butyl peroxide.

Higher initiator concentrations generally result in lower molecular weights. Apart from the actual chain-reaction, termination reactions such as disproportionation, recombination, and beta-scission can occur. In addition, depending on the initiator choice and reaction conditions, graft reactions as a result of H-abstraction and back-biting as a result of transfer reactions can also take place [26]. These reactions generally result in wider molecular weight distributions in the polymers (Figure 7.10).

It again becomes clear at this juncture that characteristic product features such as the molecular weight and molecular weight distribution can be defined in the solution polymerization process as a result of various process parameters that can be specifically managed (e.g. type of initiator, quantity of initiator, reaction temperature).

The solution polymerization process described above reaches its limits if certain functionalities that are desirable for the lateral chains of the polymer backbone cannot be introduced via corresponding acrylic monomers. It has already been mentioned earlier that the OH functionality and epoxy functionality for cross-linking with isocyanates, melamine resins, or polycarboxylic acids can be introduced via appropriate monomers (Figure 7.11).

The situation is different with respect to carbamate functionality. Its introduction into polyacrylics is not so easy since appropriate carbamate-functional monomers, though already described [27, 28], are yet to become commercially available.

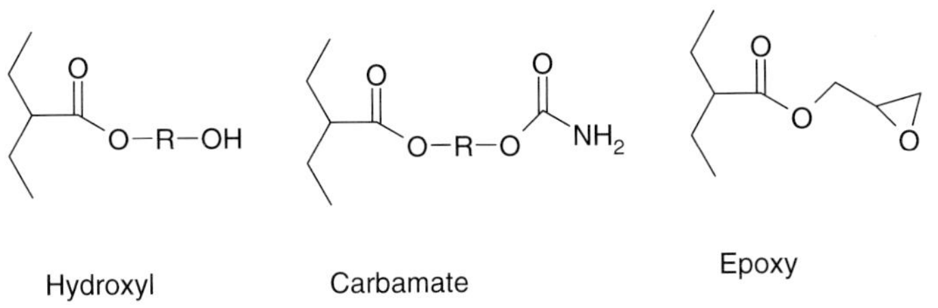

Fig. 7.11 Cross-linkable groups incorporated in a polymer backbone.

Labels: Hydroxyl, Carbamate, Epoxy

It would be attractive nonetheless to achieve such a lateral chain functionalization since it opens up the possibility of cross-linking acrylic polymers without the use of isocyanates and in this way to obtain networks that are not based on semiaminal structures but instead are based on carbamate bonds.

Various processes have therefore been developed that enable such polymers to be manufactured. In one process, the *in situ* reaction of hydroxyalkyl carbamates with (meth)acrylic anhydride is exploited to obtain the desired polymers [29] (Figure 7.12).

In another process, a hydroxy-functional polyacrylic is first manufactured, which is then converted – like a polymer – with methyl carbamate in a catalytic reaction [30, 31]. It may also be noted in this regard that aqueous polyacrylic dispersions can also be manufactured from solution polymers [32]. This requires the polymers to have an ionogenic or hydrophilic treatment. For example, monomers containing carboxyl groups such as (meth)acrylic acid and/or monomers containing ether groups can be used to manufacture polyacrylics (Figure 7.13).

Usually this is insufficient, as a result of which complex, multistage polymerization processes are used [33]. The subsequent modification with acid anhydrides is in principle another means of introducing the carboxy functionality into the polyacrylics (Figure 7.14).

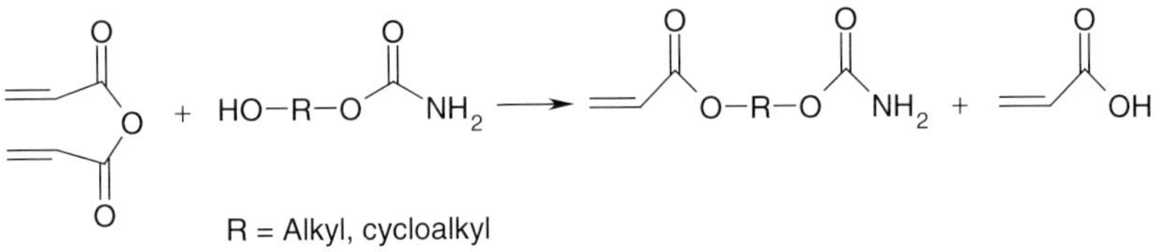

R = Alkyl, cycloalkyl

Fig. 7.12 *In situ* reaction of methacrylic anhydride with hydroxyalkyl carbamates.

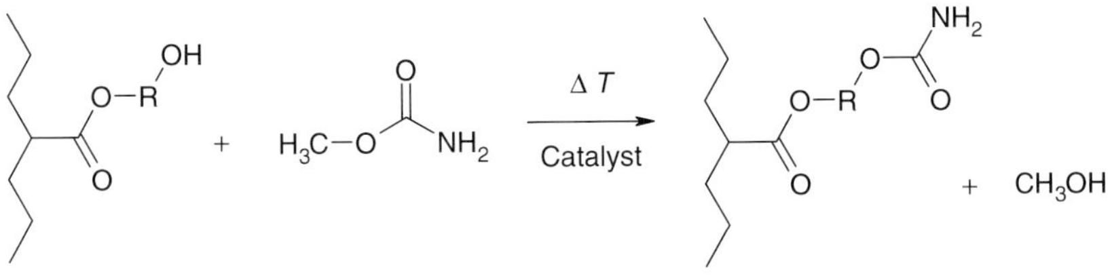

Fig. 7.13 Transesterification of polyols with methyl carbamate.

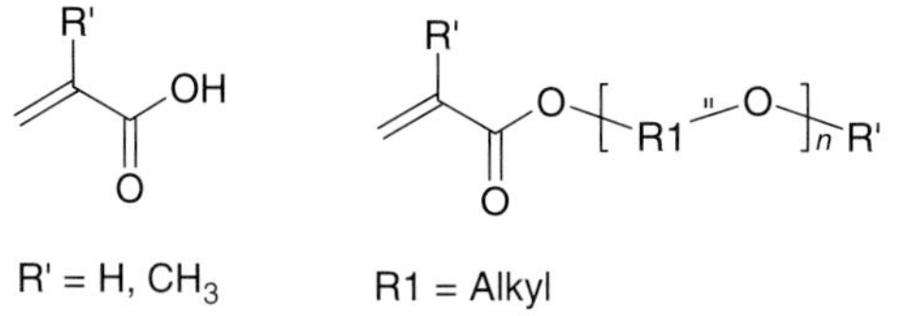

R' = H, CH$_3$ R1 = Alkyl

Fig. 7.14 Key intermediates for waterborne secondary polyacrylic dispersions.

The organic initial products are neutralized with bases and then dispersed with water. Depending on the process and solvents in question, a temporary rise in viscosity can occur, which is attributable to the formation of hydrates. Excess solvent is removed from the dispersions, depending on the desired solvent content, using vacuum processes. Such products can then be used in the manufacture of waterborne coating materials.

Cross-linking agents and customized polyacrylic resins are intensively mixed in the organic phase for the production of especially eco-efficient coating materials – solvent-free slurries. After neutralization of a small proportion of ionic groups and the addition of sterically stabilizing components, the process continues in the same way as in secondary dispersions. A further advantage here is that relatively high organic amine contents are not required.

7.2.2.2 Polymerization in an Aqueous Environment

The standard method for producing polyacrylics in water is by emulsion polymerization [34].

A characteristic feature of this manufacturing process – seen in idealized form – is that the polymerization takes place in micelles. Such micelles are formed by the addition of a characteristic quantity of an emulsifier to the reaction medium of water, because the emulsifier begins to form micelles above a certain emulsifier concentration, termed the critical micelle concentration (CMC). The nonpolar monomers diffuse into these externally polar and internally nonpolar micelles (Figure 7.15).

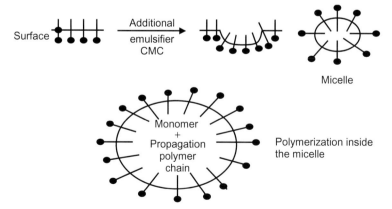

Fig. 7.15 Steps of the emulsion polymerization.

In the aqueous environment, an initiator forms radicals from the still available residual monomers, and these radicals start the polymerization process after diffusion into the micelles present there. The chain begins to grow, and rediffusing monomers, which can be very specifically defined thanks to the special reaction management process, gradually build the polymer.

Inorganic peroxides – primarily ammonium peroxodisulfate or hydrochlorides of aminic azoinitiators – are used as initiators to form radicals.

The use of peroxodisulfates yields polymers with sulfate groups, which results in good stabilization of the high-molecular polymers in water [35, 36]. However, the peroxodisulfates mainly decompose to form sulfuric acid and the corresponding sulfate salts. The resulting strongly acidic dispersions are usually neutralized with bases to deliver better usability and storage stability [37].

The polymers produced by means of emulsion polymerization are fundamentally different from waterborne systems, which are manufactured using secondary dispersions. Emulsion polymers have significantly higher molecular weights than solvent polymers as the polymerization process takes place in the micelle spaces, which are largely protected against termination reactions. The mean molecular weights are of the order of 100 000 to well over 2 million. Such molecular weights cannot be achieved using solution polymerization (Figure 7.16).

Furthermore, the emulsion polymers have only slightly cross-linkable groups in comparison with secondary dispersions, though their high molecular weights imply that they feature good physical drying properties.

Emulsion polymerization in particular, when compared with solution polymerization, also offers a raft of options for producing structured polymers (particles) with certain property profiles [38, 39] by, as already indicated above, adding monomers with a different property profile to the reaction medium at staggered times, for example, meaning – expressed very simplistically – that they are also incorporated at different growth phases of the chain.

Such structuring measures enable the minimum film-forming temperatures, hardness, and block resistance of the particles to be managed.

One special polymer type is represented by so-called polyacrylic microparticles [40, 41], which are also manufactured using the emulsion polymerization process. In this process diacryloyl-functional (meth)acrylics are also used as further comonomers. Such microparticles are used, for instance, to formulate waterborne base coats [42, 43] or as an organic initial solution for formulating solvent-borne base coats and clear coats [44]. They assist, for example, the alignment of effect pigments, increase the drying speed of solvent-borne top coats, and help to manage rheological properties in waterborne base coats.

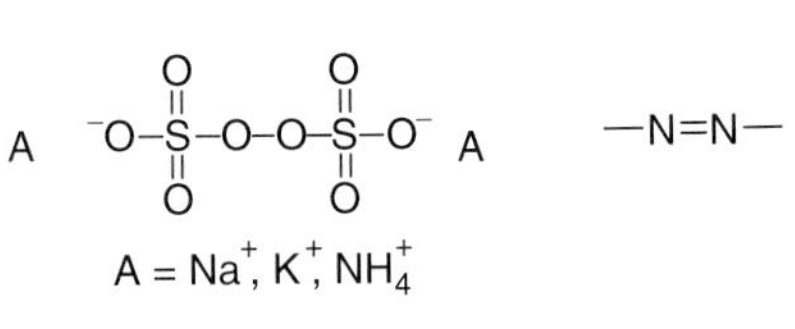

$$A^- \; {}^-O-\underset{\underset{O}{\|}}{\overset{\overset{O}{\|}}{S}}-O-O-\underset{\underset{O}{\|}}{\overset{\overset{O}{\|}}{S}}-O^- \; A^- \qquad -N=N-$$

$$A = Na^+, K^+, NH_4^+$$

Fig. 7.16 Typical initiators for emulsion polymerization.

When it comes to manufacturing polyacrylic microgels for use in organic coating materials, redox systems, which produce polymers that are low in salt groups, are generally used as initiators. As a result, these polyacrylic microgels can be converted more easily to the organic phase by salting out and also offer good surface properties.

The fact that emulsion polymers generally require low-molecular emulsifier levels of approximately 0.5–3% is definitely to be regarded as a drawback of the emulsion polymerization process. These emulsifiers are extremely polar components, which would preferably be avoided since they can be responsible for making coatings sensitive to moisture. For this reason, new processes that avoid this disadvantage are being developed. These involve carrying out polymerization under pressure in an aqueous environment without the use of low-molecular emulsifiers in the presence of certain polymers [45] or solvents [46].

7.2.2.3 Mass Polymerization

The manufacture of polyacrylics in mass is particularly eco-efficient if these are to be further processed into powder coatings or slurry coating materials. The particular challenges of these processes lie in managing high viscosities and avoiding the Norrish–Trommsdorf effect [47]. This is a self-accelerating phenomenon in the polymerization process as a result of the increasing viscosity of the reaction medium. A resulting heat buildup and possibly a shutdown of the reactor agitator would represent a serious potential danger.

Only a small number of technical processes that promise to be economically viable have therefore been developed to date [19].

7.3
Polyester for Coating Materials for the Automotive Industry

Polyesters are polycondensation products of diols and polyols with di- or polycarboxylic acids or their anhydrides and, possibly, saturated monocarboxylic acids or alcohols. Since the esterification reaction is an equilibrium reaction, the alcohol component is usually added to excess, resulting in a detectable proportion of this excess component remaining in the end product.

Polyesters are generally manufactured in batch and semibatch processes [48]. However, there are also approaches in which polyesters are manufactured continuously [49, 50].

7.3.1
Managing the Property Profile of Polyesters

The property profile of polyesters is significantly managed by the choice of polyester components. It defines the use of polyester resins for conventional and waterborne, as well as powder (see Chapter 5), primer surfacer. Here too, the key structure/effect relations, which are of decisive importance when synthesizing customized polyesters for the different coating technologies, will be given.

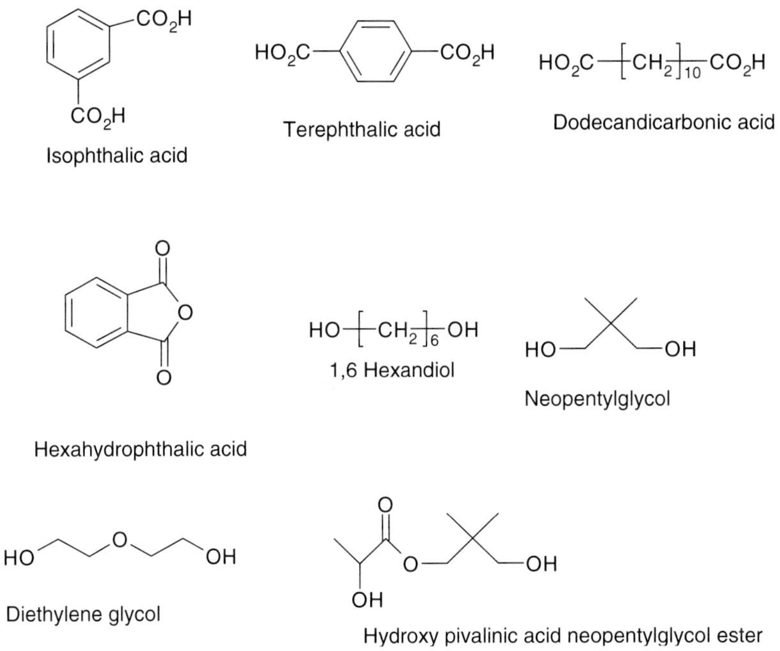

7.3 Polyester for Coating Materials for the Automotive Industry

Fig. 7.17 Typical raw materials for polyester resin composition.

Just as with the acrylics, the glass transition temperature with the polyesters is a characteristic variable that correlates with the hardness of the later coating based on these polyesters (Figure 7.17).

As a result of the use of relatively rigid aromatic dicarboxylic acids or carboxylic acid derivates such as terephthalic acid, isophthalic acid, or phthalic anhydride, the glass transition temperature of the later coating increases, whereas the use of more flexible aliphatic dicarboxylic acids or carboxylic acid derivates such as adipinic acid, dodecandicarboxylic acid, and hexahydrophthalic anhydride tends to reduce the glass transition temperature. The same applies by analogy when selecting the diols. Rigid diols such as hydroxypivalic acid neopentyl glycol ester or neopentyl glycol have the effect of increasing the glass transition temperature, whereas diols such as hexanediol or diethylene glycol have the opposite effect.

Managing the hardness is especially important, for example, when manufacturing polyesters for powder coatings. Glass transition temperatures of at least 40 °C, or preferably 50 °C or more, are required to prevent later blocking of the fine-particle powder coating. This is primarily achieved by using aromatic dicarboxylic acids such as isophthalic acid and terephthalic acid.

The flexibility of the polyesters is also significantly determined by the synthesis components, though the degree of branching and the molecular weight are also crucial factors. Thus the flexibility of the polyesters increases in the following sequence of dicarboxylic acids used: isophthalic acid replaced by hexahydrophthalic acid replaced by adipinic acid (Figure 7.18).

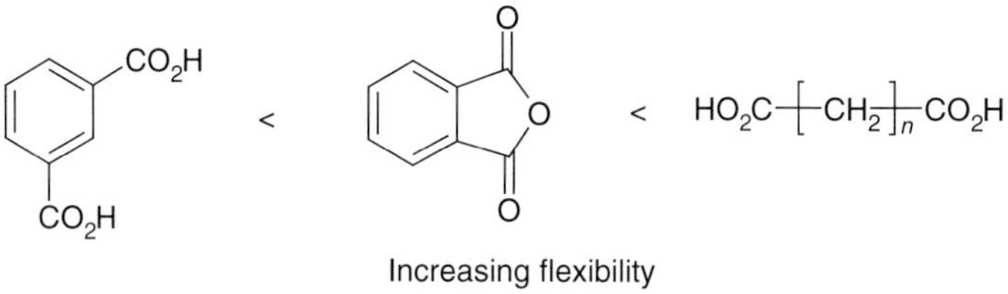

Fig. 7.18 Influence of the raw materials on increasing flexibility of the polyester (from left to right).

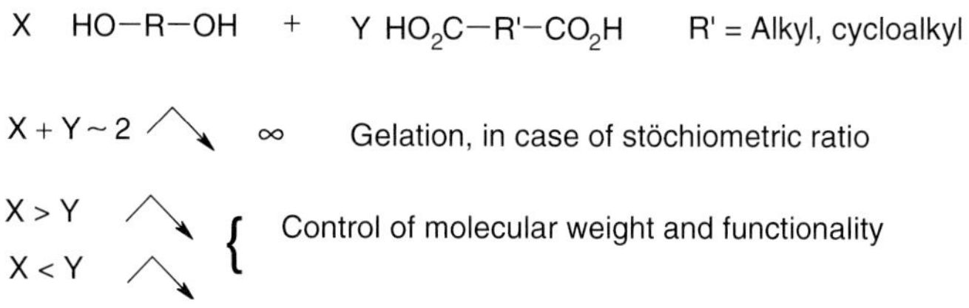

X HO—R—OH + Y HO$_2$C—R'—CO$_2$H   R' = Alkyl, cycloalkyl

X + Y ~ 2 ↘ ∞   Gelation, in case of stöchiometric ratio

X > Y ↘
X < Y ↘ { Control of molecular weight and functionality

Fig. 7.19 Ratio of the diol to the acid functionality in polyester processing and its consequences.

Linear polyesters have greater flexibility than branched products, and lower-molecular polyesters have lower flexibility compared with higher-molecular materials. The molecular weight of polyesters is managed via the ratio of diol to dicarboxylic acid. A ratio of close to 1 : 1 results in very high molecular weights. An increasing excess of one component results in a decrease in the molecular weight (Figure 7.19).

The mean molecular weights M_n of polyesters for powder coatings and polyurethane production, for example, are generally of the order of 800 to 1600. Depending on the coating material, however, molecular weights up to means of 4000 are possible.

A further important factor is the hydrophobic property of polyesters. Since the polymer backbone is formed of ester groups, there is greater sensitivity to acid or alkali media compared with polyacrylics. This weakness can be counteracted by using sterically more complex, cyloaliphatic or longer-chain acids or polyols for polyester production. Examples of such compounds include 2-butyl-2 ethyl propane 1.3-diol, bis(4-hydroxycylohexyl)propane, and dimeric fatty acids, as shown in Figure 7.20. Good solubility of the polyesters in common paint solvents is also usually desirable in order to suppress the polymers' tendency toward crystallization. To this end, several different parent components are generally used in the synthesis process, as a result of which excessively high symmetry and the possibility of the polymers becoming attached to one another is curbed.

The provision of customized polymers must also address the desired cross-linking chemistry. Here too, the synthesis components used play a key role.

The number of reactive groups per theoretical average polymer available for cross-linking reactions is limited to two in the case of linear polyesters, which are based only on bifunctional synthesis components. These products are therefore mostly used only as intermediates for polyurethane resin production or in powder

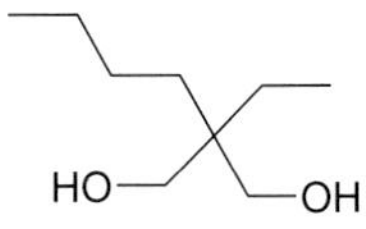

Fig. 7.20 Special raw materials for increasing the etch resistance of polyesters.

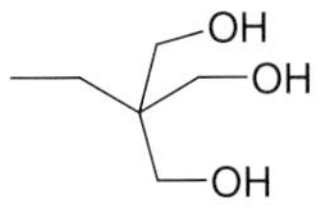

2-Butyl-2 ethyl-propane-1,3-diol

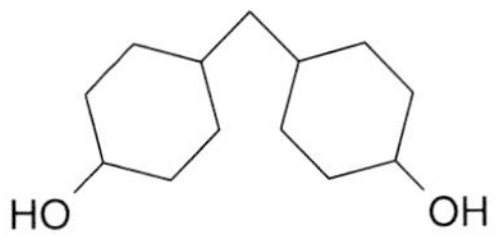

Bis(4-Hydroxycylohexyl)methane

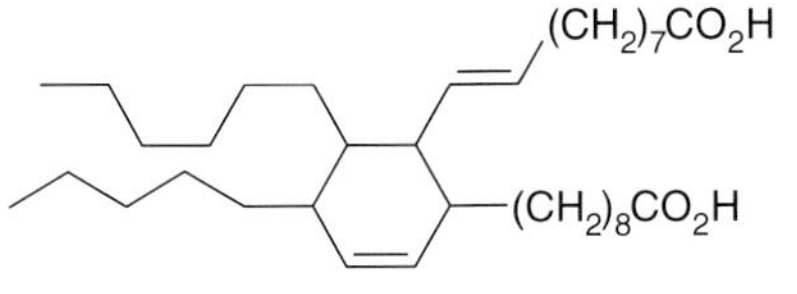

Dimeric fatty acid

Fig. 7.21 Polyols for the synthesis of multifunctional branched polyesters.

Trimethylol propane

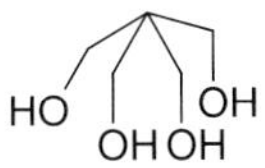

Pentaerythritol

coatings owing to their appropriate melting behavior. A higher number of reactive groups is achieved by using polyols such as trimethylol propane or pentaerythritol for polyester synthesis (Figure 7.21).

These hydroxyl groups can be cross-linked with polyisocyanates, blocked polyisocyanates, melamine resins, or tris(aminocarbonyl)triazine. Section 7.7 deals with the cross-linking agents in greater detail.

However, acid groups rather than hydroxyl groups are required for certain applications like powder or waterborne resins. Either an excess of dicarboxylic acids must be used or acid anhydrides must be added at the end of the polycondensation processes to convert the product (Figure 7.22).

In the case of UP (unsaturated polyester) resins, polyesters that cure radically, the reactiveness is achieved via double bonds, which are part of the polyester. To this end double-bond-functional, that is, unsaturated components such as maleic acid/maleic acid anhydride or endomethylene tetrahydrophthalic acid anhydride must be used as the acid component.

The proportion of these unsaturated acids to the overall acid content that is used for polycondensation controls the reactivity of the polymers and the network properties of the coatings. The longer the double-bond-free sequences in the polymer, the more elastic (plastic) the coatings. A certain proportion of phthalic acid in these polyesters increases their compatibility with the reactive thinner styrene. As far as high-gloss polyesters are concerned, compounds such as trimethylol propane or glycerine monoallyl ether must also be used to achieve the desired property profile in the coating. These allyl lateral chains result in supplementary oxidative drying through hydroperoxide formation with the oxygen on the surface

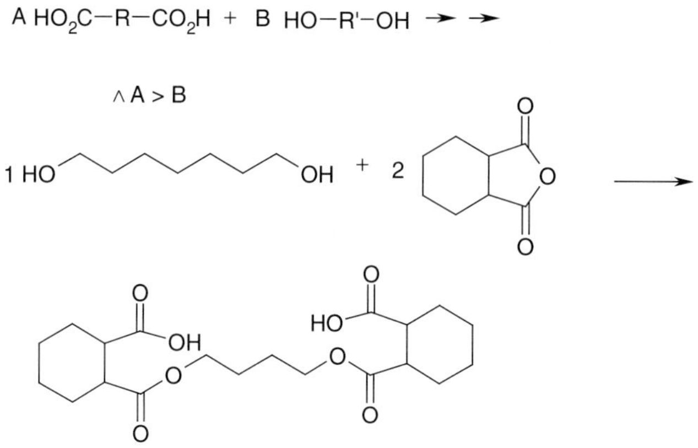

7 Polymeric Engineering for Automotive Coating Applications

$$A\ HO_2C-R-CO_2H\ +\ B\ HO-R'-OH \rightarrow \rightarrow$$

∧ A > B

1 HO~~~~~~OH + 2 [cyclohexane dicarboxylic anhydride] →

[polyester intermediate structures]

Fig. 7.22 Synthesis scheme for acid-functional polyesters.

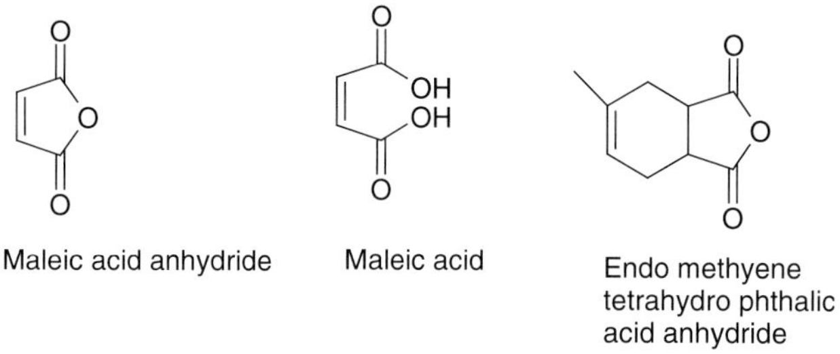

Maleic acid anhydride   Maleic acid   Endo methyene tetrahydro phthalic acid anhydride

Fig. 7.23 Raw materials for unsaturated polyesters.

of the coating film. This reduces oxygen inhibition of the polymerization process at the surface as the coating gets formed (Figure 7.23).

7.3.2
Manufacturing Polyesters

As already mentioned above, the polycondensation reaction in which polyesters are formed is an equilibrium reaction. To shift the equilibrium to the product side (polyester), the low-molecular reaction products such as water or alcohols – approximately 3–12% – have to be efficiently removed from the reaction mixture.

Steel reactors ranging in size from 5 to 60 m^3 with a column and condenser are therefore used for the production of polyesters. The polyester raw materials that are prebatched into the steel reactor are continually heated by means of an electric heater or thermal oil until the material is in the form of a mixable reaction mass. The condensation reaction is then continued at end temperatures between 200 and 300 °C while removing the low-molecular reaction products until the desired degree of conversion has been achieved. This process is termed a *melt process*. Figure 7.24 shows a schematic layout of a plant for the production of polyesters.

The conversion during the production process is monitored by measuring the residual acid content by means of the acid value [51], the OH value [51], and the

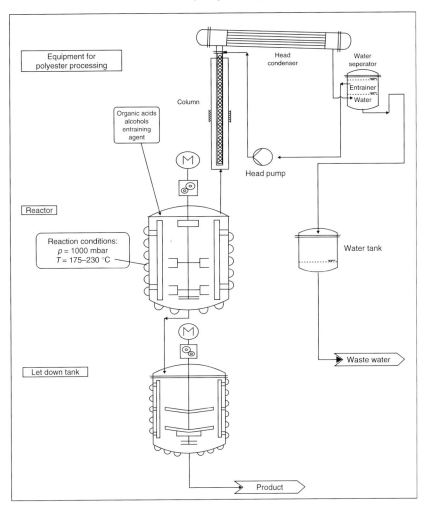

Fig. 7.24 Scheme of a production plant for polyester processing.

viscosity. Apart from wet chemical titration, more and more near infrared (NIR) spectroscopic methods are being used [52].

Since the viscosity of the reaction mixture rises during production, inert gas can also be passed through the reaction material to remove the resulting condensation. Vacuum processes are also started at the end of the condensation process, especially for the production of higher-molecular polyesters so that even relatively small volumes of water can still be removed economically at the higher viscosity. In principle, it is important to keep the process time as short as possible with an eye to product quality, product control, and environmental and economic aspects.

The thermal stress can result in marked, undesirable yellowing, especially with sensitive products. Such oxidation processes can be countered with antioxidants,

Table 7.1 Azeotrope-Forming Solvents for Water Removal

Water removal agent	Boiling point of the azeotropic mixture with water, composition of the azeotropic mixture
m-xylene	94.5 °C, 60% m-xylene + 40% water
Cyclohexane	69.8 °C, 91.5% cyclohexane + 8.5% water

usually organic phosphites, though these then remain in the product as such or in an oxidized form, and/or by blanketing with inert gas.

The azeotropic process is an attractive process that promises more rapid removal of the reaction water from the reaction mixture. In this the reaction, water is removed from the system by an entraining agent, which forms an azeotropic mixture (Table 7.1) with the reaction water. The usual practice here is to add 1–5% cyclohexane or xylene to the reaction mixture from the very start.

A side effect of the azeotropic process is a very constant production process for the products in comparison with the pure melt process. Further shortening of the process time can be achieved by means of polycondensation under pressure in combination with the azeotropic process [53].

Achieving equilibrium in the esterification reaction can be accelerated by using catalysts such as $TiCl_4$, stannic acid, or other Lewis acids. The use of protonic acids, as used in the synthesis of distillable products, is not common practice as such acids remaining in the product can have an adverse affect on the coating quality.

Only the azeotropic process with an upper temperature limited to 200 °C is used for the production of unsaturated polyesters. In addition, this process takes place strictly in an inert gas – oxygen-free – atmosphere. A further special feature is that the unsaturated polyesters are initially dissolved not in solvents but in reactive thinners such as styrene and allyl compounds at temperatures below 80 °C. Stabilizers such as methyl hydroquinone are added to the reactive thinners before the initial dissolving process to ensure storage stability in the products. These additives remain in the later paint film.

Waterborne polyester dispersions can be manufactured using various processes.

For example, ionizable compounds can be condensed directly into the polymer backbone. Dimethylolpropionic acid or dimethylolbutanoic acid [54] or compounds containing sulfonate groups [55] are primarily used for this process (Figure 7.25).

The typical method, however, is a two-stage process in which a polyester polyol manufactured in the first stage in accordance with the standard process is reacted with trimellitic acid anhydride in the second stage at a lower reaction temperature of approximately 150–180 °C. This process permits excellent management of the acid value (Figure 7.26).

These products are then diluted in water-miscible solvents, neutralized with environment-friendly bases, and finally dispersed in water. Correspondingly, these products have a higher pH value and contain amines, of course.

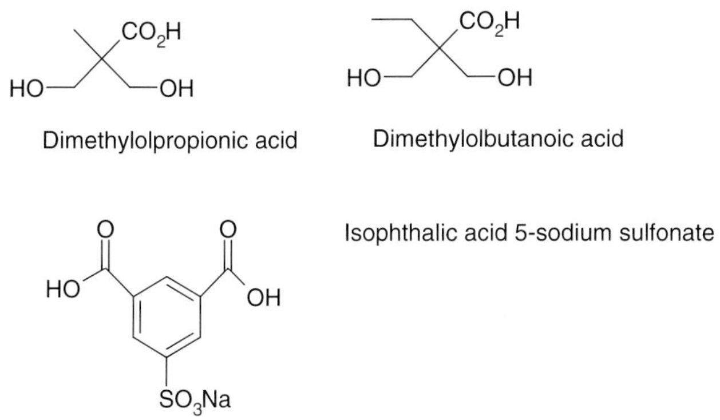

7.4 Polyurethane Dispersions in Coating Materials for the Automotive Industry

Fig. 7.25 Raw materials for the processing of polyester dispersions.

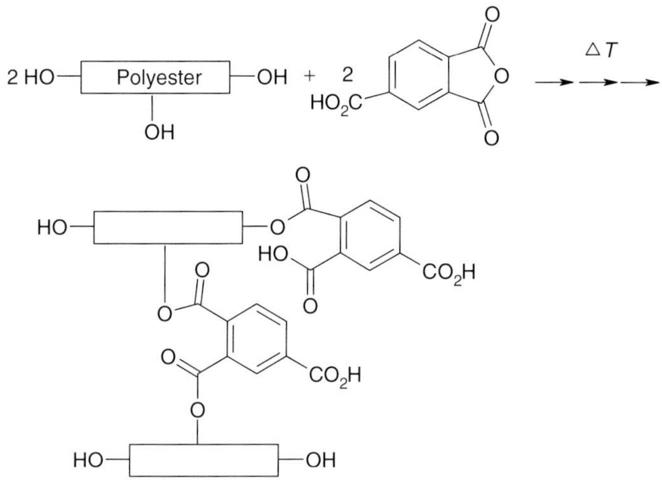

Fig. 7.26 Standard process for the production of waterborne polyester resins.

7.4
Polyurethane Dispersions in Coating Materials for the Automotive Industry

Polyurethanes are manufactured in addition reactions in process solvents. This usually involves reacting diisocyanates with diols, triols, and mainly linear polyols or especially polymeric diols such as polyether or polyester diols (Figure 7.27).

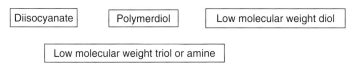

Fig. 7.27 Typical raw materials for processing polyurethanes.

Therefore this gives rise to segmented polymers – block polymers. Semibatch processes are mainly used for this application. However, continuous processes have also been described [56].

7.4.1
Managing the Property Profile of Polyurethane Resins and Polyurethane Resin Dispersions

Polyurethanes consist formally of hard and soft segments. The hard segments determine the stiffness of the polymers. They are created by the reaction of the diisocyanates with low-molecular diols. The soft segments are responsible for the elastic properties. These are generally polyesters. Even at this early stage it is clear that the paint chemist has an enormous repertoire at his disposal for managing the overall property profile of the polymer since both the chosen diisocyanate and the hard and soft segment constituents can be adapted to the desired profile. Practically all low-molecular diols, such as neopentyl glycol, hexamethylene diol, hydroxypivalic acid neopentyl glycol ester, butane diol, or dimethylolpropionic acid and polyols such as trimethylol propane or 1,2,4-butane triol can be considered for forming the hard segment. The glass transition temperature can be modified depending on the choice of the hydroxy-functional component. Furthermore, the ratio of hard segment to soft segment, which is also termed the *block ratio*, is also particularly important for the property profile (Figure 7.28).

As already mentioned above, polyesters are the most commonly used materials for the soft segment, generally with a proportion of 40–65% relative to the overall polyurethane polymer. Because of the wide choice of raw materials for the polyester synthesis process, it is possible to tailor the property profile particularly effectively – hardness, elasticity, hydrolysis stability, affinity to pigments, and flow properties. Polycarbonate diols, which offer increased UV stability and hydrolysis stability over simple polyester soft segments, can also be used as an alternative to the polyesters. If the UV stability is of subordinate importance and the demand is for a lower glass transition temperature combined with elasticity, it is possible to revert to polyether as the soft segment. Such polyether-based polyurethanes exhibit more pronounced phase separation than polyester-based alternatives since hydrogen bridges can be more easily formed into esters, which have a planar structure, than into polyethers.

The choice of the isocyanate used in the polymer has an impact on UV stability, viscosity, and drying properties.

The production of especially light-stable polyurethanes calls for the use of aliphatic isocyanates such as 1,6-hexamethylene diisocyanate (HDI) or trimethyl

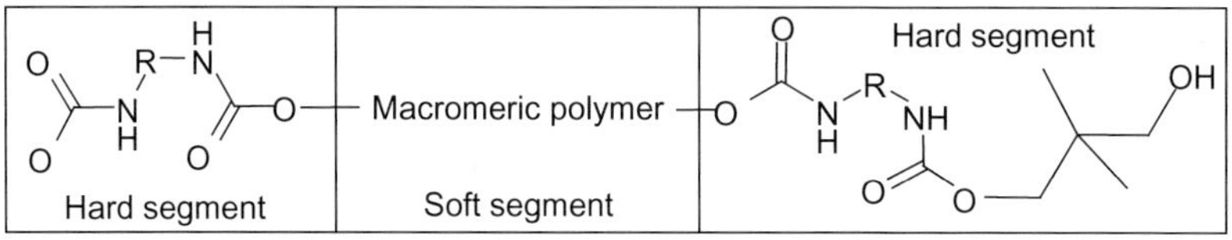

Fig. 7.28 Segment structure of polyurethanes.

7.4 Polyurethane Dispersions in Coating Materials for the Automotive Industry

hexamethylene diisocyanate, araliphatic isocyanates such as tetramethyl xylylene diisocyanate (TMXDI) or cycloaliphatic diisocyanates such as isophorone diisocyanate (3,5,5-trimethyl-1-cyanato-3-isocyanatomethyl-cylohexane) (IPDI) or H12 MDI (4,4'-dicyclohexylmethane diisocyanate) as parent compounds, as shown in Figure 7.29. However, the choice of diisocyanates also determines the viscosity level of the corresponding resin solutions with otherwise the same polymer composition. TMXDI yields the lowest-viscosity resins, IPDI and H12 MDI are more or less the same as each other, and HDI yields the highest viscosity because of its symmetry. The extent of physical drying is controlled by hydrogen bridge bonds between and within the polymer chains, on the one hand, and by particularly hydrophobic constituents, on the other. Polyurethane resins based on cycloaliphatic diisocyanate elements provide faster physical drying than those based on TMXDI.

Higher proportions of urethane groups also result in a greater degree of drying by hydrogen bridges.

The water dispersibility of polyurethane dispersions can be achieved by incorporating ether groups located terminally or in lateral chains (nonionically stabilized polyurethane dispersions) or ionizable components such as hydroxycarboxylic acids [54], or sulfonic-acid-based compounds [55]. The lateral incorporation of polyether segments results in rheologically active polyurethanes, which are used as thickeners in dispersions [57]. Cationogenic polyurethanes can also be produced. To do so, tertiary amino groups are produced in the polyurethane. However, such polyurethanes play no role in exterior applications in the automotive sector (Figure 7.29).

The visual appearance of polyurethane dispersions can vary to a large extent. Thus, milky white to yellowish-white and nontransparent to almost transparent appearances are possible. This depends primarily on the polarity of the polyurethane resins and any possibly present cosolvents and residues of process solvents. For example, rising levels of neutralized carboxylic acid cause the product to move from being nontransparent to opaque.

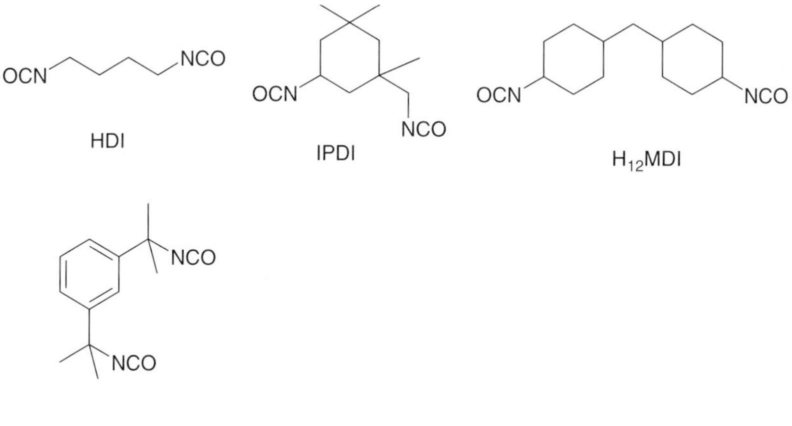

Fig. 7.29 Diisocyanates for nonyellowing polyurethanes.

Nonionically stabilized polyurethane dispersions differ significantly from the properties of their ionic counterparts. For example, nonionically stabilized dispersions are freeze-stable and are unaffected by salts, though they tend toward thermocoagulation. The latter is triggered by the thermally induced collapse of the hydrogen bridges to the ether segment, immediately causing the polyurethane to become insoluble in water. The ionically stabilized dispersions do not exhibit such behavior.

7.4.2
Manufacturing Polyurethane Resin Dispersions

Polyurethane dispersions can, in principle, be manufactured by the external emulsification of organic polyurethane resins or their initial solutions by means of special shear systems. However, this requires very large quantities of emulsifier, which make the resulting coatings very water-sensitive. Self-emulsifying systems are therefore the standard method used in the automotive industry. The self-dispersibility of polyurethanes is achieved by incorporating ionizable components such as dimethylolpropionic acid or dimethylolbutanoic acid. Longer-chain hydroxy-terminated ethers (M_n 1000–3000) can also be used to obtain such water dispersibility (see Figure 7.30).

A wide range of processes is available to form self-emulsifying polyurethanes. A common feature of these processes is that an ionizable isocyanate-terminated prepolymer is manufactured in suitable process solvents in the first stage.

To this end, as shown in Figure 7.31, the hydroxy-functional synthesis elements such as the soft segment and low-molecular diols are heated to the desired reaction temperature in an inert gas environment with an excess of diisocyanate in a process solvent, and the reaction is continued until the hydroxy-functional components have been fully used up. The standard means of monitoring the progress of the reaction is via titrimetric measurement of the isocyanate content. A more efficient and higher-quality method of monitoring the reaction is offered by IR-in-line spectroscopy [58]. Reaction temperatures between 50 and 120 °C are usual, depending on the system, reaction conditions, and process solvent. Ketones,

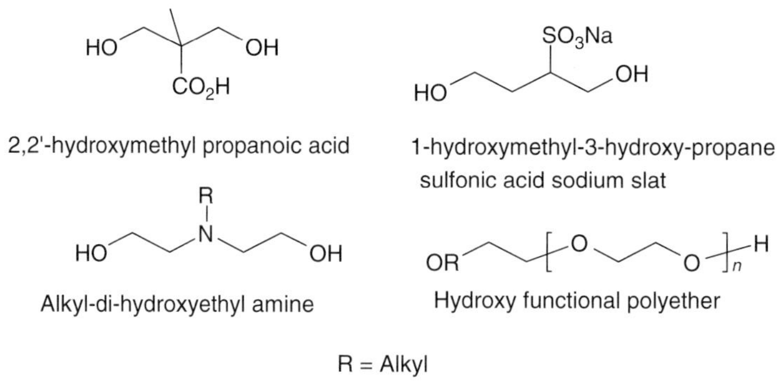

Fig. 7.30 Raw materials to design water dispersibility of polyurethanes.

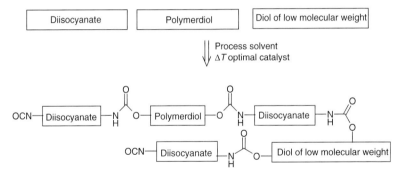

Fig. 7.31 Synthesis of an isocyanate-functional polyurethane prepolymer.

esters, ethers, N-methyl pyrrolidone, and N-methyl caprolactam can be used as the process solvent. Whereas the low-boiling process solvents can be removed, the high-boiling material have to remain in the product and therefore have to be suitable for use as coating material.

Starting from such prepolymers, three different processes for manufacturing polyurethane dispersions are described in the literature:
- the acetone process
- the prepolymer mixing process
- the melt-emulsification process.

In the acetone process [59], as in Figure 7.32, the prepolymer is preferably manufactured in acetone, other ketones, or N-methylpyrrol (NMP). If acetone is used, a higher reaction temperature can be set by increasing the reactor pressure. The molecular weights, starting from such a prepolymer, are then increased by extending the chain with polyols such as trimethylol propane, trimethylol ethane, or 1,2,4-butane triol in the organic phase.

The resulting viscosity increase is weakened by the polar solvent, which prevents the formation of a hydrogen bridge between the polyurethanes. In this process, the increase in molecular weight is very controlled and reproducible, and relatively tight molecular weight distributions can therefore be achieved. However, only molecular weights of less than 100 000 can be economically manufactured using this process. The resulting polyurethanes are then converted into ionomers by neutralization with organic amines. Once this stage is complete, the reaction mixture can be dispersed in water. Wherever possible, the process solvent is removed at the end by passing steam or using a vacuum distillation method known as a *strip process*. The dispersions obtained thus have particle sizes of between 10 and 1000 nm.

The acetone process is not ideal from an eco-efficiency perspective since strip processes are obviously energy- and capital-hungry, especially as the process solvents often cannot be reused economically. From an economic perspective, the acetone process is also restricted to the manufacture of polyurethanes with molecular weights of less than 100 000 as a huge volume of process solvent would otherwise be required in the chain-extension stage.

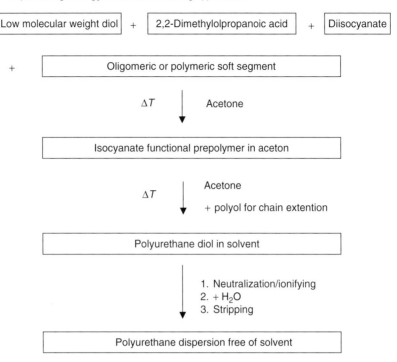

Fig. 7.32 Acetone process.

In the prepolymer mixing process, [60], as in Figure 7.33, for the production of polyurethane dispersions, an ionogenic prepolymer is produced in solution or melt by neutralizing with tertiary amines that cannot react with isocyanates. This is followed by dispersion in water to which aminic chain-extension agents such as ethylene diamine or ethylene triamine may already have been added.

A subsequent addition of polyamines to the dispersed prepolymer is also possible. However, additionally this results in reactions between the isocyanates and water, which generate carbon dioxide and an amine, which can in turn react with isocyanates to produce polyurea (see Figure 7.34).

The use of less-reactive prepolymers, such as those based on tetramethyl xylylene diisocyanate, permits discrimination in the variety of reactions available in favor of the polyamine/isocyanate reaction over the isocyanate/water reaction [61]. The chain-extension process itself can be carried out as a semibatch or continuous method [62].

Since the molecular weight increase in this process takes place in the aqueous phase, higher molecular weights can be achieved, more branched polymers constructed, and dispersions with higher solids contents manufactured than in the acetone process. This provides greater support for the physical drying and solvent resistance of coatings than by means of polyurethanes manufactured in the acetone process. However, these dispersions have significant urea content in the polymer and usually a very much wider molecular weight distribution. Because

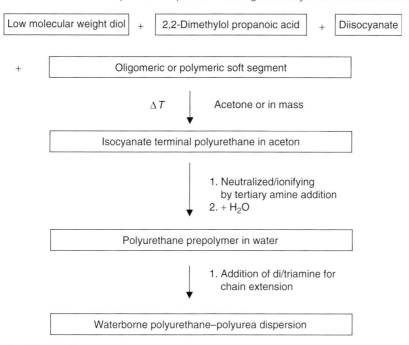

Fig. 7.33 Prepolymer mixing process.

Reaction during chain extension with water

〰〰〰NCO + H$_2$O ⟶ 〰〰〰NH$_2$ + CO$_2$

〰〰〰NH$_2$ + OCN〰〰〰 ⟶ 〰〰〰NHCONH〰〰〰

Reaction during chain extension with amine

〰〰〰NCO + H$_2$NCH$_2$CH$_2$NH$_2$

↓

〰〰〰NHCONH–CH$_2$CH$_2$–NHCON〰〰〰

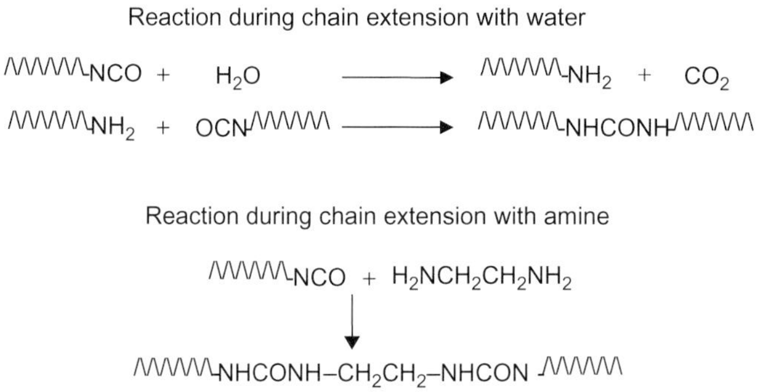

Fig. 7.34 Reactions during chain extension in the polymer mixing process.

of the additional urea functionality, these dispersions are also termed *polyurethane polyurea* dispersions. This increased polarity can cause greater swelling of coatings to occur when exposed to water.

The ketazine and ketamine processes [63, 64] are variants of the prepolymer mixing process. In these processes, the chain-extending polyamines are mixed with the organic phase in blocked form as ketazines or ketamines. The polyamine protective groups are hydrolytically separated and the chain extension is initiated only when the reaction mixture has been mixed with water. Consequently, the chain extenders are evenly distributed in the emulsion droplets, which is the ideal

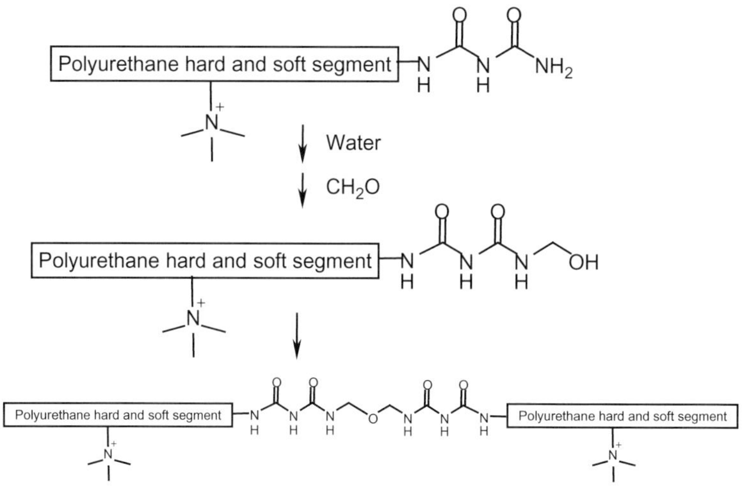

Fig. 7.35 Melt-emulsification process.

scenario. This procedure provides better monitoring of the chain-extension reaction and thus more constant molecular weights and molecular weight distributions.

In the third process, the melt-emulsification process [65], the same methodology is used as in the prepolymer mixing process, and chain extension is carried out in the aqueous phase. In this case, however, no isocyanate-functional prepolymers are dispersed in water; instead, formaldehyde-reactive urea-blocked isocyanate prepolymers are used. Figure 7.35 shows an example in which a cationic, urea-blocked prepolymer (biuret structure) is dispersed in water and reacted with formaldehyde. The resulting N-hydroxymethylol groups can then condense among themselves under appropriate conditions – acid catalysis or heat – and form higher-molecular polyurethanes.

7.5
Polyurethane Polyacrylic Polymers in Coating Materials for the Automotive Industry

7.5.1
Introduction

A relatively new class of resins that has gained in importance is that of the polyacrylic polyurethane polymers. These polymers are used in particular in the form of their dispersions for formulating automotive top coats and primer surfacers. The emphasis here is on providing low-emission coating materials with the properties of polyacrylic and polyurethane systems [66]. Hybrid binders of this type combine the good properties of polyurethanes, such as very good pigment wetting and excellent mechanical properties such as elasticity and wear resistance, with those of polyacrylics, such as UV stability and excellent management of the hydroxyl group content.

Purely physical blends of polyacrylic and polyurethane dispersions often exhibit thermodynamic incompatibilities caused by the decrease in the mixing entropy with increasing molecular weight of the polymers [67]. It is very often the case with physical blends that not even an addition of some positive properties can be observed [68, 69]. In the case of interpenetrating networks (IPN), such effects are mitigated by looping the different polymer classes together. No chemical links are created. The interpenetration of the polymer classes results in a reduction in the tendency toward phase separation, making better coating properties achievable [70]. In order to achieve the optimum coating properties, however, a chemical link between the polymer types is required.

7.5.2
Managing the Property Profile of Polyurethane Polyacrylic Polymers

Rules analogous to those for polyacrylics and polyurethanes are applicable to the management of the essential elements of fundamental properties.

In addition, properties can be managed by means of the location and number of linkage points. For example, polyurethane polyacrylic polymers with only terminal and/or lateral linkage points can be generated (Figure 7.36).

Furthermore, the physical drying properties can be influenced via the molecular weights. A linking reaction in an aqueous environment tends to produce higher molecular weights.

By varying the ratio of polyacrylic to polyurethane and the process – batch to semibatch – it is possible to make specific changes to the properties of the hybrid polymer dispersions [71]. Batch processes yield large particles than semibatch processes, and the particle size increases with increasing polyacrylic ratio. This causes a decrease in the Koenig hardness of the systems studied.

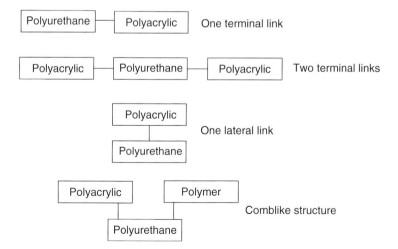

Fig. 7.36 Variation in the linkage points of polyacrylic polyurethane polymers.

7.5.3
Manufacturing Polyurethane Polyacrylic Polymers

A very wide range of methods can be used to manufacture such hybrid polymers. A common feature of most of the processes, however, is that compounds that enable radical graft reactions are incorporated in the polyurethane backbone. For example, vinyl cyclohexanoles [72], thiodiethanol [73], trimethylol propane monoallyl ether or glycerine monoallyl ether [74], and hydroxyalkyl (meth)acrylic monomers are used (see Figure 7.37).

With such polyurethane resins as the starting point, a process can be mentioned in which a polyurethane precursor is fed in during the organic phase, generating the desired hybrid polymer in this phase via polymerization [75]. As seen in Figure 7.38, dispersion process is then carried out and the process solvent is removed, resulting in the production of a solvent-free secondary dispersion.

In another process (see Figure 7.39), a polyurethane resin produced in a process solvent is first dispersed in water, and the hybrid polymer is produced in this aqueous phase by a polymerization process [76]. These dispersions are termed *primary dispersions* here. The polyurethanes dispersed in water do not exhibit a

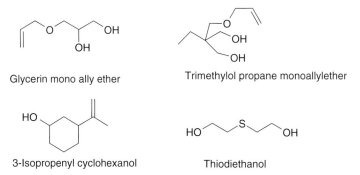

Fig. 7.37 Raw materials for introducing a grafting position into the polyurethane resin backbone.

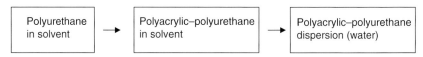

Fig. 7.38 Secondary polyacrylic polyurethane hybrid dispersion.

Fig. 7.39 Primary polyacrylic polyurethane hybrid dispersion.

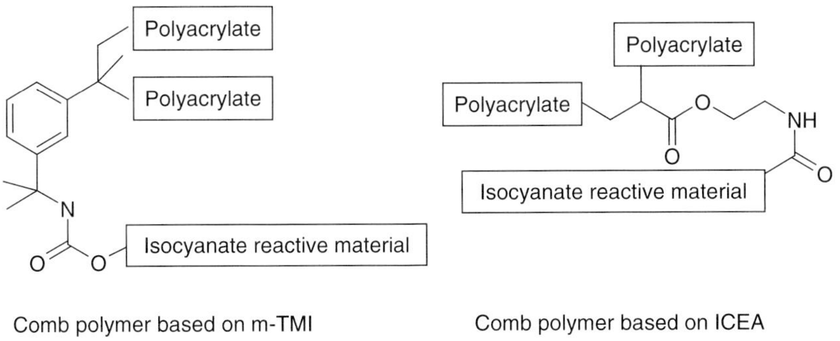

Comb polymer based on m-TMI   **Comb polymer based on ICEA**

Fig. 7.40 Comblike grafted polyurethane polyacrylic hybrids by inverse processing.

CMC according to studies by Jahny [77]. Accordingly, polymerization takes place in a particle swollen to a greater or lesser extent by monomers.

Brand-new processes are available for the manufacture of hybrid polymers in mini emulsions, though the formation of the two polymer types and the graft reaction can take place alongside each other here [78]. In a process that is the inverse of the manufacturing strategies used to date, a polyacrylic is first manufactured by radical polymerization in solution – as described in Section 7.1 – before being transformed into a hybrid polymer by polymer-analogous reactions. However, this process yields differently structured polymers, that is, with comblike structures. For example (see Figure 7.40), isocyanate-functional polyacrylics are synthesized by using, among others, m-TMI (3-isopropenyl-a,a-dimethylbenzyl isocyanate) or isocyanatoethyl (meth)acrylate (ICEA) as copolymers to build the polyacrylic polymers [79]. It is also possible to manufacture polyacrylics low in hydroxyl groups, which are reacted with di- or polyisocyanates [80].

IPN can also be manufactured using polyurethane dispersions as the seed in an emulsion polymerization process without low-molecular emulsifiers [81]. However, no chemical link is established here between the polymer types. Rather, catena-like physical loops and penetrations of the different polymer types occur at best.

7.6
Epoxy Resins

Epoxy resins are the principal binder component of modern cationic electrocoat paints, which offer outstanding anticorrosion protection even in very thin films and are extremely eco-efficient since they contain virtually no solvents and can be processed with a high application efficiency of almost 100% (see Chapter 4).

The epoxy resins for these cationic electrocoat paints are polyaddition products of epichlorhydrin and bisphenols or other generally OH-functional components, which can be transformed into cationic polyelectrolytes by incorporating, for example, primary or secondary amines.

7.6.1
Managing the Property Profile

As with the other binder classes, the property profile of the epoxy resins is substantially determined by the components used to build the polymer.

Those structures that have a positive impact on anticorrosion protection are of particular interest.

First, mention must be made of the classic components of epoxy resins, bisphenols such as bisphenol A (BPA) and bisphenol F (BPF) (see Figure 7.41). Their nonpolar structures and the large planar aromatic systems exert an outstanding barrier effect.

However, these are very rigid structural elements. In order to generate satisfactory elasticity for the automotive coatings sector, components that increase flexibility must be incorporated in the epoxy resin.

This takes place primarily by modification of the epoxy resin by means of ether-modified bisphenol A, aliphatic epoxy resins based on polyethers, or ethoxilated phenols (Figure 7.42).

Alternatively, elasticity can be increased via a hydrophobic aliphatic lateral chain in the polymer matrix [82].

The stability of the epoxy resin dispersions or the pigment-free components and precipitability of the cathodic electrocoat are of crucial importance to the quality of the cathodic electrocoat. Both properties are controlled via the proportion of ammonium groups, which can also be quaternary, if necessary [83].

The quality of the coating depends to a significant degree on the cross-linking, with the isocyanate-OH reaction generally being the standard cross-linking chemistry for the cathodic electrocoat. The strength of the networks formed by this

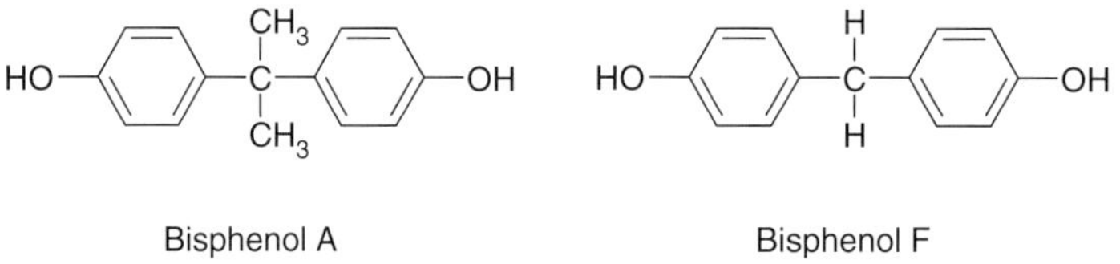

Fig. 7.41 Chemical structure of bisphenol A and bisphenol F.

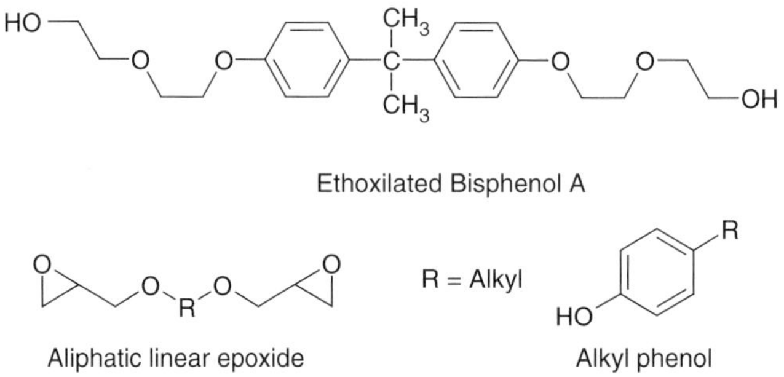

Fig. 7.42 Key intermediates for elastic polyepoxides.

cross-linking reaction is determined by the number of hydroxyl groups in the polymer matrix and the nature of the aromatic blocked polyisocyanates (see also Section 7.7.2).

Apart from these fundamental structure/effect relationships, there are very special modifications to the polymer structures, for example, with regard to pigment wetting, edge protection, throwing power, and so on, particularly in the case of the cathodic electrocoat. Interested readers are advised to consult the extremely extensive patent literature, as the amount of information available and required is far beyond the scope of this work.

7.6.2
Manufacturing Polyepoxy Resins

Classic epoxy resins can be obtained from the reaction of BPA or BPF with epichlorhydrin (shell process). It is possible to manage the molecular weights and their distribution as a function of the stoichiometric ratio of the reactants and the process used. The epoxy resins produced in this way have molecular weights of approximately 400–4000 g mol^{-1} and 1–2 epoxy terminal groups per molecule (Figure 7.43).

The polyepoxy resins for a dip coating application are constructed via an addition reaction using the advancement process in which liquid epoxies - low-molecular epoxy resins with terminal epoxy groups with alcohols or polyether diols as soft segments and/or phenols such as BPA – are transformed into higher-molecular polymers (see Figure 7.44).

These precursors are then further transformed using secondary or primary amines, giving rise to polymers with tertiary amino groups (Figure 7.45). To

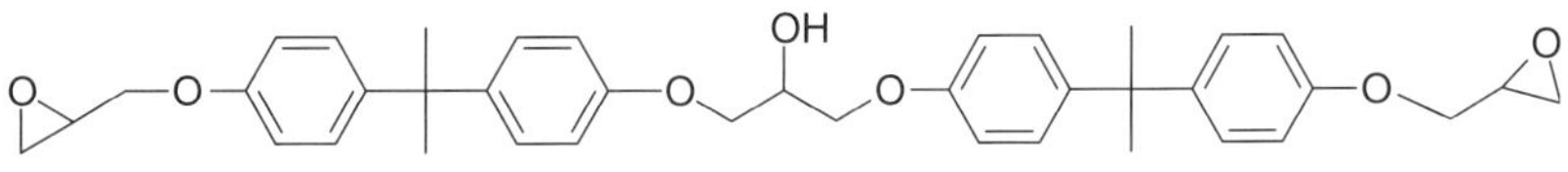

Fig. 7.43 Epoxy resin based on bisphenol-A (BPA) and epichlorhydrine.

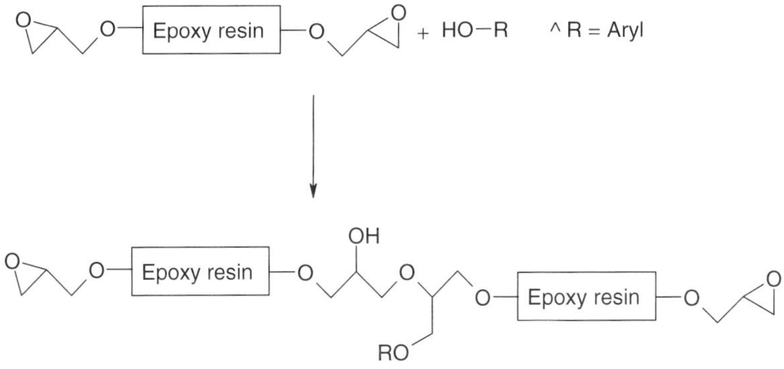

Fig. 7.44 Chain extension to epoxy-functional polymers of higher molecular weight.

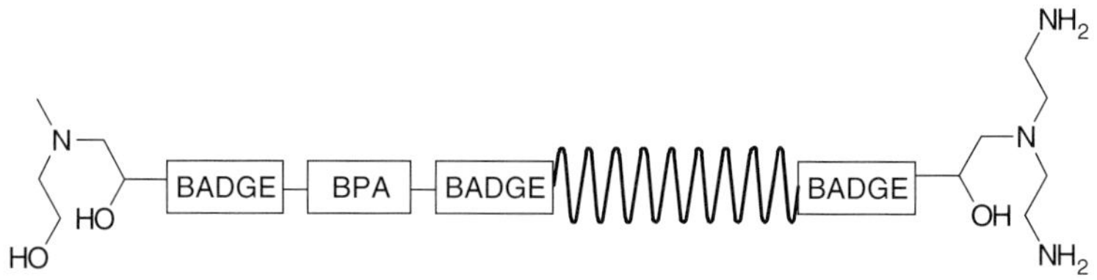

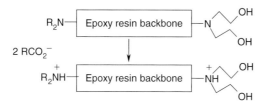

Fig. 7.45 Amino-functional epoxy polymer.

```
         R₂N—[ Epoxy resin backbone ]—N       OH
                                          OH
2 RCO₂⁻
    +                                        OH
         R₂NH—[ Epoxy resin backbone ]—NH
                                          OH
```

Fig. 7.46 Water-dispersible epoxy resin.

prevent gel formation as a result of primary amino groups, these amines are transformed into ketoximes using ketones. In the subsequent dispersion process, the protective groups are removed again by means of hydrolysis.

These amino-functional polymers are neutralized with low-molecular carboxylic acids to yield water-dispersible products (Figure 7.46).

These already amine neutralized binders are then completed in the solvent-borne phase with cross-linking agents additives, and so on, to form a pigment-free mixture and transformed into a cationic secondary dispersion with water as in Figure 7.47. The process solvent is removed in a strip process prior to further processing of the dispersion.

The same resin is used, in addition, to manufacture a pigment paste that is also cationically stabilized. A dip bath is prepared from the two components, with the dispersed constituents electrocoagulating on the workpiece, which is connected as the cathode.

7.7
Cross-Linking Agents and Network-Forming Resins

7.7.1
Introduction

The cross-linking reactions of coating materials are usually thermally induced and carried out under a thermal load.

Even with highly reactive coating systems, such as in the case of two-component paints in which the reaction is initiated by mixing the components using two-component lines, the activation energy of the cross-linking reaction has to be additionally reduced by the use of catalysts such as Lewis acids. Only then it is possible to achieve effective cross-linking at temperatures between 60 and 140 °C.

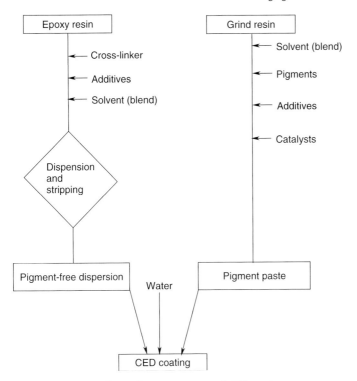

Fig. 7.47 Overview of manufacturing process of CED coatings.

The cross-linking agents are presented in detail in Section 7.7.2 and 7.7.3, with Section 7.7.2 concentrating on the cross-linking agents for the very large class of liquid coating materials, while Section 7.7.3 deals with the cross-linking agents for powder coatings. Within these sections the chemical structure of the cross-linking agents is selected as the classification principle – as was the case with the binders above.

It should also be mentioned at this juncture that networks can also be induced by UV or electron beam radiation by exciting photoinitiators and sensitizers [84]. It is also possible to combine different network-forming mechanisms such as UV and thermal network formation [85]. At present, however, there is no standardized application for this in the automotive industry.

7.7.2
Cross-Linking Agents for Liquid Coating Materials

7.7.2.1 Melamine and Benzoguanamine Resins
In the automotive industry, amino-plast resins, especially melamine resins and, to a lesser extent, benzoguanamines, are primarily used for the condensation reactions as the cross-linking reaction of polymeric polyols and carbamate-functional polymers for the production of one-component coating materials (see Figure 7.48).

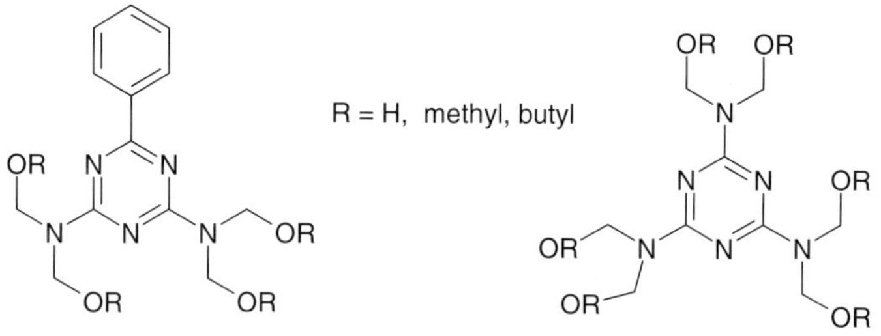

Benzoguanamine   Hexamethoxymethyl melamine

Fig. 7.48 Benzoguanamine and melamine as ideal monomeric examples for cross-linkers.

Benzoguanamines are used in automotive applications as cross-linking agents only in surfacers/primers where they contribute to particularly plastic networks. These cross-linking agents contain very small quantities of formaldehyde from the production process and also release very small amounts of formaldehyde during the cross-linking reaction [86].

The broad pallet of commercial melamine resins differs in the multinuclearity (molecular weight distribution), the degree of methylolization, the degree of etherification, and the nature of the monoalcohols or monoalcohol blends used for the etherification reaction (Figure 7.49). Only those melamine resins manufactured using a melamine–formaldehyde ratio of less than 1 : 3 and an alcohol–melamine ratio of greater than 2 : 1 are suitable for formulating coating materials.

Fully alkylated melamine resins, which are low in NH groups, are mainly used to formulate coating materials in waterborne and solvent-borne automotive OEM coatings. Such types are also often used for manufacturing high-solids paints because of their low viscosities [87]. Methoxymethyl-functional melamines are more reactive in this application than, for example, butyloxymethyl-functional melamines since methanol is very easy to eliminate during formation of the coating. Furthermore, the highly methoxymethyl-functional melamines permit trouble-free dilution in water (Figure 7.50).

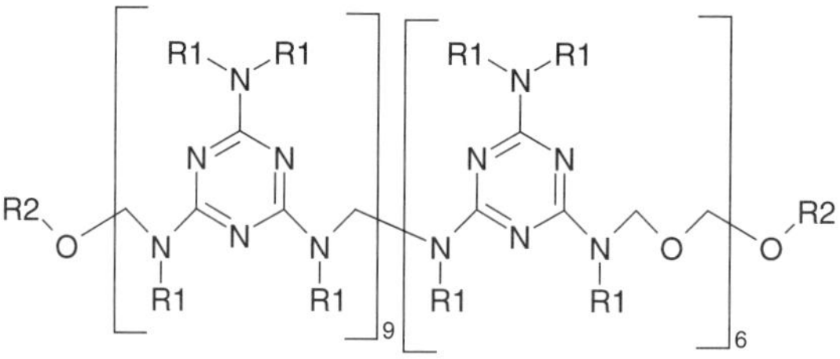

R2 = CH_3, *n*-butyl , iso-butyl

R1 = H, CH_2-OH, CH_2OR2

Fig. 7.49 Basic chemical structure of melamine resins.

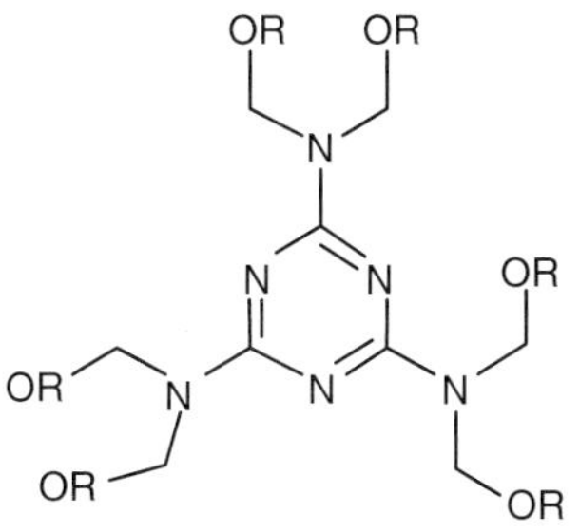

7.7 Cross-Linking Agents and Network-Forming Resins

R = Methyl

Fig. 7.50 HMMM hexakis (methoxymethl) melamine.

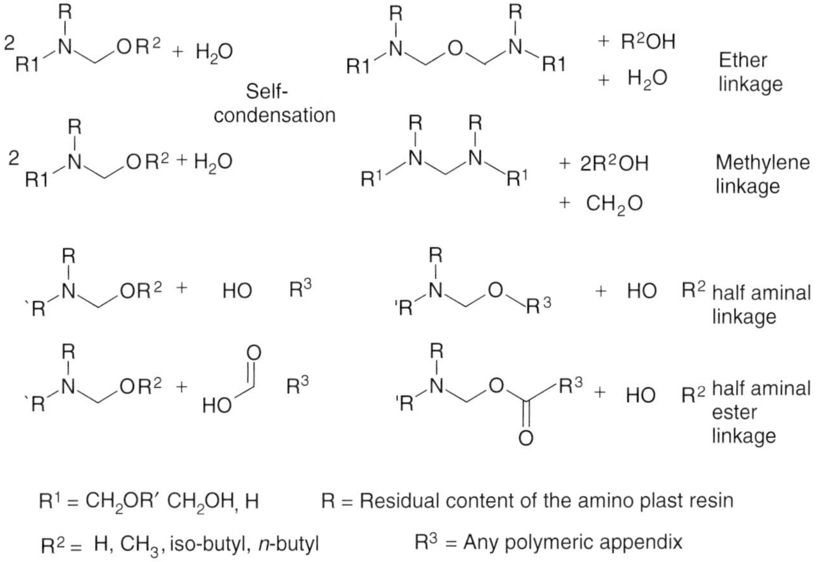

R^1 = CH$_2$OR′ CH$_2$OH, H   R = Residual content of the amino plast resin
R^2 = H, CH$_3$, iso-butyl, n-butyl R^3 = Any polymeric appendix

Fig. 7.51 Overview of amino-plast resin reactions.

However, the cross-linking reaction of hexamethoxymethyl melamine (HMMM) with polyols in solvent-borne coating materials has to be assisted via catalysis with acids and the use of relatively high temperatures. This catalysis is unnecessary in waterborne coating materials. Sulfonic acids such as para-toluenesulfonic acid, dodecylbenzenesulfonic acid, naphthalenesulfonic acid, or dinonyl-naphthalenemono- or disulfonic acid are used for this catalytic process. If the requirements relating to storage stability are particularly stringent, the corresponding amine salts of these sulfonic acids are used. At the same time self-condensation is suppressed in the HMMM types, resulting in more homogeneous coatings being obtained than would be the case with partially alkylated melamine resins. Figure 7.51 provides an overview of the reactions.

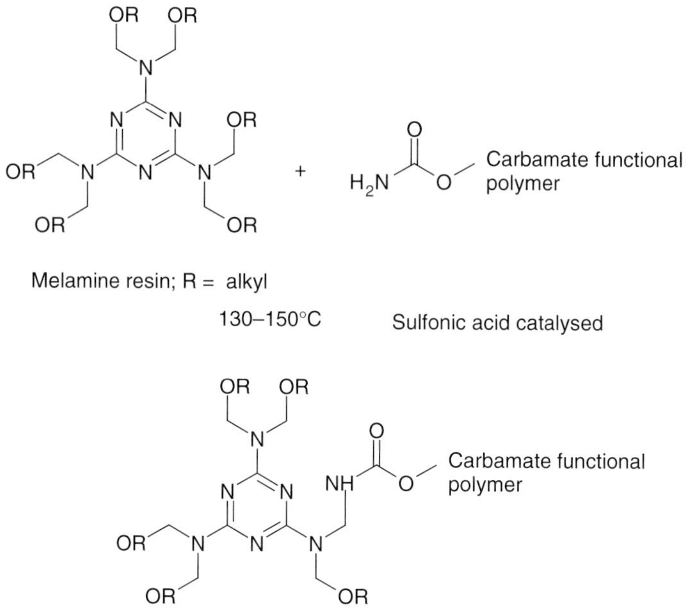

Fig. 7.52 Network functionalities of the OH and carbamate-melamine cross-linking.

The manufacturers of melamine resins now offer a broad range of products. Mixed etherified types and other modifications are available, thereby enabling customized products to be selected for special coating formulations.

If melamine resins are used as the sole cross-linking agents with hydroxy-functional polymers, coatings with very good surface hardness [88, 89] and good gloss can be obtained. However, these coatings are sensitive to acid action since the network is primarily produced by the formation of acetal esters [90].

This weakness can be compensated for by the use of carbamate-functional film-forming agents [91, 92] instead of hydroxy-functional ones since urethane aminals [93] are produced in the cross-linking reaction here, which requires slightly higher stoving temperatures (see Figure 7.52).

7.7.2.2 Tris(Alkoxycarbonylamino)-1,3,5-Triazine

Tris(alkoxycarbonylamino)-1,3,5-triazine (TACT, see Figure 7.53), which has been commercially available for some time, is also based on a triazine structural element with excellent UV stability [94].

However, this cross-linking agent produces polyurethane rather than ether networks, since the cross-linking reaction is a trans-carbamate reaction that is initiated at approximately 130–150 °C. To date, this cross-linking agent has only been used for waterborne [95] and solvent-based high-quality coating materials [96] because of its high cross-linking temperature, high price, and availability issues.

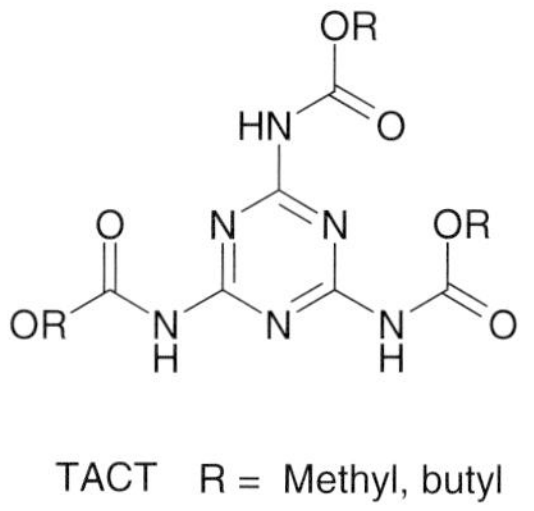

TACT R = Methyl, butyl

Fig. 7.53 Tris(alkoxycarbonylamino)-1,3,5-triazine (TACT).

7.7.2.3 Polyisocyanates and Blocked Polyisocyanates

Polyisocyanates, which are generally used for formulating two-component paints and reversibly blocked polyisocyanates that are available for formulating storage-stable one-component paints, react with hydroxy-functional polymers to form urethane groups (see Figure 7.54).

Coatings based on such urethane group-functional networks are of excellent quality – for example, in terms of UV stability, elasticity, and resistance to chemical exposure. This is, of course, only possible where the appropriate cross-linking agents and binders are specifically chosen, resulting in customized systems for the particular specification profile. Aliphatic polyisocyanates are used exclusively for this purpose. At the other end of the scale we have functional layers with a high barrier effect and very good elasticity, based on cross-linking with aromatic polyisocyanates (see Section 7.5). MDI (methylene diphenyl diisocyanate) and its oligomers as well as TDI (toluylene diisocyanate) are used as the parent compounds for this (Figure 7.55).

The term polyisocyanate covers a broad spectrum of polymers. At the basis of this broad spectrum are, on the one hand, different diisocyanates from which the polyisocyanates are formed in oligomerization processes. The most commonly used polyisocyanates are based on HDI (hexamethylene diisocyanate), followed by IPDI (isophorone diisocyanate) and H12 MDI (bis{4-isocyanatocyclohexyl}methane) (see Figure 7.29).

On the other hand, there are the different linkage structures among the isocyanates themselves, which are achieved by the process design and management during production. Polyisocyanates can be manufactured with uretdione, allophanate, iminooxadiazindione, isocyanurate, and biuret structures, for example; see Figure 7.56. Both the type of diisocyanate used and the type of oligomerization have a very marked impact on the performance of the polyisocyanates. HDI-based

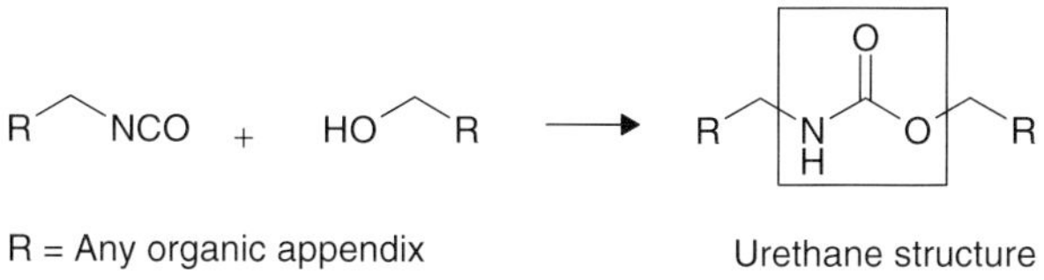

R = Any organic appendix   Urethane structure

Fig. 7.54 Cross-linking reaction of free polyisocyanates with polyols.

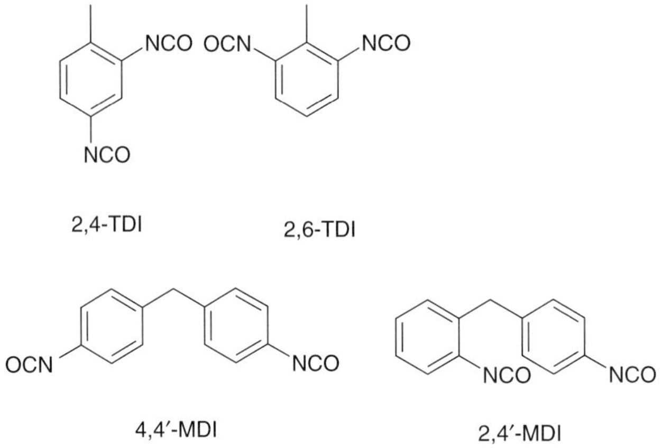

7 Polymeric Engineering for Automotive Coating Applications

Fig. 7.55 Aromatic polyisocyanates for good elasticity and barrier effects of coatings.

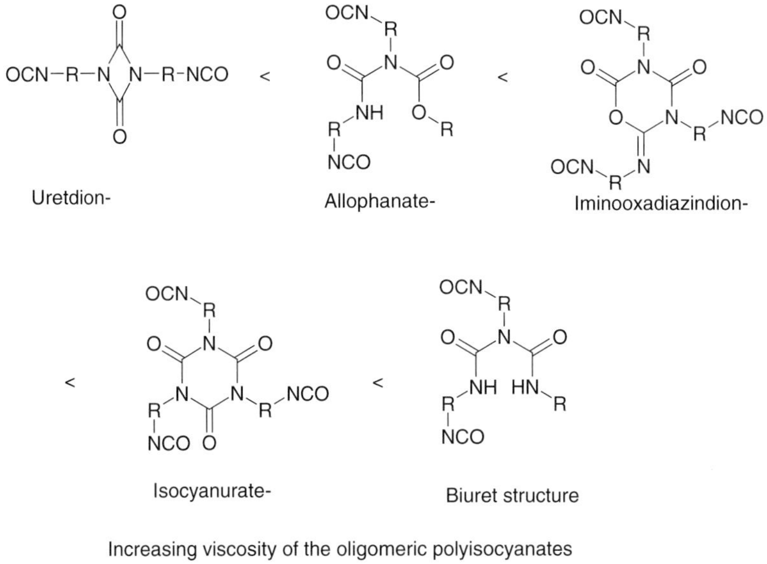

Fig. 7.56 Ideal structures of different polyisocyanates cross-linkers.

polyisocyanates with the same structure type, for example, result in lower glass transition temperatures (film hardnesses) in the coatings than those based on the corresponding cycloaliphatic diisocyanates. The reactivity of the HDI-based polyisocyanates is higher than that of those based on IPDI.

Low viscous polyisocyanates with an allophanate and iminooxadiazindione structure are the products of choice for formulating high-solid solvent-based coating materials. A very recent innovation is a partially blocked polyisocyanate that can, as it were, react with itself [97]. Coatings based only on such low-viscosity polyisocyanates often exhibit outstanding reflow but have weaknesses in their chemical

7.7 Cross-Linking Agents and Network-Forming Resins

resistance at the same time. For this reason, it is generally necessary to use blends of polyisocyanates in order to achieve the required property profile in the coating.

The reaction of polyisocyanates and hydroxy-functional polymers has to be catalyzed with Lewis acids in order to ensure an economical design for the automotive coating process and to guarantee good curing. Lewis acids such as dibutyl tin dilaurate and similar compounds are generally used for this purpose.

The formulation of low-solvent two-component waterborne materials with commercially available hydrophobic polyisocyanates is extremely difficult since the cross-linking agents are incompatible with water. Up till now it has been necessary to use extreme mixing processes to obtain approximately homogeneous coating materials [98].

Hydrophilized polyisocyanates [99] have also been available for some time to formulate waterborne two-component paints that promise better water compatibility and homogeneity in the coating materials. However, coating materials based on these materials have yet to make a breakthrough in automotive OEM coatings. The reason for this is subsidiary reactions with water, which can adversely affect the appearance of the coatings. For example, carbon dioxide is formed in the isocyanate–water reaction. The resulting amino groups then react with isocyanate to form urea. Pinholing and gloss haze can then occur. Also, there is no successful method for producing solvent-free waterborne two-component paints that guarantee outstanding film appearance. This has only been achieved to date with slurry-based paints, which are not classic emulsions [100].

Reversible blocked polyisocyanates are very suitable for manufacturing extremely storage-stable one-component coating materials. Compounds such as butane oxime [101], dimethylpyrazole [102] or acetoacetate and malonic ester [103] can be considered as blocking agents for automotive OEM paint applications (see Figure 7.57). The activation energy of the deblocking reaction is usually reduced by catalysts such as Lewis acids (bismuth, tin, and zinc compounds) or by certain amines such as diazabicyclooctane (DABCO), for example, to achieve lower stoving conditions and a better conversion.

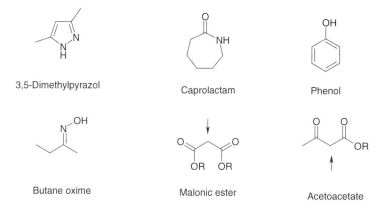

Fig. 7.57 Blocking agents for isocyanate groups.

Dimethylpyrazole-blocked [104, 105] and ketoxime (methyl–ethyl ketoxime) blocked polyisocyanates are also highly suitable for formulating waterborne coating materials.

The blocking reaction of isocyanates with malonic ester is scarcely reversible. Here cross-linking takes place mainly by means of transesterification of the malonic ester with the hydroxy-functional polymers, which gives rise to coatings that do not quite match the high property level of polyurethane-based coatings with regard to chemical resistance and yellowing.

Depending on the type of elimination product, it remains in the coating or is released in the stoving process.

7.7.2.4 Other Cross-Linking Agents for Liquid Coating Materials

In special markets such as Japan, clear coatings of high visual quality are also achieved via the cross-linking reaction of epoxy-group-functional polyacrylics with blocked di- and polycarboxylic acids. To this end, the dicarboxylic acids are transformed into their vinyl esters. This type of reversible blocking guarantees excellent storage stability in such a one-component coating material and uniform curing, which results in very good leveling of the coating. Since the network is based on pure ester linkages, pronounced hydrophobing of the coating must be achieved by means of the binder in order to guarantee good chemical resistance.

7.7.3
Cross-Linking Agents for Powder Coatings in the Automotive Industry

Powder coating materials gain importance and increasing market share in primer surfacers (see Chapter 5) and clear coats [106].

They require solid cross-linking agents, whose glass transition temperatures are substantially above room temperature in order to guarantee the material's blocking strength for manufacturing, storage, and application. Correspondingly, only a small number of cross-linking agents that are suitable for applications in the automotive sector are commercially available. Figure 7.58 provides a view of the most important agents. With regard to the manufacture of powder primers, epoxies are primarily used as cross-linking agents in combination with acid polyesters or polyacrylics [107]. These systems are inexpensive and offer excellent flexibility in the coating, have a wide application window, and deliver very smooth coatings, though these have only moderate light resistance and recoatability. Triglycidyl isocyanurate (TGIC) can also be used as a cross-linking agent for acid polyesters. The coatings achieved with this agent also have very good flexibility with significantly better light resistance and high stability against overbaking. However, TGIC is classified as teratogenic, which means that its use is restricted for occupational health and safety reasons. In addition, hydroxy-functional polyesters with blocked polyisocyanates are also used for the production of automotive primers. In contrast to systems based on TGIC or N-beta-hydroxy alkyl diamides (primide types), such powder-coating materials can be stoved at temperatures from 140–160 °C. This means that the maximum plasticity of the coating can be achieved under moderate stoving

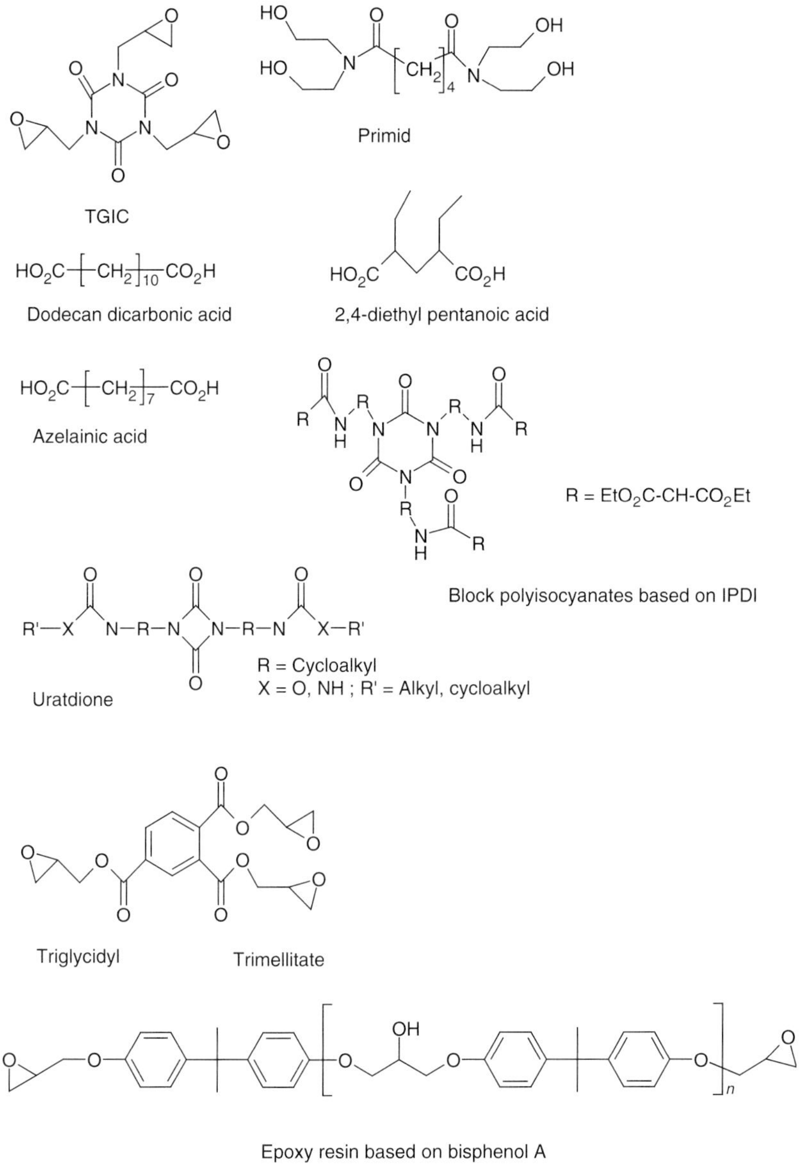

7.7 Cross-Linking Agents and Network-Forming Resins

TGIC

Primid

Dodecan dicarbonic acid

2,4-diethyl pentanoic acid

Azelainic acid

$R = EtO_2C\text{-}CH\text{-}CO_2Et$

Block polyisocyanates based on IPDI

Uratdione

R = Cycloalkyl
X = O, NH ; R' = Alkyl, cycloalkyl

Triglycidyl Trimellitate

Epoxy resin based on bisphenol A

Fig. 7.58 Cross-linkers for powder coatings.

conditions. Powder coatings based on hydroxy-functional polyesters or polyacrylics can be cross-linked with certain blocked polyisocyanates to form top coats [108].

High-priced products with high-end properties such as powder clear coats and powder primers for coating wheels are based on epoxy-functional polyacrylics as film-forming agents and dicarboxylic acids as cross-linking agents [109]. Dodecandicarboxylic acid is generally used, but other acids such as azelainic acid or

2,4-diethylpentanoic acid-1,5 are also suitable in principle. These combinations guarantee good chemical resistance, if a correspondingly high cross-link density is achieved, and high UV stability in the coating. Furthermore, these powder reaction systems can meet challenging demands in terms of appearance because of their very good flow during formation of the coating. However, the mechanical properties of these coatings are weaker than those based on epoxy cross-linking agents and polyester resin. In addition, such powder coatings have to be carefully tempered to counteract agglomeration. Overall, however, the level of polyurethane networks cannot be matched as yet.

References

1 Betz, P., Bartelt, A. (1993) *Progress in Organic Coatings*, **22**, 27.
2 Frey, T. (1995) *Farbe+Lack*, **101**(12), 1001.
3 Bosch, W., Schlesing, W., Buhk, M. (2001) *European Coatings Journal*, **10**, 60.
4 Malzbender, J., Toonder, J.M.J., Balkenende, A.R., de With, G. (2002) *Materials Science and Engineering*, **R36**(2-3), 47.
5 Satguru, R., MacMahon, I., Padget, J.C., Coogan, R.C. (1994) *Journal of Coatings Technology*, **66**(830), 55.
6 DE 4320969. (1995), Verfahren zur Herstellung von Polyurethanharzen und deren Verwendung sowie die Verwendung von Ethoxiethylpropionat zur Herstellung von Polyurethanharzen, BASF Coatings AG.
7 Wang, T.-L.D., Lyman, D.J. (1993) *Journal of Polymer Science Part A: Polymer Chemistry*, **31**, 1983.
8 Osterhold, M., Vogt Birnbricht, B. (1995) *Farbe+Lack*, **101**, 835.
9 DE 3606513. (1987), Dispersionen von vernetzten Polymermikroteilchen in wässrigen Medien, Verfahren zur Herstellung dieser Dispersionen und Beschichtungszusammensetzungen, die diese Dispersionen enthalten, BASF AG.
10 DE 2931044. (1991), Wässrige ionische Dispersionen von vernetzten Polyurethanteilchen, Wilmington Chemical Corporation
11 Lückert, O. (1992) *Prüftechnik bei Lackherstellung und Lackverarbeitung [Testing in Paint Manufacture and Processing]*, Vincentz, Hannover.
12 Fox, R., Flory, P.J. (1950) *Journal of Applied Physics*, **21**, 581.
13 O'Hara, K. (1988) *Journal of the Oil & Colour Chemists Association*, **71**, 413.
14 Neumann, C., Schmitt, G., Schmitt, B., Arndt, T. (2002) *Farbe+Lack*, **108**, 20.
15 Zimmermann, R. (1976) *Farbe+Lack*, **82**, 383.
16 Nakane, Y., Ishidoya, M. (1997) *Progress in Organic Coatings*, **31**, 113.
17 Kehlen, H., Rätzsch, M.T. (1985) *Solution Properties, Polydispersity Effects, Encyclopedia of Polymer Science and Engineering*, 2nd edn., John Wiley and Sons, New York, USA.
18 Budde, U., Reichert, K.-H. (1989) Polymer reaction engineering, in *Proceedings of the Third Berlin International Workshop on Polymer Reaction Engineering* (eds K.-H. Reichert, W., Geiseler), Wiley-VCH Verlag GmbH, Weinheim, p. 140.
19 DE Patent 19828742. (1999), Taylorreaktor für Stoffumwandlungen, bei deren Verlauf eine Änderung der Viskoität des Reaktionsmediums eintritt, BASF Coatings AG.
20 Conrad, I., Kossak, S., Moritz, H.-U., Jung, W.A., Rink, H.-P. (2001) *7th International Workshop on Polymer Reaction Engineering*, Dechema Monographien, Vol. **137**, Wiley-VCH Verlag GmbH, Weinheim, Hamburg, p. 117.
21 Webster, O.W. (1999) *Polymer Materials Science and Engineering*, **80**, 280.

22 Georg, M.K. (1999) *Polymer Materials Science and Engineering*, **80**, 283.
23 Wang, J.-L., Grimanua, I., Matyjaszewski, K. (1997) *Macromolecules*, **30**, 6507.
24 Matyjaszewski, K. (ed.) (2000) *Controlled/living Radical Polymerisation: Progress in ATRP, NMP, and RAFT*, ACS Symposium Series, Vol. **768**, ACS.
25 Bremser, W., Räther, B. (2002) *Progress in Organic Coatings*, **45**, 95.
26 Peck, A.N.F., Hutchinson, R.A. (2004) Secondary reactions in the high temperature free radical polymerisation of butyl acrylate, *Macromolecules*, **37**, 5944–5951.
27 US 3674838. (1971), Vinyl Carbamyloxy Carboxylates, Ashland Oil Inc.
28 EP 0675141. (1995), Copolymer compositions containing hydroxyl functional (meth)acrylates and hydroxyalkyl carbamate(meth)acrylates and mixtures thereof, Union Carbide Chemical.
29 EP 0675141, (1995), Copolymer compositions containing hydroxyl functional (meth)acrylates and hydroxyalkyl carbamate(meth)acrylates and mixtures thereof, Union Carbide Chemicals.
30 US 5356669 (1994), Composite color plus clear coatings utilzing carbamate functional polymer composition in the clearcoat, BASF Corporation
31 WO 01/14431. (2001), Process of the preparation of carbamate functional polymers, PPG.
32 DE 4322242. (1995), Wässriges 2-kcomponenten-Polyurethan-Beschichtungsmittel, Verfahren zu seiner Herstellung und seine Verwendung in Verfahren zur Herstellung einer Mehrschichtlackierung, BASF Coatings AG.
33 Rink, H.-P., Lettmann, B. (2002) *Farbe+Lack*, **108**(4), 22.
34 Gerrens, H. (1995) *Polymerisationsverfahren – Kapitel "Emulsionspolymerisation"*, DECHEMA-Kurs "Polymerisationstechik".
35 Goodwine, J.W., Ottwill, R.H., Pelton, R., Vianello, G. (1978) *British Polymer Journal*, **10**, 173.
36 Kolthoff, I.M., Miller, I.K. (1951) *Journal of the American Chemical Society*, **73**, 3055.
37 DE 10213229. (2003), Wässrige 2K PUR-Systeme, Bayer AG.
38 Kirsch, S., Pfau, A., Leuniger, J. (2002), Control of Particle Morphology and Film Structure of Carboxylated Composite Straight Acrylic Particles, *FSCT Mid Year Symposium in Waterborne Coatings: Meeting the Challenge*, Orlando.
39 Kirsch, S., Pfau, A., Stubbs, J., Sundberg, D. (2001) *Colloids and Surfaces A: Physicochemical and Engineering Aspects*, **183**, 725.
40 EP 1163304. (2004), Überzugsmittel und deren Verwendung bei der Mehrschichtlackierung, Du Pont de Nemours Company Incorporated.
41 Funke, W. (1988) *Journal of Coatings Technology*, **60**(767), 69.
42 EP 0234361. (1987), Dispersionen von venetzten Polymermin-lerotiden in wässrigen Medien, Verfahren zur Herstellung dieser Dispersionen und Beschichtungszusammensetzungen, die diese Dispersionen enthalten, BASF Coatings AG.
43 WO 03/ 089477. (2003), Highly crosslinked polymer particles and coatings composition containing the same, PPG.
44 EP 0029637. (1980), Acrylic resin composition and method of making it, Cook Paint & Varnish Company.
45 DE 10010405. (2001), Wässrige (Meth)Acrylatcopolymerisat-Dispersionen, Verfahren zu ihrer Herstellung und Verwendung, BASF Coatings AG.
46 EP 0841352. (2003), Verfahren zur Herstellung von Polyacrylat-Sekundär-Dispersionen, Bayer AG.
47 Rempp, P., Merill, E.W. (1986) *Polymer Synthesis*, Hüthig&Wepf, Basel, p. 87.
48 Kaska, J., Lesek, F. (1991) *Progress in Organic Coatings*, **19**, 283.
49 WO/ 068498. (2002), Method and device for the continuous condensation of polyester material in the solid phase, Bühler AG.

50 EP 0963396. (**1999**), Verfahren zus Durchführung von Polykondensationsreaktionen, Bayer AG.
51 Goldschmidt, A., Streitberger, H.-J. (**2003**) *BASF Handbook on Basics of Coatings Technology*, Vincentz, Hannover, p. 272.
52 http://www.rxeforum.com/FP18-13_page2.htm
53 EP 0409843. (**1993**), Verfahren zur Herstellung von Polyesterharzen, BASF Coatings AG.
54 U.S. Patent 3412054. Water dilutable polyurethanes, (**1968**), Union Carbide Corporation
55 DE 2446440. (**1974**), Verfahren zur Herstellung von Polyurethanen, Bayer AG.
56 EP 0303907. (**1990**), Verfahren zur Herstellungen von wässrigen Polyurethandispersionen, Bayer AG.
57 Karunasena, A., Brown, R.G., Glass, J.E. (**1998**) *Advances in Chemistry Series 223*, American Chemical Society, pp. 496.
58 Reimann, H., Joos-Müller, B., Dirnberger, K., Eisenbach, C.D. (**2002**) *Farbe+Lack*, **108**(5), 44.
59 Dietrich, D., Uhlig, K. (**1992**) Polyurethanes, *Ulmann's Encyclopedia of Industrial Chemistry*, 5th edn, Wiley-VCH Verlag GmbH, Weinheim, Vol **A21**, p. 665.
60 Dietrich, D. (**1981**) *Angewandte Makromolekulare Chemie*, **98**, 133.
61 Cody, R.D. (**1993**) *Progress in Organic Coatings*, **22**, 107.
62 WO Patent 01/14441. (**2001**), Process for preparing polyurethane polymers, Henkel Corporation
63 Ger.Offen. 2,725,589. (**1978**), Verfahren zur Herstellung von wäßrigen Polyurethan-Dispersionen und -Lösungen, Bayer AG.
64 US Patent 4,269,748. (**1981**), Process for the preparation of aqueons polyurethane dispersions and dilutions, Bayer AG.
65 Ger.Offen. 1913271. (**1969**), Verfahren zur Herstellung von Polyurethanen, Bayer AG.
66 DE Patent 4010176. (**1991**), Verfahren zur Herstellung einer mehrschichtigen Lackierung und wässrigem Lack, BASF Coatings AG.
67 Vollmert, B. (**1988**) *Grundriss der Makromolekularen Chemie*, E.Vollmert-Verlag, Karlsruhe.
68 Hegedus, C.R., Kloiber, K.A. (**1996**) *Journal of Coatings Technology*, **68**, 39.
69 Kukania, D., Golob, J., Zupancic-Valant, A., Krjink, M. (**2000**) *Journal of Applied Polymer Science*, **78**, 67.
70 Frisch, K.C., Xiao, H.X. (**1989**) *Journal of Coatings Technology*, **61**, 51.
71 Sebenik, U.B.S., Golob, J., Krajnc, M.B.S. (**2003**) *Polymer International*, **52**(5), 740.
72 US 2004/0102595. (**2004**), Polyurethane and graft copolymers based on polyurethane, and their use of producing coatings materials, adhesives, and sealing compounds, BASF Corporation
73 DE 10053890. (**2002**), Sulfidgnuppenhaltige Polyurethane und Polymergemische auf dieser Basis sowie ihre Herstellung und ihre Verwendung, BASF Coatings AG.
74 WO 97/29874. (**1997**), Aqueons polyurethane resin and grafted polymer thereon, BASF Coatings AG.
75 DE 4414032. (**1995**), Als Beschichtungsmittel geeignete Polyurethane, BASF AG.
76 EP 1124871. (**2003**), Polyurethane-Polymer-Hybrid-Dispersion mit hoher Filmhärte, Verfahren zu ihrer Herstellung sowie deren Verwendung, Degussa AG.
77 Jahny, K., Adler, H.J.P., Vogt-Birnbrich, B. (**1999**) *Farbe+Lack*, **105**(10), 131.
78 Barrère, M., Landfester, K. (**2003**) *Macromolecules*, **36**, 5119.
79 EP 0661357. (**1995**), Polymeric pigment dispersants for use in coating compositions, BASF Corporation
80 EP 0588794. (**1995**), Waterborne acrylourethane pigment dispersant polymer, E.I. Dupont de Nemours and Company
81 Chen, S., Chen, L. (**2003**) *Colloid and Polymer Science*, **282**, 14.

82 DE 3518732. (**1986**), Wasserverdünnbare Bindemittel für kationische Elektrotauchlacke und Verfahren zu ihrer Herstellung, BASF Coatings AG.
83 US 3936405. (**1974**), Novel pigment grinding vehicles, PPG.
84 Shukla, V., Bajpai, M., Singh, D.K., Singh, M., Shukla, R. (**2004**) *Pigment and Resin Technology*, **33**(5), 272.
85 Oaka, M.O., Ozawa, H. (**1994**) *Progress in Organic Coatings*, **23**, 323.
86 Goldschmidt, A., Streitberger, H.-J. (**2003**) *BASF Handbook on Basics of Coating Technology*, Vincentz, Hannover, p. 88.
87 Goldschmidt, A., Streitberger, H.-J. (**2003**) *BASF Handbook on Basics of Coating Technology*, Vincentz, Hannover, p. 647.
88 Pourdeyhimi, B., Wang, X., Lee, F. (**1999**) *European Coatings Journal*, **10**, 100.
89 Courter, J.L., Kamenetzky, E.A. (**1999**) *European Coatings Journal*, **10**, 24.
90 Gregorovich, B.V., Hazan, I. (**1994**) *Progress in Organic Coatings*, **24**, 131.
91 US Patent 5, 356, 669. (**1994**), Composite color plus clearcoatings utilization carbamatfunctional polymer composition in the clearcoat, BASF Corporation
92 WO Patent 94/10211. (**1994**), Aminoplast-curable filmforming compositions providing films having resistance to acid etch, PPG.
93 Higginbottom, H.P., Browers, G.R., Ferrell, P.E., Hill, L.W. (**1999**) *Journal of Coatings Technology*, **71**(894), 49.
94 Essenfeld, A., Wu, K.J. (**1997**), A new formaldehyde-free etch resistant melanine crosslinking, *Proceedings 24th Waterborne, High Solid, and Powder Coatings Symposium*, New Orleans, p. 246.
95 US 6300422. (**2001**) Tris(alkoxycarbonylamino)triazin crosslinked waterborne coating system, Lilly Industries Inc.
96 DE Patent 19725187. (**1998**), Beschichtungsmittel und Verfahren zu dessen Herstellung, BASF Coatings AG.
97 Renz, H., Bruchmann, B. (**2001**) *Progress in Organic Coatings*, **43**, 32.
98 Schultz, S., Wagner, G., Ulrich, J. (**2002**) *Chemie Ingenieur Technik*, **74**(7), 901.
99 Pires, R., Laas, H.J. (**2001**) *European Coatings Journal*, 16.
100 Woltering, J. (**1998**) *Journal für Oberflächentechnik*, **38**(9), 14.
101 EP 0576952. (**1997**), Einbrennlacke aus in Wasser löslichen oder dispergierbaren Polyisocyanatgenischen, Bayer AG.
102 EP Patent 0159117. (**1993**), Blocked isocyanates, The Baxenden Chemical Company.
103 DE 4405042. (**1995**), Verfahren zur herstellung von blockierten Polyisocyanaten, die erhatten blockierten Polyisocyanate und deren Verwendung, Du Pont Performance Coatings.
104 DE Patent 19810660. (**1999**), Wäßrige Polyisocyanatrernetzer mit Hydroxypivalinsäure und Dimethylpyrazol-Blockierung, Bayer AG.
105 WO Patent 97/ 12924. (**1997**), Waterdispersible blocked polyisocyanates, Baxenden Chemicals Ltd.
106 Klemm, S., Svejda, P. (**2002**) *Journal für Oberflächentechnik*, **42**(9), 18.
107 Moran, D.K., Verlaak, J.M.J. (**1993**) *Modern Paint and Coatings*, **83**(6), 23.
108 Wennig, A., Weiß, V., Grenda, W. (**1998**) *European Coatings Journal*, **10**, 244.
109 DE 4227580. (**1994**), Verfahren zur Herstellung von Pulverlacken und nach diesem Verfahren hergestellte Pulverlacke, BASF Coatings AG.

8
Paint Shop Design and Quality Concepts

Pavel Svejda

8.1
Introduction

The paint shop is one of the most complex production areas of vehicle manufacture. From today's perspective, the most important paint application procedures can be found here, involving a relatively long process chain. In the paint shop, the highest demands are made on the functional and visual quality of the painting, on the productivity of the painting installations, and on the environmental compatibility of the processes. These are responsible for the high degree of automation that can be found in automobile painting.

In most paint shops, the individual coating processes are classified into coherent functional fields. They are arranged in such a way in the layout of the painting installation that a material flow results that is as simple and logical as possible, in relation to the connection of the paint shop to the neighboring production areas, the body shop, and the assembly line.

A standard coating line (see Figure 8.1) for painting 60 units per hour is about 2 km long. The dwell time of a body is between 6 and 11 hours.

About 30–50 people are employed per shift in a fully automated paint shop, mainly for maintenance, process control, and trouble shooting. The process chain includes value-adding and non value-adding scopes of work. Non value-adding jobs are typically manual jobs, for instance, repairs of body shop faults, sanding and polishing, cleaning, smoothing, and repainting. A future objective is to eliminate non-value-adding jobs completely, or at least reduce them to the minimum extent possible. Value-adding processes have reached a high degree of automation today, and it is expected that full automation will be achieved in the future [1].

The increasing pressure for reduction in costs is reflected in the effort to reduce the cost per unit (CPU). This has led to innovations in the customer–supplier relationship (see Section 13.2), and in the painting process. The standard painting process, which has been used for years by all Original Equipment Manufacturers

8 Paint Shop Design and Quality Concepts

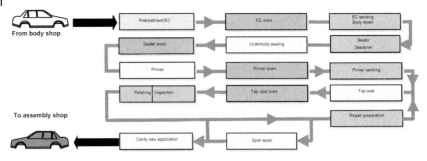

Fig. 8.1 Process chain of a typical automotive OEM paint shop.

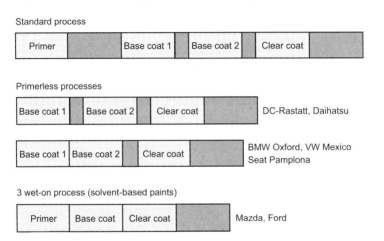

Fig. 8.2 New processes.

(OEMs), consists of the steps primer, base coat 1, basecoat 2, and clear coat. Consolidated processes are now being introduced which involve shorter process times, where either the primer application is dispensed with, or where all coats are applied wet-on-wet, without intermediary drying (see Figure 8.2).

Surface coating technology is going through an exciting time. The purpose here is clear – cost reduction, environmental compliance [2], and improved quality.

With this as the background, the state-of-the-art processes of automotive coating as they have actually evolved till now are described in the following section.

8.2
Coating Process Steps

The material flow in the paint shop, which has been described in the introductory chapter, is characterized by a process chain that includes a host of different coating processes. There are dipping processes as well as painting and coating material atomizing applications. These are briefly explained here according to the material flow or the process sequence as far as is necessary. The chapter on coating installations then elaborates on the corresponding installation technology.

8.2.1
Pretreatment

Pretreatment consists of the steps of precleaning, degreasing, purging, and phosphating. Precleaning removes the rough contaminations. Degreasing solubilizes grease, for example, deep-drawing greases, oil, wax, and other contaminations acquired from the earlier working processes. Phosphating following after a purging process, serves as a temporary corrosion protection, and improves the adhesiveness of the paint film when it is applied.

8.2.2
Electrocoating (EC)

Electrocoat paints are water soluble (suspensions of binders and pigments in d.i. (deionized) water) with only low proportions of organic solvents (approximately 3%).

Electrocoating covers all dip painting processes, where the paint precipitates on the workpiece owing to chemical conversion and associated coagulation of the binder. These conversions are caused by an electric current flow from an external electrode via the conductive paint, to the workpiece. The advantages of the electrocoating process are as follows:
- complete and uniform coating, also in cavities;
- very good corrosion protection;
- no formation of drops and runners;
- very low paint losses with the corresponding purging techniques;
- very good possibilities of process and quality control;
- electrocoat installations are fully automatic, and are continuously operated.

The process is faced with some disadvantages, namely,
- high installation-specific expenditure with correspondingly high investment cost;
- relatively high material cost;
- considerable expenditure for the monitoring of the paint bath.

In the vehicle series painting, the cathodic dip painting has been in use since about 1975. Compared to the anodic dip painting it provides advantages such as an excellent corrosion protection, a uniform film thickness distribution, a better wraparound, a good edge covering, and a lower current consumption. The disadvantages existing in the beginning, such as lower film thickness and high baking temperatures have been reduced over a period of time by means of further improvements in paint materials. Epoxy resins are used as a binder for the cathodic electrodeposition paints because of their known characteristics, such as good corrosion protection and good adhesion to the underground and to further coatings.

Normally, the epoxy resins are not water soluble and must therefore be modified to be able to convey into a 'water-soluble' state in a so-called neutralization reaction with organic acids like acetic acid. At the electrodes, and at the workpiece, the water is electrolyzed by the electric current, and gases (H_2 and O_2) as well as ions are released [18]. The hydroxyl ions (OH^-) created at the interface to the cathodically poled workpiece cause a reversal of the neutralization reaction and lead to a coagulation of the binder (see Chapter 4).

The layer of coagulated paint particles that is loose in the beginning is pushed toward the electrode by the paint particles which are wandering because they are still electrically charged. At the electrode, the coagulated material with high water content, is fixed and compressed by the material that follows later. At the same time, the electric field causes removal of ions that carry along hydrate shells. This leads additionally to a dehydration and solidification of the film by means of electro-osmosis. Within a short time, the nonvolatile portion of the film increases to more than 90%, although the paint bath itself only has about 20%. In this manner, the film reaches a firmness that allows the rinsing processes that follow.

$$\begin{aligned}
&\text{P-NR}_2 + \text{R-COOH} &&\rightarrow&& \text{P-N}^+\text{R}_2\text{H} + \text{R-COO}^- &&\text{Neutralization reaction}\\
&\text{Water insoluble} &&&& \text{Water soluble} &&\\
&2\text{H}_2\text{O} + 2\text{e}^- &&\rightarrow&& 2\text{OH}^- + \text{H}_2 &&\text{Water electrolysis}\\
&\text{P-N}^+\text{R}_2\text{H}^+\text{OH}^- &&\rightarrow&& \text{P-NR}_2 + \text{H}_2\text{O} &&\text{Coagulation}\\
&\text{Water soluble} &&&& \text{Water insoluble} &&
\end{aligned}$$

P: Long chain of film forming polymer
R: Organic rest (1)

It is essential for the understanding of coating processes that the precipitation is not realized by the attraction of the paint particles to the workpiece surface in the electric field and the attachment and discharge thereafter. Rather, the decisive factors for the coating are the electrochemical processes in the diffusion controlled layer at the cathodic surface and the reactions of the binders, as already described. The mass transport is exclusively determined by convection. Therefore, a sufficient recirculation of the dipping bath is indispensable.

8.2.3
Sealing and Underbody Protection

The overlapping, spot-welded metal sheets must be sealed in such a way that no humidity can penetrate between the metal sheets and water in the vehicle interior, which may lead to corrosion there. On the weld seams, high viscous

8.2 Coating Process Steps

Fig. 8.3 Application of under body protection material.

polyvinylchloride (PVC) material is mostly sprayed as paths with airless application or extruded by flat stream nozzles (see Figure 8.3). The underbody protection also serves as protection from corrosion, mostly for areas exposed to a high strain because of stone chips. It is applied partially two-dimensionally, for instance, in wheel arches and in the rocker panel area (see Chapter 10).

8.2.4
Paint Application

The primer surfacer applied on top of the electrocoat protects the cataphoretic electrocoating film from ultra violet (UV) radiation, serves as a surface smoothing primer for the following top coat film, and reduces the risk of damage to the layers below, in case of stone chips. Bumps and faults stemming from the body shop like grinding remains can be repaired by sanding the primer coat. The primer is applied with the high-speed-rotating application with electrostatic charging of the paint material. For reasons of volatile organic compounds (VOC) emission, hydroprimer materials are mostly used in Europe and powder, to a certain degree, in North America (see also Chapter 5). A further reduction of emissions has been achieved with the development of two different processes.

8.2.4.1 Function Layer and Primerless Processes
The function layer combines the characteristics of the primer and the base coat material. It is matched in color to the following base coat material, which is then only applied in one coat on the wet or flashed off function layer. This is also valid for the metallic effect material. In this process, the entire process chain consists only of four paint applications – cathodic electrocoating, function layer, base coat,

and clear coat, compared to five layers for the conventional metallic painting – cathodic electrocoating, primer surfacer, basecoat 1, base coat 2, and clear coat. Apart from the emission reduction, this process has the advantages of the reduced installation investment and overheads owing to the omission of one painting line (see Chapter 13). A disadvantage is the fact that faults from the body shop and the cataphoretic electrocoat process that are not eliminated after the cataphoretic electrocoat, can only be processed after the top coat painting.

8.2.4.2 Powder

The primer surfacer application with powder material is practiced mainly in North America. The powder primer application is distinguished by the known advantages of a powder coating – no solvent emissions, possibility of material recirculation, and the resulting high transfer efficiency of more than 98%, simple installation and application technology. The disadvantages are as follows:
- relatively high film thickness
- high material usage
- limited possibilities for a color change in connection with powder recirculation.

The top coat is applied after a thorough cleaning of the entire car body. The prevalent process for the top coat application is the application of a waterborne base coat, followed by a clear coat. One- and two-component high-solid clear coats are mainly used as clear coats (see Chapter 6). For waterborne clear coats, powder clear coats (see Figure 8.4), and water-dispersed powder clear coat systems, so-called powder slurries are formed and so these could not capture any significant market share.

Fig. 8.4 Application of powder clear coats by robots and bells.

Solid base coats are applied as a single coat, using high-speed-rotating atomizers. Metallic effect paints are applied in two coats, the first coat with high-speed-rotating atomizers, the second, usually pneumatically. The reason for the second pneumatic application lies in the desired effect of the painting, which can be reached to the required extent only in certain application cases with high-speed-rotating atomizers. Also, the repairing process is more easily accomplished owing to the fact that repair of top coats, especially in the field, is carried out manually by pneumatic guns. It can be expected, however, that the electrostatically supported application of both coats will prevail in the near future.

With the clear coats, there has been an increase in development activities towards achieving high scratch resistance. These, and other equally important features, are achieved by paint formulations that are linked with a higher degree of polymerization by means of UV radiation.

The top coat application is followed by a quality control check. Here, the paint film is examined for faults like dirt inclusions, wetting disturbances, runners, and other such defects. Additionally, the film thickness and the visual parameters like color shade, gloss, and leveling are measured regularly.

8.2.5
Cavity Preservation

The corrosion protecting measures are finalized with the sealing of the cavities with wax materials. For this, two procedures are usually followed – spraying and flooding. For spraying, special nozzles are inserted in the cavities, and an exactly measured quantity of material is sprayed inside each cavity. For flooding, the cavities are filled with flooding wax, under pressure.

For the application of the wax, two automation procedures are used. The procedure that is clearly more flexible with regard to the position of the bores, is based on the introduction of the nozzles by robots (see Figure 8.5). Here, several pneumatically fold-out nozzles are arranged in a nozzle-exchange head. The robot retrieves the required nozzle-exchange head from a magazine, applies it to the corresponding cavities and replaces it, to seize the next nozzle-exchange head. In

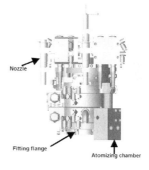

Fig. 8.5 Cavity preservation step with robots and nozzle-exchange head.

the other automation procedure, numerous nozzles are arranged in a frame. After the car body is positioned above the nozzle plate, the plate is lifted, and all the nozzles are inserted simultaneously into the corresponding bores in the vehicle underbody and applied. With this procedure, many cavities can be sealed within a short cycle of time, provided the cavities are reachable.

8.3
General Layout

Several levels of the building are usually involved in the arrangement of the different process steps of a paint shop. The bottom level has the dipping tanks for pretreatment, cataphoretic painting, and the entire technology for waste water and exhaust air. On the level above that are the spray booths and the work decks (see Figure 8.6). The next higher level has the dryers and the car body storage area. The air-supply units are located on the top level, in the so-called *penthouse*.

Installation areas like the primer and the top coat area, where cleanliness is of prime importance are often designed according to a clean room concept. The car bodies are moved into this area after a thorough cleaning. Entry into this area is restricted and requires special clothing [3].

The automatic cleaning of the body surface can be done by dry or wet methods. In either case, the process consists of two stages. In the first step, the inner areas and seams are blown out for removing dirt particles, which may have been incorporated into the surface mainly during pneumatic-spray applications. The body is driven through an installation with horizontal, vertical, and other direction-oriented air nozzles through which airstreams are pumped at more than $40\,\text{m s}^{-1}$. The dirt particles are carried by the airstreams to the end of the zone, where they are trapped by filter units. After filtration, the clean air is recycled and reused to feed the nozzles. The second step is the cleaning of the body surface by rotating rollers equipped with feathers. The feathers pick up the dirt particles which are sucked away from the rollers. The rollers are programmable and rotate at the right distance from the body surfaces.

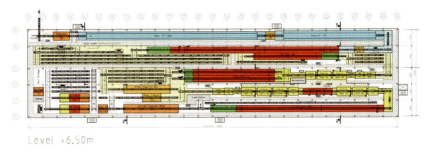

Level +6.50m

Fig. 8.6 Typical layout of the paint shop functions.

Fig. 8.7 3D layout of a transportation segment in a paint shop.

Another method for removing dirt from the body surface is simple washing. In this case, the body is dipped and treated with rotating brushes, similar to a washing station, or cleaned with high-pressure water jets. Dirt in the inner sections is removed with the high-pressure water jets. For best results, the nozzles are guided in the required direction by robots. After the washing step the wet body has to be dried completely in a dryer. This is to prevent any wetting marks by water droplets, which can be blown out during paint application.

An optimum layout must meet the following requirements:
- short lead times
- straight-line material flow
- compact, space-saving design
- consistent implementation of dust-protection measures
- concentration of installation technology and work decks
- short and easy access for operating and service personnel
- simplified maintenance and service conditions.

The layout and the design are finalized increasingly with the help of 3D computer aided design (CAD) systems (see Figure 8.7). Without doubt, the expenditure involved here is worthwhile. Not only is safety increased owing to planning, but examination and safeguards can be put in place, well in advance. This will reduce cost outflows, which may emerge at a later stage. These are, for example, collision observations of installation components or space constraints at the time of installation of large-volume assemblies. Simulations of the material flow or the robot stations can be integrated, ergonomic studies can be made. Lastly, various virtual-reality tools offer the possibility of inspecting the paint shop long before the first ground-breaking ceremony [4].

8.4
Coating Facilities

8.4.1
Process Technology

A painting line is divided into various zones. Paint is applied under the necessary conditions of temperature and humidity in the spray booth (see Figure 8.8). The overspray that is generated during painting is absorbed by the air that flows vertically through the booth and is directed to the washout.

The booth is supplied with air by an air-supply installation. The fresh air that is aspirated by fans, is filtered in several stages, heated up, humidified, and pumped into the booth uniformly through a filter from the ceiling via a pressure room, that is, *plenum*, with a downdraft speed of 0.2–0.5 m s^{-1}, depending on the application procedure.

The outgoing air, which is loaded with the overspray, flows through a washout system for example, a Venturi gap, where the particles in the overspray are absorbed by water. The outgoing air is guided through a rotating air-to-air heat exchanger or the so-called *absorption heat exchanger*. Here, the heat and the humidity are exchanged with inward air supply that flows in the opposite direction. The paint

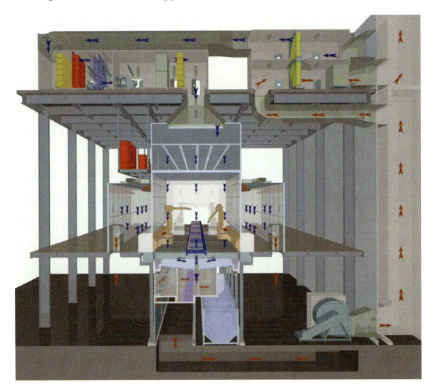

Fig. 8.8 Cross section of a spray booth.

material that has been carried in the booth water, is flaked by the addition of a coagulation agent and removed from the water with the using various technologies. The booth water is guided in a closed loop. The evaporated water and the water removed from the paint sludge are added to the booth water.

8.4.2
Automation in the Paint Application

The evolution of the fully automatic paint application has passed through numerous stages. In the early phases, the exterior areas of the car bodies were coated by painting machines. The individual paint coats were at that time already being applied by high-speed-rotating atomizers, except for one – the second metallic base coat. For this, pneumatic atomizers that are mainly characterized by one feature, namely, low transfer efficiency, were necessary. The machine concept developed further to the robot concept. The driving forces behind this move were the striving for higher flexibility and effectivity.

Transferred to the painting technology, this means that robot installations provide more flexibility in regard to the vehicle design, and at the same time, they allow for a higher effectivity owing to the reduction of the number of atomizers, the increase in transfer efficiency, and the corresponding reduction in the overheads. The pneumatic atomizer for the second metallic application is currently being superseded by the high-speed-rotating atomizer. In the future, the bell–bell application with robots will, in all likelihood, be used for exterior painting. The fully automatic interior painting which has always been a domain of robot technology, has also developed away from pneumatic application during the last few years. Interior application with rotating atomization is the latest stage of evolution as of now [5].

There are two different techniques of movement or transportation of the body for painting by robots. Both have some common features. The first technique is the 'painting in tracking', which is characterized by continuous movement through the robot station, and the second is the painting in the 'stop and go' method. The latter needs a fast drive into the robot station, where the body is then fixed and positioned for the robots.

The choice of the technique has significant impact on the design of the station. The 'painting in tracking' preferably uses robots with six axes. This concept is characterized by less complex technology and lower investment costs. Furthermore, it allows for a higher painting capacity because no additional time is needed for the special conveying and positioning techniques. However, the 'stop and go' technique also has its advantages. The positioning of the body is more accurate, and the application process is not influenced by variations in the body position caused by the conveying system. The number of robots can be reduced, especially with a longer dwell time. In case of technical failure on the part of the robots, the remaining robots can take over the painting step, as long as two robots are installed on either side of the body.

The painting of the inner sections can be realized by both techniques. However, fixed robots cannot be used for this purpose.

8.4.2.1 Painting Robot

A painting robot is a freely programmable automatic painting machine in bend-arm style, which has six rotatory axes – three main and three hand axes. It can be equipped with an additional traveling axis. With a painting robot, all painting processes in the interior and exterior areas of automobile bodies can be carried out. A painting robot mainly consists of the components displayed in Figure 8.9.

A painting robot has six movement axes (see Figure 8.10):

1. axis 1 (horizontal turning movement of the housing on the basic body)
2. axis 2 (pivoting movement of arm 1 on the basic body)
3. axis 3 (pivoting movement of arm 2 at arm 1)
4. axis 4 (rotatory movement of the hand axis directly at arm 2)
5. axis 5 (rotatory angularly offset movement of the middle hand axis component)
6. axis 6 (rotatory movement of the hand axis at the atomizer flange).

To be able to adjust the working range of the robot optimally to the painting task, the necessary robot kinematics can be realized by using arms with different lengths. A sufficient carrying weight, a high dynamics, and high path accuracy are other important features of a robot system. For the painting of interiors, it may be advantageous from the point of view of the available painting time, if so-called *handling robots* (door and hood openers) complement the system. However, if the

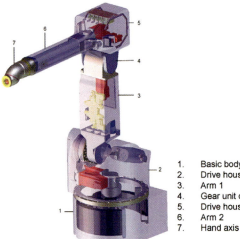

1. Basic body
2. Drive housing of the axes 1 and 2
3. Arm 1
4. Gear unit of axis 3
5. Drive housing of the axes 4, 5 and 6
6. Arm 2
7. Hand axis

Fig. 8.9 Design of painting robot.

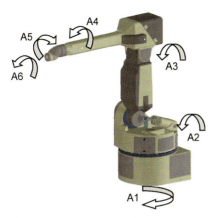

Fig. 8.10 Movement axes of a painting robot.

painting time is assessed properly, door and hood handling by the painting robot may reduce the investment cost.

Painting robots can carry various atomizer systems like pneumatic atomizer, high-speed-rotating atomizer, and powder atomizer. It is important to note here, that the painting robot is suited for electrostatics without restrictions. Today's standard for painting robots is the layout of all media lines to the atomizer through the robot arm. The application equipment must not restrict the high dynamics of the robot. Therefore, dosing pumps, air volume sensors, and proportional valves are installed as close as possible to the atomizer.

Robot control must be particularly efficient that is, a robot must have an operating system capable of multitasking. The hardware and software components of the control should have a modular architecture. Painting-specific functions such as for instance, a convenient main-needle control for the atomizer, a function for the exact repositioning and continuing of the painting process after a conveyor stop, or a function enabling the painting robot to automatically move backwards on the path after an emergency stop – especially important for interior painting – make the robot operation safer for process and also increase its availability.

8.4.3
Application Technology

The term *application technology* in a paint shop means the overall equipments namely, the atomizer, the dosing, and the color-change technology. In automotive OEM painting, the following are the three different atomization principles mainly used:
- the airless atomization
- the pneumatic atomization
- the high-speed-rotating atomization.

8.4.3.1 Atomizer

The main requirements on an automatic atomizer and its periphery are the reproducibility of all atomizer parameters and the spray pattern. The latter depends not only on the characteristics of the control systems, but also on the manufacturing quality of the components used for atomization like the nozzle, air cap, bell disk, and so on.

The atomizer functions are mostly controlled pneumatically by solenoid valves. The control system also process the feedback signals of the atomizer, like the position of main needle and speed of the bell disk. For automatic application, it is important to define a very fast response behavior for the atomizer. Both the switch-off and the switch-on of the atomizer and the required value changes of the paint and air volumes must be realized exactly at the programmed points. This requirement is not easily met, especially because pneumatic systems, which use compressed air, have longer reaction times, whereas, today's path speeds of the automatic painting machines lie in a range above $1000\,\text{mm}\,\text{s}^{-1}$. Breathing of the hose lines additionally aggravates this problem. If the atomizer switching time is for instance, 50 milliseconds, which is quite a good value, a painting robot moves 50 mm on its path during this time. Therefore, if the atomizer like a pneumatic gun has to be turned on at a certain point, the control signals for the atomizer needle, the dosing pump, for the atomizing, and the shape of the spray cloud regulating air – the so-called *horn air* – must be set correspondingly earlier.

The availability of automatic painting installations may not be impaired by unnecessarily long downtimes for maintenance and servicing work. Therefore, it must be possible to exchange each component to be serviced in as short a time as possible. For an atomizer, this means that all media, *control air*, and signal lines are conducted on one connection plate on which the atomizer with its counter flange is fastened. The attachment is realized by means of a screw or a bayonet coupling.

8.4.3.1.1 Airless Atomization
Airless atomization converts the energy of the fluid under pressure into kinetic energy. The geometry of the nozzle determines the efficiency of the conversion [6]. A turbulent fluid jet with a high flow speed emerges from the nozzle into the inactive environment atmosphere. The inert forces generated by the internal turbulence of the jet are higher than the surface force, so that the jet atomizes [7]. Shearing forces arising from the jet delay support this process.

8.4.3.1.2 Pneumatic Atomization
In pneumatic atomizers, the energy necessary to dissipate the fluid jet leaving the nozzle at low speed is delivered by the atomizing gas (compressed air). The fluid is accelerated a little beyond the nozzle port by the atomizing air. The high relative speed (in case of pneumatic high-pressure atomizers, the atomizing air leaves with the speed of sound) between the air and the fluid jet effects an atomization of the fluid by the generated shearing forces. Internal turbulences in the fluid jet do not appear owing to its low speed [6, 8, 9].

The air cap and the paint nozzle are the parts essential for atomization and for achieving the paint-film quality (see Figure 8.11). The atomizing air required for

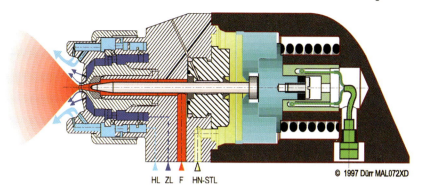

Fig. 8.11 Scheme of a pneumatic atomizer.

the atomization of the paint jet emerging from the nozzle, flows out of an annular gap that is formed by the bore in the air cap and the air nozzle that is arranged in it centrically. Further air jets from surrounding bores support the atomizing process and prevent the formation of a low-pressure area with negative pressure in front of the air cap that would lead to a soiling of the atomizer head. Fan jet (horn air) bores regulate the shape of the spray jet. The horn air flowing out of these bores can form a round jet with an almost circular spray pattern to a fan jet with an elliptic spray pattern. Both the atomizing air and the horn air are fed and controlled separately in automatic atomizers. The paint nozzle is closed and opened by means of a nozzle needle that is activated pneumatically with the control air. The paint flow volume is adjusted with dosing pumps. A return line serves to purge the atomizer during the paint color-change process.

8.4.3.1.3 **Rotating Atomization** In the painting technology, two models of rotating atomizers, the disk atomizer, and the bell atomizer are used.

The centrally fed fluid distributes on the atomizer as a film, the film thickness normally reducing toward the edge of the atomizer. When the throughput is increased with a constant speed of the atomizer, the decomposition mechanisms, that is, dripping off, filament decomposition, lamella decomposition, and atomization appear consecutively [8].

Basically, rotating atomization is similar to airless atomization, where a fluid jet with interior turbulences emerging into an inactive environment, is delayed and atomized. The atomization energy provides the rotation and the centrifugal forces connected to it. Accordingly, the droplet diameter mainly depends on the speed and the diameter of the atomizer.

High-speed-rotating atomizers (see Figure 8.12) consist mainly of an atomizer bell disk and its drive. Air turbines are mostly used as drives. Depending on the paint material and throughput quantity, the atomizers are operated at speeds between 15.000 and 60.000 min^{-1}. The speed is usually recorded optically and controlled. The bell disks have a diameter of magnitude in the order of 50 mm, and have different geometries and edge surfaces.

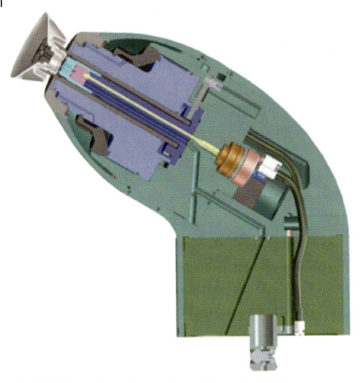

Fig. 8.12 Scheme of a rotating atomizer – high-rotational bell.

The paint is centrifuged tangentially off the bell disk edge and atomized. It is then deflected by a shaping airflow that leaves the air bores. These bores are arranged so that they are annular and coaxial to the atomizer axis behind the bell disk. The atomized paint is formed into a spray jet and deflected by electrical field forces to the grounded substrate to be coated. In the middle, the circular spray pattern has principally a field with a lower film thickness. The forming depends on the operating parameters and the bell disk geometry.

The paint material is charged by contact with the bell disk that is connected to high voltage potential of 50–120 kV by contact charging, or by agglomeration of air ions to the paint droplets, that is, ionization charging. The air ions are generated by external electrodes. Contact charging is used for electrically low-conductive paints based on organic solvents, and for electrically conductive waterborne paints, together with a potential separation system (see Section 8.4.4). Electrically conductive paints are normally processed with ionization charging because of lower investment costs (see Figure 8.36).

Besides the shaping airflow, electrical field forces are also involved in the transport of the paint droplets to the grounded substrate, so that a high transfer efficiency of 85–90% is achievable in this application.

8.4.3.2 Paint Color Changer

Automatic painting installations, especially in the automobile industry, must be flexible with respect to colors. For this, a purgeable paint color-change and dosing technology is required. The color change must be carried out in a short time, and the purging losses must be minimal.

The paint supply of a painting installation is located in a paint-mix room, the so-called *paint kitchen*. Here, a paint-circulation container is set up for each paint. The containers are linked to circulation lines which are supplied by means of a pump. The paint-circulation pipe, which leads the paint to the painting line, supplies the automatic painting machines mostly through stubs, and ends in the paint-circulation container again (see following chapter). The individual stubs are led to the color-change equipment (paint color changer) inside the automatic painting machine and connected to the paint control valves there. A dosing system arranged after the color changer supplies the atomizer.

The color-change system is conceived in such a way that the color-change can be carried out in the gap between two successive car bodies. Depending on the distance of the vehicles and the conveyor speed, only few seconds are available for the color change. To achieve the required short color-change time, and to reduce the paint and solvent losses generated during a color change to a minimum, the color changer is usually installed directly in front of the atomizer (see Figure 8.13).

The color changer (see Figure 8.14) consists of several function modules that allow any number of paint connections to be strung together. The central component is the connection strip into which the paint control valves are screwed in. The paint control valves have air connections and are activated pneumatically by means of solenoid valves. The connection block that serves to attach the paint hose lines by means of hose connection plates, is located on the connection strip.

On the paint side, the color changer ends at the paint pressure regulator. In addition to the connections for the various colors, one connection each for solvent and pulse air is available.

The color change is carried out as follows:
- Close the paint control valve for applied paint.
- Blow the paint line to the atomizer free with pulse air.
- Purge the paint line to the atomizer alternately with solvent and air pulses; with the last air pulse the line is blown dry.
- Open the paint control valve for the next paint color, prepaint paint up to the atomizer.

The paint and solvent losses that result during the color change can be reduced by the use of special color-change systems. The most efficient reduction of paint losses during color change is however, the assembling of car bodies with the same color to larger so-called color blocks that are moved through the painting installation without a color change. A prerequisite for this is the installation of a color sorting stack next to the paint shop.

Fig. 8.13 Color changing installation in the robot arm.

1. Connection strip
2. Paint control valve
3. Connection block

Fig. 8.14 Color changer unit.

8.4.3.3 Paint Dosing Technology for Liquid Paints

For a good painting result an exact dosing of the required paint volume is very important. Basically, it is possible to set the paint volume by means of a dosing pump or a paint volume control. Control systems based on the paint pressure without a paint flow meter are unsuited for waterborne paint systems that show a shear thinning behavior.

Paint volume controls dose the paint material with a paint pressure regulator, that is activated by a proportional valve depending on the paint volume that has been set. A flow meter supplies the information about the current paint volume. Gear wheel gauge heads, inductive gauge heads, and Coriolis force flow meters can be considered as measuring transducers. For fast changes of required values, as this is required in the use of painting robots, the paint volume control is less suited, because the entire system is relatively sluggish (pneumatic actuators) and requires a longer setting time. For this, dosing pumps, mainly gear wheel dosing pumps are much better suited.

The gear-wheel dosing pump (see Figure 8.15) consists of a pump housing, and a wheel set that is driven by a servo motor via a shaft. The medium arriving at the inflow (suction room) is transported toward the outlet into the pressure room via the tooth gaps. The engagement of the teeth into the tooth gaps displaces the filling toward the outlet. The flanged-on valve block enables a fast purging of the dosing pump during a color change.

8.4.3.3.1 Pigging Technology

In configurations which need to manage a high number of colors it may be advantageous to install the dosing and color-change system outside the spray booth. However, this results in the atomizer being connected to long paint feeding lines that have to be purged during a color change. To avoid rejection of the material contained in the lines, it is possible to push it back into the circulation line by means of a pig. The pig, a cylindrical body adjusted to the diameter of the paint pipe, moves between the atomizer and the dosing

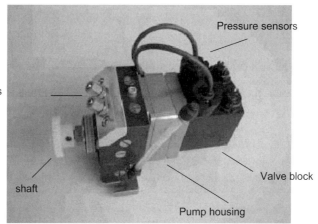

Fig. 8.15 Dosing pump unit.

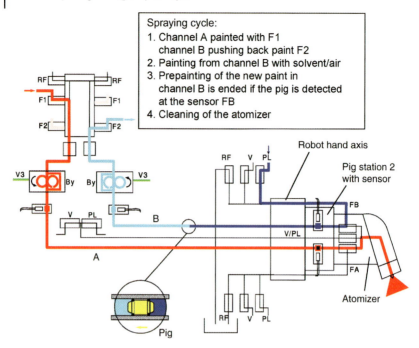

Fig. 8.16 Pigging system for paint recovery in circulation lines.

pump (see Figure 8.16). Together with an A/B color changer, extremely short paint color-change times can be achieved with the use of pigging technology.

8.4.3.4 Paint Dosing Technology for Powder Paints

The painting powder must be fluidized for dosing, transportation to the spray gun or bell, and for spraying (see Section 5.5). This is achieved either by using respective containers having porous bottoms through which air is blown with eventual supporting vibration, or by local fluidization in the area of the powder sucking pipe. The latter technique does not need a fluidization container and the powder can directly be taken from the delivery bag, usually called *big bags*.

The powder injector sucks the material under low pressure out of the powder container or bag. The powder flow is determined by the amount of conveying air. The addition of air guarantees a sufficient air speed of more than $10\,\text{m s}^{-1}$. The design and the parameters of the injector, in combination with appropriate powder materials and pipe diameter, together with straight piping lines have significant impact on the reproducibility and uniformity of the powder output. The output can be controlled by weighing systems beneath the fluidization container, or by sensors within the pipes.

The injector technique is not appropriate for precisely adjusted, reproducible powder outputs. A powder dosing pump with a new level of accuracy has been developed recently. A pneumatic driven, double piston pump (see Figure 8.17) sucks alternately the powder from the fluidization container and conveys it to the

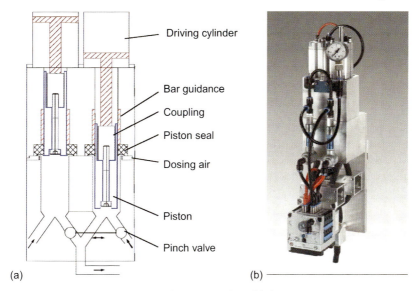

Fig. 8.17 Powder dosing pump – scheme (a) and model (b).

application unit via a hose. Pinch valves close according to the piston movement, the sucking, and the output pipes or hoses. The powder flow is determined by the volume of the piston chambers and the pumping frequency. The essential advantages of this powder pump are as follows:
- high dosing accuracy
- powder out put up to 300 g min^{-1}
- change of powder flow via pumping frequency
- length of hose lines to the application units up to 25 m
- hose diameters are 50% less in comparison to the injector
- minimal air usage of about 10 Nl min^{-1}.

Powder transportation is carried out with these pumps in the almost complete absence of air. Therefore, it is more suitable for feeding the rotational bell application of powder, where the powder cloud is achieved only by the rotation of the bell.

8.4.4
Paint-Material Supply

The task of a paint supply system is to supply the application technology with the required quantities of material at the required pressure. For this, to achieve a high quality of the paint application, all paint parameters must be imperatively kept within the process window provided. For example, variation in material temperature leads to viscosity changes in the paint materials, resulting in structural differences or sagging, in case of deviations in the coating. In most paint systems there is a tendency of sedimentation of the contained binder and pigment components.

Therefore, prevention of these sedimentation movements is high-priority task of a paint supply system. Inhomogeneous paint material inevitably leads to an inadequate painting result. Settled paint components may lead to blockages and to the gradual blocking of lines. Flaked paint components result in soiling of the paint film.

8.4.4.1 Paint Supply Systems for the Industrial Sector

In the industrial sector with painting installations for a higher parts throughput and correspondingly high paint-material consumption, paint supply systems with paint circulation are frequently used to supply the atomizers. Such paint supply systems can be classified into two areas – in the paint-mix room with the container groups and all components required for the delivery and provision of the paint materials and in the circulation line system with the takeoff points.

8.4.4.2 Paint Mix Room

In the paint-mix room or 'paint kitchen' (see Figure 8.18), the container groups for the supply of the painting installation with paint material and with purging agent and solvent are set up. An area of the paint-mix room is for the delivery of the paint containers separated. In this container room, the paint containers are docked to the container groups for filling and additional containers are stored. Here, the container is prepared to be connected to the container groups, that is, the agitating of the paint material and the setting of the processing viscosity. The part of the paint-mix room with the container groups is separated by open ground from the other areas and is frequently conceived as a clean room to ensure optimum cleanliness of the paint supply systems. A paint-mix room with the required sensor and actuator equipment requires a control system. This is located in a separate room. The monitoring of all functions and the operation of the paint supply systems is carried out from an operation room by means of a viewing system.

8.4.4.3 Container Group

All paint materials required for supply to the application units are stored in the container groups (see Figure 8.19). The material is conveyed from the circulation tank with a compressed air operated piston pump through a filter into the supply line of the circulation line. A pulsation damper smothers the pressure surges that are caused by the piston pump. The required material pressure in the circulation

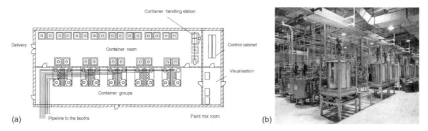

Fig. 8.18 Layout and picture of a paint-mix room ('paint kitchen').

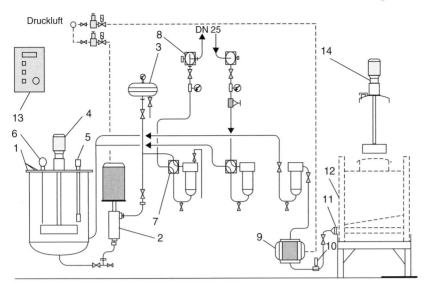

Fig. 8.19 Design of a container group.

1. circulation tank 300 l
2. circulating pump
3. pulsation damper
4. agitator
5. level monitoring
6. thermometer
7. change-over ball valve
8. pig connection
9. filling pump
10. container filling level
11. fill stub
12. delivery container
13. operating panel
14. agitator for container

line is set with a pressure controller in the return line. The pressure must have a certain value so that enough paint is available at the takeoff points. Guided back through the return line, the paint is filtered again and gets back into the circulation tank. The circulation tank has an agitator, so the material cannot settle. A thermometer records the paint temperature, the level monitoring records the filling level. If the filling level falls below a programmed value, the filling pump (usually a compressed air operated diaphragm pump) is activated, that conveys material from the supply containers to the circulation tank, until the specified maximum filling level is reached. The minimum and the maximum filling level of the circulation tanks are kept fairly close to each other. This prevents the material from drying up on the inside wall of the tank, which would lead to contamination.

Pig stations are installed in the supply and return sections of the circulation line. To clean the ring pipe, a cylindrical body (pig) can be inserted into the circulation lines through the pig station, pushed through with solvent and compressed air, and removed in the return line. This achieves an optimum cleaning of the ring pipe. For cleaning and maintenance of the supply and return line filters, four way change-over ball valves are installed at the filters.

In the design of the container groups, it must be ensured that they are easy to clean and that paint deposits on the components are avoided. This is achieved by the following design principles:
- no dead storage volumes
- smooth surfaces (inside polished tank walls), no undercuts
- gap-free connections.

All paint-carrying components of the container group are made of stainless steel like 1.4302 according to DIN. This is also valid for the total circulation line system.

8.4.4.4 Circulation Line System

This system supplies the application stations of the painting installation with paint material. The pipeline route is laid out in the area of the booth washout system. This provides for stable operating conditions owing to a uniform temperature. The consumers can be connected from below, without any impairment of the clean room concept in the booth area. Various constructions exist for the circulation line systems.

The takeoff points of the circulation lines are designed as T-outlets that can be closed with a ball valve. A hose, usually steel-encased, leads from the ball valve to the bulkhead plate in the painting machine or at the takeoff points of the booth.

The pipes, according to DIN 11 850, are blank inside with a surface roughness of approximately 1 µm, material for example, 1.4404, and are made almost gap-free with the use of so-called *aseptic* screw-pipe connections and orbital-welded connections. Fittings are designed in such a way that clearance volumes are reduced to a minimum. The advantages of circulation line systems with such a design are as follows:
- low pressure losses because of a gap-free pipe and fitting connection, which reduces paint shearing and energy consumption;
- no paint deposits in the circulation line system;
- easy to clean, easy to purge and hence, a reduced consumption of the purging agent during color changes;
- piggable, paint recovery during color change possible.

To exclude the influence of the temperature fluctuations on the paint material, the lines are insulated. Under extreme temperature conditions, heat exchangers are used for material tempering.

8.4.4.5 Basic Principles for the Design of the Pipe Width for Circulation Lines

The pipe width is designed in such a way that a flow speed of between 0.1 and 0.4 m s^{-1} is achieved at the required tapping quantity. If the flow speed is any lower, there is a danger of depositing and of a gradual blocking up of the pipe profile. Too high a flow speed would result in an unnecessary shearing stress of the paint materials. Sensitive paint systems can be destroyed by an excessive flow speed. The calculation bases summarized in the following assume a Newtonian behavior of the paints. This is not the case, particularly with water based paints, where, however, it is not important for the design of the line dimension.

For a stationary, laminar flow in pipes with circular profile (Reynolds number $Re = v_m\, d/v < 2320$) the Hagen–Poiseuille equation is valid for the pressure loss Δp:

$$\Delta p = 32 l v_m \eta / d^2 \tag{2}$$

Here, l is the length of the line, v_m the medium flow speed, η the dynamic viscosity, and d the inside diameter of the line.

The medium flow speed v_m can be calculated from the material flow Q, using the following relation:

$$v_m = 4\, Q/(\pi d^2) \tag{3}$$

The dynamic viscosity η is linked with the kinematic viscosity v by the relation:

$$\eta = v\rho \tag{4}$$

where ρ is the density of the paint materials.

8.4.4.6 Paint Supply Systems for Small Consumption Quantities and Frequent Color Change

For smaller consumption quantities and a more frequent color change, the supply systems so far described are not suited. The purging of such a system is time-consuming despite an easy to clean design and involves a relatively high consumption of the purging agent. Also, the system volume is often so high that a material quantity required for the filling is too high relative to the actual paint-material requirement. For those 'low color' requirements, small circulation systems and supply systems for special colors (individual color shades) are therefore used [10].

8.4.4.7 Small Circulation Systems

The small circulation system is set up as close as possible to the painting line, to reduce the length of the circulation line. The profile of the pipe is also kept small and therefore, the system volume is low. Hard to purge components like for instance, pulsation dampers are omitted. The paint containers used are hoops with volumes of approximately 25 l. They find simultaneous use as circulation tanks. A diaphragm pump is used as a circulating pump because it is easier to purge. Owing to the smaller line profile, the number of the simultaneously used takeoff points is limited. If, for instance, such a small circulation system is used for the supply of a vehicle top coat line with special colors, this means that only individual vehicles and not vehicle groups can be painted with the special color.

The circulation line is only operated for purging, filling, and in case of no painting operation, for circulation. During the painting process, a blocking valve built in at the end of the circulation line is closed, so that the ring pipe is used as a stub. The pressure necessary for painting is built up in this manner.

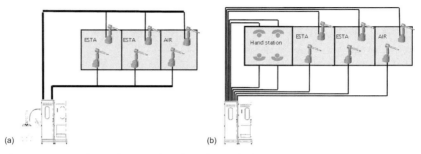

Fig. 8.20 Main stub supply (a) and single stub supply (b) for special ('Low') color application.

8.4.4.8 Supply Systems for Special Colors

Supply systems for special colors are conceived in such a way that a short color-change time, a minimum purging agent consumption, and low paint losses are achieved.

The special color-supply system is built up close to the spray booth. A system volume increasing circulation line is omitted. The atomizer is supplied from the delivery container through individual stubs or through main stubs with outlets leading to them (see Figure 8.20). The stubs are either filled with the exact amount of paint that is required for the painting process that is, 'push-out procedure', or the paint is pushed back into the paint tank after the painting process that is, 'reflow procedure'. In both cases, very small amounts of paint and solvent are lost.

The paint supply cabinet of these systems (see Figure 8.21) contains a purging container, a diaphragm pump, a gauge head, a color changer with return valves, pig-sending stations, and the delivery container that stands on a traveling device and which is closed by a lid. The stubs connect the pig-sending stations of the paint cabinet to the pig-receiving stations of the takeoff points. In the stubs, the pigs move between the sending and receiving stations and push the paint in the line ahead

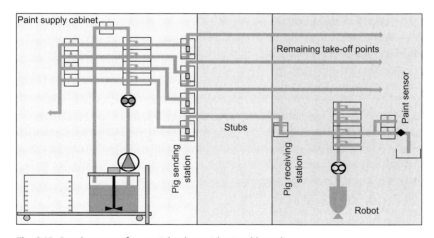

Fig. 8.21 Supply system for special colors with piggable stubs.

of them toward the atomizer. In addition to the pig-receiving station, each takeoff point has a color changer that is placed as close as possible to the pig-receiving station, a dosing pump, and an atomizer.

The functions of the system during the paint filling, the painting process, and the purging are almost fully automated. The only the task of the worker is to change the paint container and to release the system by pushing the necessary buttons. After release, the paint delivery container standing on the traveling device moves into the paint cabinet. Once the lifting door closes, the hoop lid with the built-in agitator and the suction and return line pipe gets lowered. Then the pump circulation line and the stubs up to the tapping points are filled. Here, two different cases are possible.

In the first case, the paint volume required for the painting process is lower than the hose contents. The pig is therefore started during the prepainting process. In the second case, the paint volume that is required is larger than the hose contents. The pig is then started only when the paint volume contained in the stub is just sufficient for the painting scope.

After the painting process, the color changer in the automatic painting machines switches over to the next paint to be applied. Following this, the special color supply is purged automatically. The purging process is divided into several steps in which all paint-carrying components like stubs, pigs, valves, container lids with agitator and suction, and return line pipe are cleaned. During the cleaning of the hoop lid in the cleaning container, the displacing device with the delivery container moves out of the paint cabinet. The hoop can now be removed and replaced by another one with the next special color.

8.4.4.9 Voltage Block Systems

Owing to their high electrical conductivity, the electrostatic application of waterborne paint materials is normally made by outside charging with electrode fingers (see Chapter 6). An electrostatic direct charging system requires a so-called 'voltage block' to avoid an electric short circuit between the charging electrode and the grounded components. There are various options for a voltage block. The most important criteria for their differentiation are the suitability for a color change and continuous and noncontinuous supply of the application technology.

8.4.4.10 Voltage Block Systems with Color-Change Possibility

The galvanic separation device displayed in Figure 8.22 is based on the use of two separable containers from live atomizers by means of a high-voltage-resistant insulating distance (piggable paint hose). The two purgeable containers with a volume of 1500 ml can also be galvanically separated from the paint supply and the color changer by an insulating distance. This ensures on the one hand a continuous painting process even when the supply container is being refilled, on the other hand, a fast color change being carried out with the system.

In addition to the above described voltage block, systems are used that carry the paint along for the painting job in the atomizer. Here, either the entire atomizer is exchanged by a filled one, or a cartridge filled with paint material is inserted

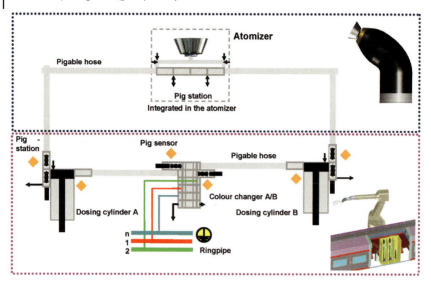

Fig. 8.22 Voltage block based on piggable paint pipes.

in the atomizer. The cartridges are rinsed at a filling station and filled with paint material. A handling device inserts the cartridge in the atomizer. The paint material is dosed by a hydraulically driven piston pump. A continuous painting process is not possible with these systems, however.

8.4.4.11 Installations for the High Viscosity Material Supply

High viscosity materials with PVC bases are used in large quantities in the automobile industry for seam sealing and for the application of underbody protection (see Section 8.2). The materials can be supplied centrally from tanks or, in case of smaller consumption volumes, locally from barrels that are arranged close to the application. In case of a central supply, a circulation line, the low-pressure circuit, is fed with pneumatically driven piston pumps and the material is guided around the seam sealing/underbody protection booths. The low-pressure circuit supplies the high-pressure pumps that are close to the booth, solidify the underbody protection material, the most commonly used material being PVC, to the required pressure of 80–150 bar and bring it into the supply lines (stubs) of the booths. From these lines, the individual takeoff points are supplied. Normally, a material tempering is required to ensure a constant material viscosity and consequently, permanent-application conditions. The width of the spray jet and the material throughput, depend on the material viscosity in the airless application.

In automatic seam sealing, applicators – so-called *flat stream nozzles* – are being used to an increasing extent. These do not atomize the PVC material, but extrude a material seam. These nozzles require dosing systems for an exact material dosing

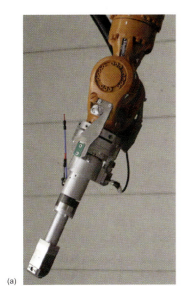

Fig. 8.23 Flat stream application head (a), and piston dosing system (b).

that is proportional to the traveling speed (see Figure 8.23). Here, the high-pressure range can be omitted in the material supply.

8.4.5
Conveyor Equipment

The various installation areas and the included coating processes, are connected to conveyor systems. The execution of these systems is adjusted to the process. Whereas for dipping processes and coatings at the underbody, overhead conveyors are mostly used, floor conveyors are used for all other working steps. The various levels in the paint shop are bridged by lifting stations. Individual installation areas are decoupled from each other by means of buffer and color sorting stacks. By means of identification systems that are connected to the car body and numerous reading points in the painting installation, the position of each car body is known at all times. This is required to supply the control of the automatic painting machines with information about paint number and vehicle model.

An interesting development has taken place in the field of conveyor technology equipment in the dipping processes of pretreatment and electrocoating (see Chapters 2 and 4). In these new processes the car bodies are dipped vertically, with a rotating movement into the bath. The inlet and outlet zone, as it was required in the so far prevalent pendulum conveyor equipment with an immersion angle of 45°, can be omitted, so that the dipping tank becomes shorter by approximately

Fig. 8.24 Dipping of a vehicle into the cathodic electro-coat tank (a): pendulum conveyor equipment; (b): RoDip technology.

20% (see Figure 8.24). Overall installation length is reduced, and this is reflected in a clearly lower investment, especially in terms of building cost.

There are other equally important advantages – lower energy requirement owing to smaller tank sizes, less bath carryover owing to rotating movement, uniform coating and reduction in coating faults, and soiling of the horizontal areas. The operational experience shows that savings in the consumption of water and energy in the magnitude of 50% are possible. In the consumption of chemicals, 30% can be saved. Owing to the better quality, the electrocoat sanding on critical horizontal surfaces can be practically eliminated.

8.5
Paint Drying

Paint application is followed by the drying and hardening of the wet paint film. In this step two different targets have to be recognized are as follows:
- drying as a flash-off step between two painting areas like the waterborne base coat and the clear coat; in this intermediate drying process, only the solvent in the paint film is removed by evaporating to a certain degree [11];
- drying for solvent removal and cross-linking of the paint film, as in the case of the cataphoretic dip painting, of the primer and of the clear coat–top coat application; here, the paint film is hardened by chemical reactions that lead to a cross-linking of the molecules of the binder [12].

The required energy is supplied in the form of heated and recirculated air in so-called *convection ovens*.

At relatively high temperatures and long process times requiring long ovens together with high plant capacities, the operation of the oven consumes lot of energy. For the oven construction, the energy losses and measures for their

8.5 Paint Drying

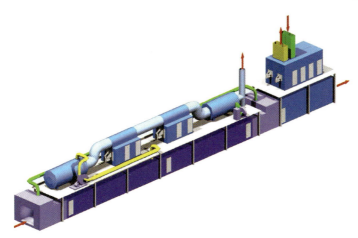

Fig. 8.25 Oven for cross-linking of paints.

reduction like heat exchanger, insulation, and air seal play an important part. Flash-off zones often have a segment with IR (infrared) radiators at the entry area that heat up the paint film by absorption of the IR rays and lead to a shortening of the oven length.

The air seal at the entry and exit of the oven prevents hot air from escaping. Cold air trying to get inside the oven is heated up to the object temperature. This effect is achieved by a specifically guiding the air into the air seal, as shown in the Figure 8.26. The type with a horizontal entry air seal allows the layout of the oven on the level of the painting line. In case of the A-type air seal, the oven is located above the process level. The car bodies are conveyed into the oven by an elevator. This design allows for arrangement of the primer and top coat lines next to each other in the layout. The car body painted in the primer line is conveyed backward through the oven to the entry of the top coat line.

Typical length and process times are summarized in Table 8.1. These data are only indicative, because they are dependent on dwell time, line speed, and the paint material itself. IR-zones are not listed.

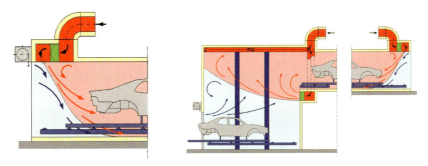

Fig. 8.26 Design of oven air seals: horizontal entry (a), A-type (b).

Table 8.1 Typical Data of Drying Stations and Ovens in Waterborne Base Coat–Clear Coat Application Step (Conveyor Speed: 3.5 m min^{-1}, 30 Units h^{-1})

	Station	Length (m)	Dwell time (min)	Temperature (°C)
Base coat	Ambient flash off	7	2	23
	Base coat dryer	17.5	5	75
	Air seal	3.5	1	23
	Cooling	7	2	23
	Air seal	1.75	0.5	23
Clear coat	Ambient flash off	28	8	23
	Oven	105	30	150
	Cooling	21	6	23
	Total	190.75	54.5	

8.6 Quality Aspects

8.6.1 Control Technology

Automatic painting installations with units like process equipment, environmental equipment, conveyors, robots, and application equipment are complex systems whose controllability is subject to the performance of the total process and the coating result. Therefore, the functionality and the operability of the control-technology equipment are essential, and high demands are made on the appropriate control technology. The most important factors are listed below:

- open, modular, and flexible architecture
- compatibility with international standards
- process-orientated
- high uptime
- convenient, clearly arranged viewing system on PC basis, uniform operating philosophy
- integrated error diagnostics and emergency strategies [13].

The modular design of the control technology is realized by means of a division into individual station controls that communicate with each other, with the periphery, and with the host computer by means of a system bus.

The station control also has a modular design and consists of the following components (see Figure 8.27):

- motion control for the control of the robot drives as well as of the valves, controllers, and sensors of the application technology;

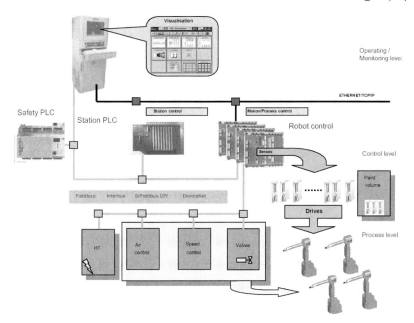

Fig. 8.27 Control configuration of a painting station.

- station control for the coordination of the individual in the station-integrated motion controls and for communication with the neighboring stations and the host computer;
- safety control for the coordination of all safety-relevant data such as for instance, EMERGENCY OFF signals, personnel protection, fire protection.

The hardware and software used for control purposes should be as uniform as possible in all the stations of a painting installation. This facilitates the operation, maintenance, and troubleshooting, and also increases the uptime of the installation.

The viewing and operating system has an especially important function within the installation control. The tasks of this system can be summarized by the functions operating, viewing, programming, and parameterizing. A clear graphical desktop with operator prompting simplifies the operation of the installation and the complex parameterizing. In case of fault and error messages, help functions and fault diagnoses support fast localization and elimination. With the potential of simulating a prepared painting program, possible program faults can be located at an early stage.

Automatic painting machines, especially painting robots, are programmed increasingly offline, that is, with simulation systems that emulate the characteristics of the robot and its control. Program modules that have been created offline are then loaded into the robot control and are optimized in the teach-in mode. This mode allows robot programming by means of a manual approach, thus saving the

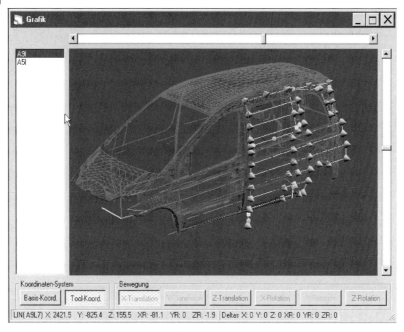

Fig. 8.28 Three-dimensional viewing of a robot path program.

movement using the teaching pendant. The requirement for offline programming is a geometric and dynamic pattern of the robot cell and the vehicle to be painted. The functionality of these simulation systems can, to some extent, be transmitted to viewing systems, so that a 3D-based viewing and editing of the programs is possible (see Figure 8.28).

8.6.1.1 Process Monitoring and Regulation

The complexity of automatic painting installations and the large number of different process parameters makes it complicated for the plant operator to maintain a high quality of production, and for the service personnel to eliminate defects without delay. Systems that support the operator in the diagnosis, optimization, and monitoring of the processes are already in use. New systems are being developed to further improve on these, by taking into consideration quality-oriented control of the painting processes and process parameters. The linking up of such systems with all levels of the process and the installation, and with the control technology necessary for doing so, makes them very effective tools [14].

- *Process Data Recording and Viewing:* All parameters and data of the installation and application technology, the paint supply, and the quality control important for the process are recorded in the form of actual values that are related to the car body. With the aid of free access to data sets of each car body and machine, process diagnoses and fault analyses can be carried out.

- *Determination of Remaining Service Life of Components:* From the load of a component and its expected service life, it is possible to arrive at its probable remaining service life. With the use of this system, plans for preventive maintenance can be drawn up. At installations that operate in 24-hour shifts for the whole week, the availability can be clearly increased by the in-time, targeted replacement of worn-out components during production-free time.
- *Teleservice:* For enterprises producing in various locations and not wanting to build-up an expert team for each location, teleservice offers interesting possibilities. With today's advances in telecommunication, it is easy to transmit data, information, language, and pictures to the main office or to service departments of the installation manufacturers, where specialists quickly localize faults and offer solutions.
- *Automatic Quality Analysis:* The reliable, objective, and reproducible measurement of the quality-relevant data like film thickness, color shade and leveling are absolute requirements for a quality-oriented process-control system. The systems required for this are available today and can be used flexibly by means of measuring robots (see Section 8.7).
- *Creating and Viewing of Management Data:* The management level of a paint shop requires constantly updated data and facts about the state of the painting installation, to be able to make decisions based on the facts available. These may be short-term decisions such as introduction of extra shifts, or changes in operations owing to a high repair rate, or long-term measures like change of a subcontractor owing to quality problems. In this connection, cost planning on the basis of consumption quantities plays an important role. The system accesses the data from the process data recording and consolidates them to the required management data.

8.6.2
Automated Quality Assurance

Fully automatic painting lines can create a constant quality of paint films. The requirements for this are given below:
- matching and optimization of paint materials, their application, and the workpiece to each another
- optimum installation layout
- optimum application and process conditions, that is, process parameters.

These requirements and conditions are necessary, but not sufficient for the final paint film quality, and so quality inspection and monitoring as quality assurance measures cannot be omitted, even if all process parameters are regulated within narrow limits.

Quality inspection and monitoring are carried out visually these days, by means of measurements by trained personnel after the painting and drying process (see Figure 5.5). Aside from the different and erratic performance of the specialist personnel, this type of quality monitoring is in the sense of a quality control circuit due to the very long reaction time (time between application and quality inspection check) only suited to a limited extent. These facts show the necessity of automatic quality measuring systems that record the paint film quality as directly as possible after the application. This requirement cannot be fulfilled for quality features like paint-film leveling or color shade, because such features can be found only during the final film creation process. However, the film thickness, which is the most important value for the paint-film quality, can be measured directly and without contact after the application, in the wet film stage. For this, measuring systems are available as shown in Figure 8.29. The displayed sensor works according to the photothermic principle, where a very short laser pulse is directed on the paint surface to be measured, and this heats up briefly by some degrees Kelvin. The progression of the reflected heat radiation is recorded by an infrared optical system and a detector. From the characteristics of the measuring curve, the measuring system determines the corresponding individual film thickness.

The measured data are immediately processed and shown on the workpiece, according to their place of measurement. They can be used both for a 100% quality inspection of individual workpieces (see Figure 8.30), and for installation and process optimization, as well as for statistical evaluations and trend analyses [15].

With the introduction of primerless processes (see Section 8.1 and Chapter 6), the online film thickness measurement assumes increasing importance. The base coat film has, in these shortened processes, the task of ensuring the UV protection of the cataphoretic electrocoat film lying beneath it, and to exclude delamination

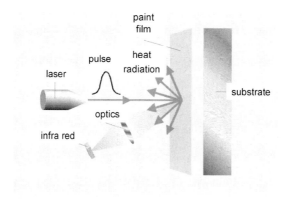

Fig. 8.29 Principle of a contactless measurement of film thickness by photothermic monitoring (source: Optisense).

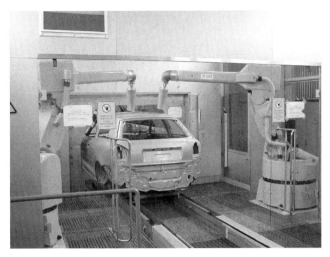

Fig. 8.30 Robot-guided contactless online measurement of film thickness.

which can appear at a later stage. For this, a minimum film thickness is required for each color shade. This must be measured immediately after the paint application at each vehicle, with the objective of a 100% quality monitoring.

The contactless color-shade measurement is carried out with multiangle spectral photometers that have been conceived for the operation, together with industrial robots (see Figure 8.31). The sensors provide the spectral-reflection values (CIE-L*a*b) and other colormetric values that are necessary for the characterization of the color shade (see Chapter 6). The surface to be measured is lighted by a xenon

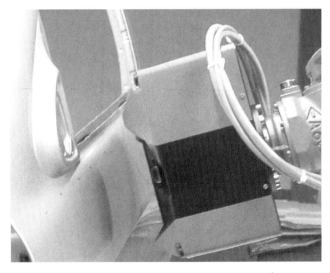

Fig. 8.31 Robot-guided, contactless online measurement of color and surface leveling (source: Optronic).

flashlight at 45° and the measuring values are recorded at angles of 15, 25, 45, and 75°. The surface temperature is measured by an IR sensor, and compensated with regard to the color shade owing to the thermochromical dependence of various color shades, especially red. The positioning of the measuring head is of special importance. Since color-shade measurement depends on the measuring angle the positioning of the sensor head must be carried out very precisely. The robot is responsible for the prepositioning. There are different procedures in the fine positioning; this can be carried out either by the measuring head itself, or by the robots. In the first case, the measuring head must have the corresponding mechatronics. In both cases, distance sensors which work on ultrasound or infrared basis are required. The measuring heads can be equipped with leveling meters for the contactless measuring of the paint-film levelling.

In addition to the measurement of the film thickness and of the optical quality features, an automated fault inspection with image processing systems can also be carried out today. Normally, all paint-film faults that lead to a change of the topography on the surface are recognized. Robot-guided camera/lighting units are used for this purpose. The recorded pictures are evaluated by computers and rated regarding customer relevance. In the next paint station, the relevant faults are marked for reworking.

8.6.2.1 Process Optimization in Automatic Painting Installations

With the described sensors, the most important measuring procedures for an automated characterization of paint film surface are available. An optimization of the painting processes or an operation of the processes in their optimal area can therefore, be carried out. The following example for a process optimization describes a case from the field, before and after the introduction of the online film thickness measurement in an automotive paint shop.

The film thickness distribution on a surface is subject to a more or less distinct fluctuation that is caused by many factors. The main factors are the application technology, the setting of the atomizer parameters, the number of overlaps of the individual spraying paths, and spray-booth conditions.

In case of a very strongly fluctuating film thickness distribution, it must be ensured that the film thickness minimum has enough hiding power to keep the color independent from that of the substrate. The medium film thickness value must be set accordingly high (see Figure 8.32b), at approximately 9 µm. Figure 8.32a shows the film thickness leveling after optimization has been carried out, where it was possible to determine the film thickness leveling directly after the application. Here the middle film thickness could be reduced to approximately 7 µm with a reduction in the degree of fluctuation.

Apart from a quality improvement in the form of reduced cloudiness of the coating, the consumption of paint material and the costs have been clearly reduced, so that the investment for the film thickness measurement pays off after a short

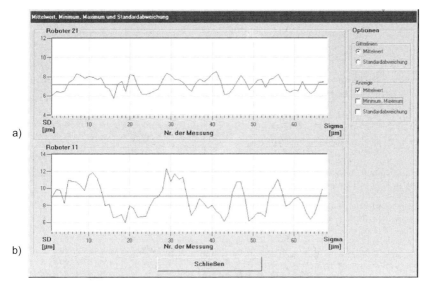

Fig. 8.32 Optimization of film thickness by on line measurement technique.

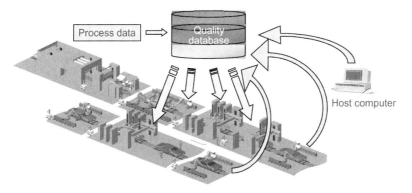

Fig. 8.33 Process-control circuit for coating quality.

time. The use of the measuring station is not eliminated because of this step, as the status reached must be maintained.

Further development of the automated quality recording is concentrating on expert systems, which should help in closing quality control circuits as much as possible (see Figure 8.33). The installation operator, then has an important tool that supports him in troubleshooting and elimination of defects. However, for the time being fully automated quality controlled painting installations remain a distant dream.

The units and equipment that have been described can reach paint shop availabilities of up to 98% and to about 85% of first run ok painting units.

8.7
Economic Aspects

Global competition between the car manufacturers has resulted in the economics of an automotive paint shop becoming more and more a competitive factor [16]. Efficiency is decisively affected by the personnel costs, the operating costs, and by the quality that is required during production. Personnel costs are influenced not only by the degree of automation, but also by expenditure on maintenance and service. Operating costs consist mainly of the energy cost and the material cost for paints, solvents, and auxiliary chemicals. The energy cost can be controlled with various measures, and the material costs mainly by an efficient application technology. Constant quality with an OK rate or no-touch rate of the coated bodies as high as possible, is mainly ensured in fully automated processes. The no-touch rate, where no repair or polishing is made on the body, reaches up to 85% in excellent paint shops.

Potential for cost reduction will be shown with the help of selected, distinct examples. In the first example, potentials of an overall layout are considered, the second example is involved with savings by the introduction of full automation, and the third example finally shows the advantages of the electrostatic application.

8.7.1
Overall Layout

The overall layout has the task to harmonize different requirements, for instance, material flow, processes, product quality, and the requirements of the adjacent production areas. If this succeeds, this may bring about considerable cost savings throughout the life span of the installations. The following are the main factors:

- energy savings by an optimized, process-oriented ventilation of the installation
- reduction of the personnel cost by an increase in productivity
- reduction of the maintenance expenditure by means of modular installation concepts, compact design, and short reaction times
- increase in efficiency with short cycle times and quality improvement.

Calculated for the service life of the installation technology, and considering the capital costs involved, the overheads saved involve an amount that is absolutely in the area of the investment. Figure 8.34 shows an exemplary estimate.

8.7.2
Full Automation in Vehicle Painting

Full automation does not mean carrying out only the external application with robots, but also the painting of interiors, i.e. engine compartment, trunk door shuts

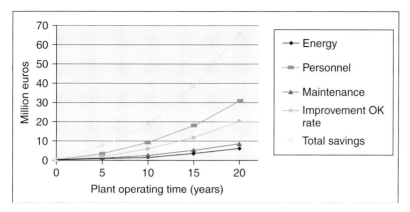

Fig. 8.34 Saving potential with overheads depending on the service life of a paint shop for 30 units h^{-1}.

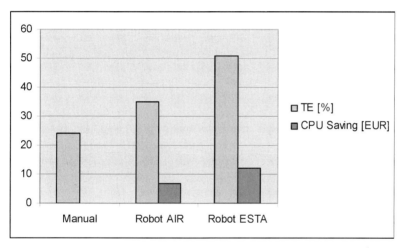

Fig. 8.35 Transfer efficiencies and cost per unit (CPU) savings by different methods of inside painting.

(see Chapter 6). As explained in the following paragraphs, rotating atomization, together with electrostatic paint charging is used increasingly for the interior painting.

Manual painting of the interiors has a transfer efficiency of approximately 25%. This can be increased to more than 50% by using the process of rotating atomization. This means that transfer efficiency is doubled and/or there is a reduction in the paint-material consumption. This is connected with a correspondingly high cost reduction (see Figure 8.35). The values are calculated for a middle-class vehicle.

A reduction of the color change losses compared to the manual application leads to additional cost savings. Furthermore, it is well known and needs no further elaboration that full automation offers advantages regarding quality improvement and first run rate.

8.7.3
Exterior Application of Metallic Base Coats with 100% ESTA High-Speed Rotation

The second metallic base coat film on the car body exterior areas continues, in most cases, to be pneumatically applied in the standard painting process (see Chapter 6). The reason for this is that the achievable metallic effect formed by metallic flakes contained in the paint material is better and easier to repair. How is this effect created? In the pneumatic atomization, a droplet spectrum is generated with a medium droplet size of approximately 25 µm, including a high level of smaller droplets. These smaller droplets do not contain any metallic flakes and are lost as overspray. The paint droplets hitting the car body surface contain, therefore, more metallic flakes. A paint film with a high portion of metallic flakes is created that are distributed evenly and are aligned parallel to the surface. The standard rotating atomization, as it is used everywhere today for the first base coat application, generates a rougher droplet spectrum. The overspray losses are very low. This leads inevitably to a lower relative portion of metallic flakes in the applied paint film as compared to the pneumatic application. In addition to this, in the larger paint droplets there are more metallic flakes that form conglomerates and as a result, prevent an even distribution and parallel alignment in the film. The metallic effect is less distinct, which means a lower flop in the form of a lower difference in the brightness with different viewing angles (see Chapter 6). The paint film is generally darker.

The answer to this problem lies in the adjustment of the rotating atomization to the pneumatic one with regard to the droplet spectrum. The solution for this is to provide for higher atomization energy. This is achieved by a higher speed of the bell disk together with its larger diameter. More atomization energy and an overall smaller droplet spectrum lead to somewhat lower transfer efficiency that is, however, still twice the pneumatic application, at approximately 70%. The result is fairly good. The portion of the metallic flakes in the paint film is higher, and the flakes are finely distributed and aligned parallel to the surface, exactly as in the pneumatic application. The cloudiness, also an important quality criterion (and a typical problem in the pneumatic application), is clearly lower. Adjustments in the paint material may only be required for certain 'difficult' color shades; however, these can often be avoided by technical measures. One of the measures is, for instance, the possibility of the high tension reduction together with larger shaping air volumes, however, with reduced transfer efficiency. An examination in the technical center should be considered here, as the case arises.

Cost savings are mainly achieved by the paint material reduction. This is approximately 0.45 l per vehicle for an upper middle-class car with an outside area of 9.5 m^2 to be painted. Additionally, there is reduced energy consumption owing to a lower booth air downdraft speed, less paint sludge, and less VOC emissions [17].

The 'Bell–Bell' application of the first and second metallic base coat has been propagated considerably. An example of this, using outside charging devices, is shown in Figure 8.36.

Fig. 8.36 Second application step of metallic base coats with high-speed-rotating atomizer and electrostatic paint charging.

8.7.4
Robot Interior Painting with High-Speed Rotation

Interior painting means application at door entrance areas, as well as on areas in the engine compartment and trunk which are hard to reach. To be able to use a rotating atomizer, it must be designed to be very compact. In the primer and base coat area, where, nowadays, usually waterborne paints are used, a voltage-block system is required, if the advantages of the electrostatics are to be utilized (see Section 8.4.4). The high-speed-rotating atomizer must allow a settling of the spray pattern width over a very wide area (see Figure 8.37), and also a high rate of paint outflow. High painting speeds are essential for interior painting.

A voltage-block system that is particularly suited for the interior painting because it is locally separated from the atomizer, and therefore, allows its compact size,

(a) (b)

Fig. 8.37 Spray pattern settings for widths between 50 (b) and 600 mm (a).

8 Paint Shop Design and Quality Concepts

has been shown in Figure 8.22. It is based on the pigging technique that provides for the required insulating distances between atomizer, dosing equipment, and color-change equipment. The atomizer is supplied with paint-material by means of piston dosing in a gentle, low-shearing way. The dosing exactness is independent of the paint viscosity. A major advantage of this concept is the possibility of painting with an unlimited paint volume and without interruption. Docking and refilling of the atomizer, which is connected with a significant time loss, is not required.

In direct comparison with pneumatic atomization, electrostatically supported interior application provides not only a much higher transfer efficiency of 40–65% compared to 30–35%, but also additional advantages like shorter painting paths in general (see Figure 8.38), a higher quality with the electrostatic wraparound, as well as a reduction of the air volumes in the spray booth owing to a lower air downdraft speed, which means energy savings. However, the question still remains – is the higher technical expenditure worth it? One just needs to consider the paint material saved by the higher transfer efficiency illustrated by the example of a vehicle of the upper-middle class in the metallic base coat application. For the interior application at door entrance areas, in the engine compartment and trunk, the paint-material reduction works out to approximately 0.5 l for a car body. Depending on the annual paint consumption, and the material price, the retrofit of the pneumatic application to ESTA (electrostatic application) high-speed rotation has an amortization time of about one to two years. The use of the ESTA high-speed rotation is much easier and more economic in the clear coat interior application, because no voltage block is required here.

There is a distinct cost saving with 100% electrostatic spraying. Considering an already fully automated painting process, but with ESTA /pneumatic application steps, the requirement for a typical car body of an upper middle-class model works out to 6.1 l of base coat material. With a 100% ESTA application, the consumption is only 4.9 l for the same painting. This is a reduction of approximately 20% [17].

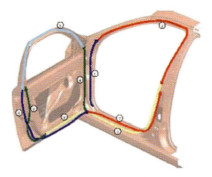 

Fig. 8.38 Spray pattern of interior applications with electrostatic high-speed rotation.

References

1 Mutschelknaus, R. (**2006**) "Staff optimisation and economy enhancement in the paint shop–an innovative conception", *Proceedings of 5th International Strategy Conference on Car Body Painting*, Berlin.
2 Challener, C. (**2006**) *Journal of Coatings Technology*, **3**(7), 28.
3 Svejda, P. (**2003**) In *Prozesse und Applikationsverfahren* (eds U., Zorll, D., Ondratschek), Vincentz, Hannover, p. 41.
4 Domnick, J. (**2003**) *Metalloberflache*, **57**(1/2), 27.
5 Svejda, P. (**2007**) *Journal für Oberflächentechnik*, **47**(3), 26.
6 Walzel, P. (**1990**) *Chemie Ingenieur Technik*, **62**(12), 983.
7 Schmidt, P., Walzel, P. (**1984**) *Physik in Unserer Zeit*, **15**(4), 113.
8 Richter, T., Wilhelm, S. (**1991**) *Kalkulierter Sprüh, Methoden des Zerstäubens von Flüssigkeiten in der Verfahrenstechnik*, Maschinenmarkt, Würzburg 97, 11 p. 26.
9 Lefebvre, A. (**1989**) *Atomization and Sprays*, Hemisphere, New York, p. 27.
10 J. Gamero, et al., (**2005**) *Journal für Oberflächentechnik*, **45**(9), 34.
11 P. Henshaw et al., (**2006**) *JCT Research*, **3**(4), 285.
12 Ortlieb, K. (**2000**) Trocknen–Verfahrensübersicht, *Jahrbuch Besser Lackieren 2001*, Vincentz, Hannover, p. 222.
13 Svejda, P. (**1999**) Steuerungs-und Simulationstechnik in automatischen Lackieranlagen, *Jahrbuch für Lackierbetriebe 2000*, Vincentz, Hannover, p. 276.
14 Leisin, O.P. (**1999**) Diagnose, Optimierung, Überwachung und Regelung von Lackieranlagen, *Jahrbuch für Lackierbetriebe 2000*, Vincentz, Hannover, p. 377.
15 Svejda, P. (**1999**) Moderne Automatisierungskonzepte für den zerstäubenden Lackauftrag, *Jahrbuch für Lackierbetriebe 2000*, Vincentz, Hannover, p. 297.
16 Wright, T. (**2006**) *Coatings World*, **11**(3), 26.
17 Svejda, P. (**2006**) 100% electrostatic application, *International Surface Technology*, Vieweg, Wiesbaden, p. 20.
18 Kittel, H. (**2004**) *Lehrbuch der Lacke Und Beschichtungen Band 9*, Hirzel, Stuttgart, p. 142.

9
Coatings for Plastic Parts

9.1
Exterior Plastics
Guido Wilke

9.1.1
Introduction

During the last 30 years, the use of plastic in automotive body construction, including interior and exterior parts, has increased to about 10–12% of the total vehicle weight. The integrated body-colored plastic parts for exterior applications have contributed significantly to the change of car design since the time when exterior mirrors and other supplied parts gave the vehicle a strange look. Adjusting these parts to suit the body was more an evolutionary rather than a revolutionary development, with some outstanding milestones like the almost total replacement today of the chromium bumper by the plastic bumper. Today, the amount of exterior plastic parts is not only much higher than in the 1970s, but also almost every exterior plastic part is coated, most of them in body color. The consumption of body-color-painted plastic parts in Europe was estimated as €270 million for 2001 [1]. In North America, the consumption of plastic coatings for exterior car body parts in 2002 was worth $364 million; the annual growth rate from 2002 to 2007 is estimated to be 4% [2]. However, when analyzed on a 'per vehicle' basis, the growth of plastic coatings in North America shows a slower rate, compared to the 1980s and early 1990s [3]. In this chapter, the main driving forces that led to the use of painted plastic parts, will be presented, and some of the challenges, which have been discussed in greater detail in Chapter 7, will be touched upon, briefly.

9.1.1.1 Ecological Aspects
Plastic substrates, with their outstandingly low density, contribute to the demand in weight reduction of all cars, although the density of some exterior plastic materials is increased by reinforcing components like glass fibers. An important factor to be considered with respect to weight reduction is the final weight of the part, as the wall thickness often has to be increased owing to the limited tensile strength of most polymers compared to steel. Although there are examples where weight

Automotive Paints and Coatings. Edited by H.-J. Streitberger and K.-F. Dössel
Copyright © 2008 WILEY-VCH Verlag GmbH & Co. KGaA, Weinheim
ISBN: 978-3-527-30971-9

reduction is not the decisive argument for selecting plastic as construction material, various legislations make it likely that plastic will continue to play an important part in the ongoing competition between all light-weight materials. In contrast to the original equipment manufacturer (OEM) painting process of the car body, the manufacturing of plastic parts offers additional options of emission-reduction techniques in surface decoration. These include the in-mold-coating (IMC) and foil techniques.

9.1.1.2 Technical and Design Aspects

Form integration was, and continues to be, one of the major driving forces for the use of plastic exterior parts. Bumpers, mirror shells, license plates, and door-handle housings are completely integrated into the car body in such a way that the shapes of assembly parts merge into the car body without or with only small gaps. Secondly, injection-molding technology, in principle, provides an almost unlimited freedom of design. It opened the door to a much larger variety of assembled parts made with thermoplastic polymers. In the early days of exterior plastic coating, very few form-integrated parts had coatings, which provided a general color contrast with a silk-finish. As the trend toward use of plastic in exterior assembly parts increased, and the surface area of plastic got larger, this portion had to be coated for color and gloss. Today, painting the assembled plastic parts in car body color is standard. Moreover, when it comes to large and central car body parts like hoods and tail gates, in the construction of which the material used, in some cases, has switched from metal to plastic, these are naturally coated in body color. Using sheet molding compound (SMC) for tail lids made it possible to integrate antennas into the body shape. New ways of designing roofs have been found by using a combination of glass and plastic, and side by side, using decorative-foil techniques.

Apart from design-generated arguments of replacing steel by plastic, there are some technical facts about polymer materials that reflect the technical advantages of providing plastic with a surface coating. Most polymer materials are not stable to ultraviolet (UV) radiation and the durability of color and gloss gains more importance with the increasing lifetime of vehicles. UV-absorber and light scavengers in the respective coatings help plastics to withstand this environmental impact economically, owing to the fact that the whole bulk of plastic need not be loaded with these additives. The majority of plastic parts consist of thermoplastic polymers, which, in general, have limited resistance to mechanical abrasion, chemicals, and solvents. This problem can be overcome by the use of cross-linked and resistant coatings. Furthermore, these improvements of coated plastics can be seen at low temperatures. Thermoplastic parts produced by injection molding often show flow lines that have to be covered by a coating. Fiber-reinforced polymers, which are popular for parts with higher demands in tensile strength and stiffness, often show a poor surface quality owing to roughness, and succumb to local failures like pinholes and shrinkholes. To make these substrates suitable for subsequent decorative painting by top coats, smoothing the surface is carried out with special coatings. In the early days of plastic coatings, initial and long time adhesion, cracks, and solvent-induced declining impact performances were real road blocks. Thanks to a much better understanding about viscoelastic properties of coating materials and painted plastic parts, although not eliminated until today, these problems did not prevent the further expansion of plastic coatings [4]. A lot of this progress has become possible with the introduction of 2K polyurethane (PUR) chemistry.

9.1.1.3 Economical Aspects

The mass production of molded parts at low cost give plastic the potential of being a less expensive substrate than metal. The workability of plastics allows cheaper production of single, integrated pieces, whereas several pieces may be needed if metal is used [5]. Whether plastic or steel is more expensive also depends strongly on the number of units produced. In any case, painting the parts raises the manufacturing costs significantly. Plastics have opened the door to more decentralized manufacturing, that is, outsourcing of add ons for the car body by the module concept. Today front ends, with form-integrated bumpers and headlamps, are state of the art. The plastic industry estimates an increase to 18% of plastic materials in car vehicles until 2008 [6, 7]. It is expected that a substantial contribution to this growth will be in the form of more car body parts with large and flat surface areas like fenders, hoods, and tail lids. Already today, special models like the VW-Beetle have a share of exterior plastic of more than 30%.

9.1.2
Process Definitions

9.1.2.1 Offline, Inline, and Online Painting

There are basically three different ways in which built-on plastic parts can be fed into the body-painting or body-assembly process, as shown in Figure 9.1 [8].

The term *online painting* is currently used if the plastic parts are mounted prior to pretreatment and pass through the entire painting process together with the steel body.

In the case of inline painting, plastic parts are attached to the car body in the coating line, either before or after application of the primer surfacer. Built-on parts being completely painted in a separate coating line, mostly at the supplier's end, are mounted on the body at the end of the painting line or in the final assembly line.

Actually, all three procedures can be found in the automotive industry.

9.1.2.2 Process-Related Issues, Advantages, and Disadvantages

9.1.2.2.1 Inline and Online Painting
The advantages of integrating plastic built-on parts for vehicles into the coating line by way of online or inline painting can be summed up as follows:
- improved color matching with the body;
- reduction of painting processes and investment costs;
- simple logistics.

When plastics are incorporated into the car body coating line, only a limited and expensive choice of heat-resistant plastics are available. Since automotive spray

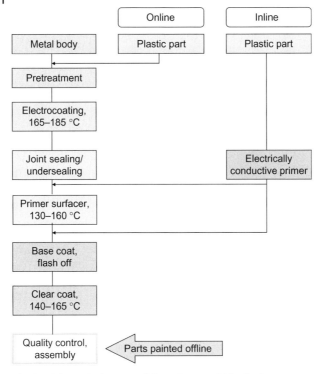

Fig. 9.1 Schematic diagram of the points at which plastic parts are incorporated in the car-manufacturing process.

paints for steel bodywork are generally cured at temperatures between 130 and 190°C, only plastics with the appropriate level of heat resistance are suitable for inline (130–160°C) and online (165–190°C) painting.

Secondly, the surface of the plastic part needs to be electrically conductive to be suitable for ElectroSTatic Application (ESTA) and the related high material transfer efficiencies. Offline application of an electrically conductive primer is state of the art. Transport and complex handling of these coated parts represent quality risks owing to surface defects. More severe and disadvantageous for the inline-process is the danger of soiling, that is, generation of dirt when parts are attached between two OEM painting steps. The danger of soiling is considerably reduced because online parts do not need to be channeled between two painting stations. Nevertheless, online painting calls for substrates that can withstand the baking temperatures of cathodic electrocoating with temperatures in the range 165–190°C. Like inline parts, to prevent film defects, they need an electrically conductive surface for ESTA-paint application [9].

Substrate materials that are equipped with electrical conductivity and therefore, do not require any precoating with electrically conductive primer, are now in the phase of development. Table 9.1 contains some examples of car models fitted with inline and online painted plastic exterior parts.

Table 9.1 Typical inline and online painted exterior plastic parts at various car manufacturers (2004)

Fender–inline	Tail lid–online
VW New Beetle	Mercedes CLK
Mercedes Vito	Mercedes S-coupé
BMW 6	BMW 6
Renault Megane	

9.1.2.2.2 **Offline Painting** In Western Europe more than 90% of all plastic parts are painted offline either by a supplier or in a separate paint shop at the automotive manufacturer. Volks Wagen, Audi, and BMW are examples of car manufacturers having offline and inhouse painting facilities. In principle, offline painting can be done in two different ways. The offline painting line is located in the same plant, close to the car body painting line, operating with the same application equipment, and being supplied with the same base-coat batches used for the OEM process. Alternatively, there could be a separate plant for plastic parts, with individual application equipment, but using the same base-coat batches as the car body painting line [9].

Most plastic parts are painted offline by an assembly-part supplier company. Offline painting by a supplier enables the use of customized painting superstructures, which can ideally match the specific characters of the substrates, geometrical-part design, application conditions, and paint shop layout. In this sense, decentralized painting by a supplier or job coater offers maximum opportunities for tailoring each paint-application process and coating-layer build up, individually. The paint-supplier–part-manufacturer relationship can lead to a mix of paint layers by different paint suppliers. This freedom of choice and flexibility may be a blessing to the plastic-part supplier, but is at times a problem for the OEM. In any case, the coating layers have to be adapted to each other to prevent failures like interfacial delamination. High demands in color and appearance matching the car body often requires a tremendous effort from the part and paint suppliers. Deviations in color shades and gloss may result in high costs for painting equipment, high-level investment, and operating costs, a high outlay on service, and logistic problems. One risk faced by the paint supplier here is the high complexity of customer-specific base-coat supply. Countermeasures can be the consolidation of pastes and 'plastic-part feasible' OEM base-coat formulas that only need to be tinted individually. It can be seen that offline coating is the most used approach to provide plastic assembly parts with car body color [1]. Figure 9.2 shows the painting process of the New Beetle bumpers at Volks Wagen's Puebla plant. The New Beetle is an example of a car with a high proportion of plastic as exterior-construction material.

Efforts to reduce emissions by the offline paint-application process have reached very different levels, owing to the individual character of the supplier market. Often, offline paint shops still perform a lot of pneumatic hand spraying for small parts. In future, more and more electrostatic and robot-assisted application techniques will be

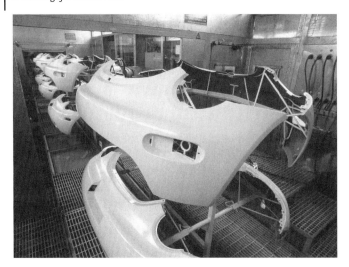

Fig. 9.2 Painting of New Beetle bumpers (source: BASF Coatings).

used for best compatibility with cost and environmental demands. Electrostatically assisted spraying is already established in bumper painting and also in the coating of small pieces, like door grip handles or mirrors.

Finally, offline manufacturing of assembly parts of plastics today, in many cases, is more than just processing and painting of the plastic part. Part suppliers have taken over a significant portion of the responsibility for complete modules. Examples for these are exterior mirrors consisting of plastic housing, glass insert, and motor and bumper modules, including front/backlights, covering, and carrier.

9.1.3
Exterior Plastic Substrates and Parts

9.1.3.1 Overview

Plastic substrates and their applications in coated exterior parts cover a large spectrum, some examples of which are shown in Table 9.2.

The highest consumption of exterior plastics counts is for bumpers. It is expected that the amount of plastics consumed for bumpers will increase to 550,000 t on a worldwide basis in 2010 [1]. Thermoplastic polyolefins (TPOs), like polypropylene (PP)+ethylene–propylene–diene–monomer (EPDM), in particular, are valued for low material costs and an excellent balance between design flexibility, mechanical properties, and recyclability [5]. Therefore, the use of TPO is growing and is expected to continue at growth rate higher than that of other plastic materials. Thermostable plastic substrates like polyphenylenether (PPE)+polyamide (PA), SMC, and polyurethane–reinforced-reaction-inmold (PUR–RRIM) are more expensive. SMC is consumed in high volumes in North America. The main applications are large body parts for passenger cars and fascias of transportation vehicles. SMC and also PUR-RRIM often result in additional costs by way of painting repair work. In

Table 9.2 Plastic substrates and their applications in coated exterior parts

Plastic substrate[a]	Coated exterior parts–examples
ABS	Mirror housings
PC+ABS	Mirror housing, registration plate supports, and pockets
PC+ASA	Coverings
PBT	Grilles
PC+PBT	Grilles, mirror housings, bumpers, front modules
PBT-GF30	Gas lids
PA6-GF30	Door handles, grip handles, hub caps, gas lids, trims
TPO (PP-EPDM)	Bumpers, ram guard strips, spoilers
PPE+PA	Fenders, tail gates
ABS+PA	
PUR-RRIM	Side member coverings, bumpers
UP-GF (SMC)	Tail lids, hoods, roofs, spoilers, fascias
EP-GF	Hoods
EP-C	Roofs
PMMA	Headlamp lenses

a Abbreviations according to DIN EN ISO 1034-1.

general, it would be more appropriate to calculate overall costs of painted parts, which would be affected to a lesser extent by the cost of the plastic raw material, and to a greater extent by painting and repair costs. From the point of view of repair work, fiber-reinforced duromeric plastics belong to the more expensive solutions in plastic-coatings technology [7].

9.1.3.2 Basic Physical Characteristics

9.1.3.2.1 Mechanical, Thermal, and Electrical Properties

In contrast to metal substrates, the variety of physical properties of plastic is much higher. Extreme firmness is found, as well as rubberlike ductility. Polymers, in general, have both elastic and a viscous character, and so they are defined as *viscoelastic* materials. The mechanical properties of polymers, in general, are investigated in various deformation experiments. Different kinds of deformations are possible: pulling, bending, compressing, and twisting. The characteristics of these deformation experiments are the fact that the results are strongly dependent on the temperature and the speed of deformation. The tensile experiment is one of most established deformation-tests, delivering a stress–strain curve with the following characteristic values:

- stiffness, expressed by the tensile modulus (MPa)
- tensile strength, expressed by either stress at yield or stress at break (MPa)
- flexibility, expressed by strain at break (% of dimension change).

Additionally, the tenacity can be obtained by determining the surface area below the stress–strain curve. Although these characteristics can only reflect a small part

Table 9.3 Mechanical data of thermoplastic substrates, obtained by the tensile experiment (ISO 527,-1-2) [10]

	Tensile modulus (MPa)	Stress at yield (MPa)	Strain at yield (%)	Stress at break (MPa)	Strain at break (nominal) (%)
PP+EPDM	500–1200	10–25	10–35	–	>50
ABS	2200–3000	45–65	2.5–3.0	–	15–20
PC+ABS	2000–2600	40–60	3.0–3.5	–	>50
PBT	2500–2800	50–60	3.5–7.0	–	20–>50
PC+PBT	2300	50–60	4.0–5.0	–	25–>50
PBT–GF30	9500–11 000	–	–	130–150	2.0–3.0
PC+ASA	2300–2600	53–63	4.6–5.0	–	>50
PA6[a]	750–1500	30–60	20–30	–	>50
PA6–GF30[a]	5600–8200	–	–	100–135	4.5–6.0

a Conditioned at 23 °C, 50% air humidity.

of the viscoelastic properties of polymers, they are often used to compare substrate materials with each other (see Table 9.3). Another reason for the importance of knowing the mechanical properties of plastics is their relevance for selecting the appropriate coating system.

With the exception of unsaturated polyester (UP), PUR, and epoxy (EP), all the plastic substrates in Table 9.2 are thermoplastics. Most of them do not withstand the temperatures used in curing OEM ovens, and are therefore painted offline. The exceptions are PPE+PA and glassfiber-reinforced polyamides (PA-GF), which can also be painted inline/online. Modified variations of most thermoplastics with fillers like talcum are also used to increase thermal and dimensional stability. Among the duromeric plastics, UPs are combined with glass fibers and transformed to SMC. Recently, carbon-fiber materials have been introduced as exterior plastic substrate for roofs. Because of their higher thermal stability and comparable low coefficient of thermic expansion (CTE), duromeric plastics are suitable for integration into the OEM coating process from the beginning (online painting). There are also applications known where parts from duromeric materials like SMC and carbon-filled plastics are coated offline. Owing to the need for dimension stability of the plastic parts, to prevent part deformation under baking and hot climatic conditions, it is advisable that thermal expansion of both the coating and the plastic are of similar value. Plastic substrates are not naturally electrically conductive. Dust adsorption induced by electrostatic charging can be the source of cosmetic film defects. In the case that plastic parts are painted electrostatically, the surface has to be equipped with electrical conductivity. The standard procedure is to apply a conductive primer, although intrinsic electrically conductive plastics have been developed as well, but have not yet been used owing to very high costs [11].

9.1.3.2.2 Solubility, Swelling, and Resistance to Solvents Finding a proper solvent balance in coating materials for plastic parts is an important task. With regard to

Table 9.4 Solubility of polymers in solvents [12]

Thermo-plast	White spirit	Benzene	Ethylether	Acetone	Ethyl-acetate	Ethanol
PS	0/−	−	−	−	−	+
ABS	+/0	0	−	−	−	0
PMMA	+	−	+	−	−	+
PC	+	0	0	0/−	0	+
PBT	+	+/0	+	+/0	0	+
PA	+	+	+	+	+	+
PP	+/0	+/0	+/0	+	+/0	+
PPE[a]	0	0	0/−	+/0	0	+
EP	+	+	+	0	0	+
UP	+	0	0	0	0	0
PUR	+	+	+	0	0	0

a Modified, + Insoluble, 0 swelling, − soluble

adhesion, a certain degree of swelling is of advantage. On the other hand, solvents should not dissolve the polymer matrix or relieve internal stress, which would result in cracks and reduced mechanical-impact resistance, especially at lower temperatures. For these reasons, some knowledge about the solubility of polymers in various solvents is useful (see Table 9.4).

Polyamides take up water to a significant degree, depending on the parts structure. This is a relevant factor not only with regard to mechanical characteristics, which vary with the water content, but more with respect to the appearance of a coating. The water of PA containing blends, painted in an OEM-integrated process, tends to evaporate in the primer surfacer or clear coat oven to generate pinholes and pops.

9.1.3.2.3 Surface Tension According to *Young*, wetting of substrates by liquids can be investigated by the contact angle between solid and liquid phase. Good wetting is achieved when this contact angle is very small, meaning that a liquid droplet has spread out and covered a large area of the solid surface (see Figure 9.3). A high surface tension of the solid as well as a low surface tension of the liquid both promote a small contact angle, as can be seen in *Young*'s equation:

$\gamma_{sg} = \gamma_{sl} + \gamma_{gl} \cos \theta$ Young equation;
γ_{sg} : surface tension solid (= interfacial tension solid/gas);
γ_{sl} : interfacial tension solid/liquid;
γ_{lg} : surface tension liquid (= interfacial tension gas/liquid);
θ : contact angle. (1)

Plastic substrates have low surface tensions compared to steel. This counts especially for the TPOs, which made it difficult to achieve good adhesion of coatings

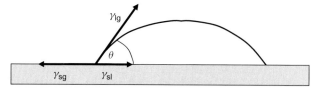

Fig. 9.3 Contact angle and surface tension according to *Young*.

Table 9.5 Surface tension of selected solvents, resins, and plastic substrates [13–15]

Substance	Surface tension (mN/m$_1$)[a]
Solvents and resins	
Water	73
Ethanol	22
Ethylenglycol	49
n-Hexane	18
n-Decane	24
Glycerin	65
Toluene	28
Alkyd-resin	33
Acrylate-resin	33
Solids	
Iron	45
Aluminum	45
POM[b]	40
PVC	39
PMMA	39
PA	38
ABS	34–38
PC	35
PPE	34
PS	33
PBT	30
PUR	30
LDPE[c]	30
PP	28

a Rounded.
b Polyoxymethylene.
c Polyethylene, low density.

on bumpers for a long time. Surface tension of liquids can be determined directly, whereas those of solids have to be determined indirectly by measurement of the described contact angles (Table 9.5).

It has been proved that the simple contact angle approach is not sufficient for explaining adhesion. It could be shown, that surface tension has two parts, a disperse and a polar part [16]. The fact that polypropylene, in contrast to polystyrene

(PS), needs to be activated for getting good paint adhesion cannot be explained by the overall surface tension value, but by the difference in polar parts of the surface energy.

9.1.3.3 Part Processing and Influence on Coating Performance

Apart from the pure polymer characteristics, coating surface properties are influenced significantly by the part manufacturing process. Moreover, the interaction is mutual, as part characteristics also can change with the painting process and depend on the coating properties. It goes beyond the scope of this article to describe a complete list of demands for a painting-friendly part molding process, so only the basic principles can be presented.

The technical quality of the polymer material itself is influenced to a great extent by part-processing parameters. For example, the resistance toward solvents is largely dependent on internal stress, which is the result of insufficient heat conductivity and local differences in cooling and solidifying at the end of the molding process. One consequence is that mechanical performance like mechanical-impact resistance can worsen at lower temperatures. Internal stress also can affect thermal part dimension stability, as does wall thickness and construction.

Visible flow lines, joints, and so on, are often transmitted through the paint layers to negatively affect the optical performance of the coating [17]. Parts containing too high an amount of recycled material can decline in thermal resistance owing to a degraded average molecular mass, resulting in cohesive failure when those parts are exposed to a hot steam jet as used in the car wash.

Molding conditions can intensely influence the morphology of TPO. Molding can increase or decrease the content of rubber close to the surface, which is suspected to be essential for interlocking with the polymer chains of the paint resin [18]. When leaving the processing tool, a plastic part might be covered with additives helping to disconnect the part from tool surface. These additives can affect paint adhesion, especially if they should contain silicones [15]. Primer to substrate adhesion or interfacial adhesion in a multilayer paint can also be affected by internal polymer additives like stabilizers, antioxidants, or slip additives. However, the cause of this failure is difficult to ascertain, as their migration to the surface and transfer through the multilayer paint film takes place only under the cure conditions for the coating.

Coating film properties depend on both paint material and the application process factors. The way a coating is applied to an assembly part also depends on the geometrical nature of the part design, so film formation might occur in different ways on different surface locations, with specific thermal conductivities and other physical properties. As a result, the performance of paint-film properties, like adhesion, are often observed to be location dependent.

9.1.4 Pretreatment

The main tasks of pretreatment are to prepare the plastic surface in such a way that coatings find the best conditions for their transformation into a smooth, cured film without optical defects derived from contamination, and for improving adhesion of the coating to the plastic surface.

The pretreatment process requires, at a minimum level, the following steps:
- cleaning,
- decharging,
- activation.

Chemical purity of surfaces is a relevant factor for both adhesion and film smoothness. Liberating the surface from dirt and additives from the injection-molding tool is achieved by cleaning. In many plastic coating lines, cleaning the surface with organic solvents is still the way of dirt removal prior to painting. In general, one should avoid the use of solvents that give rise to stress, strain, deformation, and cracking of the plastic. For most substrates, polar solvents are the least critical, and isopropanol is the most common choice.

In Western Europe, waterborne coatings are gradually replacing solvent-based materials, and as a consequence, the requirements for a reliable and powerful pretreatment have increased.

Today, in modern plastic coating lines, the standard cleaning method is a five-zone power wash procedure, starting with cleaning by detergents, followed by rinsing with water alternately, which is completed by a water-blow-away zone, followed by a drying zone.

In some cases, a temperature-dependent migration of additives from the polymer material takes place during paint curing, which can lead to interfacial delamination. As thermal-additive migration happens either after aging or under baking conditions, this kind of surface contamination cannot be removed by pretreatment.

Owing to the high electrical resistance of polymers, they can easily be electrically charged on their surface, resulting in a tendency to adsorb dust. Painting of plastic parts so contaminated, may lead to film defects, which can only be eliminated by additional rework. Therefore, behind the cleaning zones and right before the paint-application booth, a deionization step should be implemented, operating with ionized air to remove charges from the surface, and to prevent dust uptake.

Wetting is an essential precondition for adhesion. As outlined in Chapter 3, plastic substrates have a low surface energy. With the exception of TPO, normally, no activation is needed and adhesion can be provided with a carefully substrate-adapted coating formulation. In the United States and some European countries, adhesion promoters based on chlorinated polyolefines (CPOs) are used to modify TPO surfaces for better interaction with the subsequently applied coating material. This can be regarded as a kind of chemical treatment, giving the polymer surface a better ability to physically bind the applied coating. The mechanism by which CPOs adhere to the substrate is still the subject of intensive research. Entanglements of coating and substrate polymer chains could be proved by x-ray photoelectron spectroscopy (XPS), secondary-ion mass spectrometry (SIMS), and laser scanning confocal fluorescence microscopy [19, 20]. In this way, CPO-type adhesion promoters can achieve mechanical interlocking of top coat with the nonpolar TPO surface [21]. Although it has been stated that using CPO primers

9.1 Exterior Plastics

and dispensing with flaming on the average is cheaper than flaming plus painting a conventional primer, in Western Europe the trend is to get away from traditional solvent-based adhesion promoters. This is because of their low solid and chlorine content. Modern plastic-coating processes for TPO parts running with a high proportion of waterborne coatings and without adhesion promoters need one of the following activation methods:

- flaming
- plasma (low pressure)
- corona discharge
- fluorination.

Today in Europe, flaming is the most common method for activating TPO [22]. About 90% of all activations zones operate with flaming. Early investigations showed that flaming increases the surface energy of TPO from 23 to more than 40 mN m^{-1}, depending on the number of flaming cycles [23].

There are also lines operating with plasma pretreatment, which need a vacuum chamber. The Corona discharge method is a standard activation method for thin packaging films and expensive, when applied to car bumpers [5]. The mechanism of activation is still under investigation. The activation seems to enhance subsequent interdiffusion of polymer chains or the formation of interphases with superior mechanical properties in comparison to the bulk properties of the adhering partners. Different mechanisms of activation, which are discussed, are as follows [24]:

- elimination of weak boundary layers;
- increased surface roughness to allow mechanical interlocking;
- chemical changes at the surface owing to the formation of functional groups opening;
- the possibility for acid–base interactions and chemical reactions.

With regard to the chemical changes at the surface, it is generally accepted that flaming produces a higher content of oxygen-containing functional groups in the surface near regions of the polymer matrix as compared to the other methods. Hydroxyl, aldehyde, keto, and carboxyl groups have been identified by physical and analytical methods. The oxygen content can be determined by electron spectroscopy for chemical analysis (ESCA).

The functional groups are formed by reactive species like O_2-radical-anions, OH-radicals, or hydroperoxy radicals being created in the flaming or plasma processes, and reacting subsequently with the surface. The increased content of oxygen is limited to a few nanometers below the surface [20].

Flamed plasma, or corona-pretreated plastic surfaces, increases the surface tension only for a short time, so that painting is recommended within an hour. Fluorination, as well as flaming, can raise the surface tension up to more than 70 mN m^{-1} [15, 25]. In contrast to plasma, flame, and corona pretreatment, fluorinated surfaces show no time dependence in surface tension. In any case, the efficiency of the pretreatment is also dependant on the type of the plastic itself [26].

9.1.5
Plastic-Coating Materials

Depending on the purpose and function of plastic coating for exterior automotive parts, there are different paint buildup concepts that will be presented:
- multilayer coatings in car body color with high gloss;
- single-layer systems;
- either lusterless and in contrast color;
- or glossy and transparent.

For OEM-serial coatings, environmental compatibility is a major task in the development of modern plastic coatings. Whereas in the vehicle-OEM painting processes, low emission coating materials like the waterborne base coats (see Chapter 6) have prevailed in the market to a wide extent, in offline painting parts of plastics, these have not reached the same level of market penetration. Owing to the temperature sensitivity of most thermoplastics, the material concepts for environmental friendly coatings are limited to waterborne products and high-solid products up to 100% systems like UV coatings. Powder coatings are, owing to their high-bake temperatures, limited to SMC, PUR-RRIM, PPE+PA and glass-filled PAs. Apart from functional and ecological aspects and regional differences, the composition of a coating system for exterior plastic parts is determined by fundamental technical factors, which result in some principles that have to be considered as prerequisites for a successful coating process. They should be kept in mind even at the stage of raw-material selection.

9.1.5.1 Basic Technical Principles of Raw-Material Selection
The formulation of a plastic coating has to take into consideration the specific characteristics of the performance requirements of the coating and coated part, of the plastic substrates, dimensions of the part, type of parts pick up, application equipment, and the application conditions. The appropriate profile can be achieved only by selecting the right components from the portfolio of resins, pigments, extenders, solvents, and additives.

9.1.5.1.1 **Resins** Both the plastic substrate and the coating are polymer-based materials, and show their performance not as isolated units, but in combination as a painted part. Attention has to be paid to the influence of the resin structure and cross-linking to stiffness, tensile strength, and low temperature impact performance. In order not to weaken those substrate performances, an important precondition for adequate painting of plastic parts is to adjust the mechanical properties of the coating film to the plastic material. In early days of plastic coatings, often the coating materials applied were too hard. The result was the generation of cracks in the coating, which were transported to the substrate, and increased the chances of breaking at lower impact energies. Therefore, it is recommended and understandable to use flexible resins for flexible substrates. The cross-linked coating should have a CTE similar to the plastic, as otherwise part deformation may result. The fulfilling of this demand is a function of the polymer network's specific viscoelastic character.

As for OEM coatings, cross-linking improves chemical and mechanical resistance of the cured plastic top coat. In a similar sense, typical demands of workability like polishability of clear coats can only be fulfilled by using cross-linked clear coats with a sufficiently high glass-transition temperature.

9.1.5.1.2 Pigments/Extenders

In combination with the resin systems, pigments and extenders or fillers have a significant influence on the thermal expansion, glass-transition temperature, and mechanical properties of the coating film. Other technical criteria for selecting pigments and fillers are their influence on adhesion and sanding performance. With regard to the alkaline sensitivity of some plastics, pH-neutral pigments and fillers should be chosen for the coatings that are in direct contact with the substrate surface. Electrically conductive pigments enable primers to be coated electrostatically. The first choice for such pigments are carbon blacks with a high dibutylphthalate-adsorption (DBP) value. The color of these primers can range from medium to dark gray. To achieve a better color matching while keeping the base-coat film thickness low, light colored primers containing special electrically conductive mica pigments have been developed [27].

For the overall film performance of all automotive and industrial coatings, the pigment/binder ratio is of importance. The selection of color pigments is oriented to the OEM-master-color panel and the feasibility of color matching with the car body.

9.1.5.1.3 Solvents

Solvents, which are ingredients of plastic coatings, have to be selected to reduce physical impact to the substrate, particularly in the case of acrylonitrile-butadiene-styrene (ABS) and polycarbonate (PC)–ABS blends. On the other hand, solvents also help to remove lipophilic contaminants and swell the substrates, and this contributes to better adhesion by increased molecular entanglement. The latter, in particular, is an important precondition for adhesion to TPO substrates, and is strongly influenced by the type and amount of solvent contained within the applied paint. Conventional plastic coatings should have a solvent composition specifically adapted to the substrate, to achieve good wetting while preventing stress cracking or a strong redissolving, which can result in wrinkling.

Another important factor is the flash-off and drying situation in an offline paint shop. It is necessary to tune the solvent balance of the applied coatings, depending on the design and part suspension on the conveyor skid, for the specific layout of the application and drying zone.

9.1.5.1.4 Additives

For all performance problems, the standard range of additives for OEM base coats and fillers should be the first choice.

To achieve sufficient adhesion, one has to take into consideration the ability of coating materials to wet the substrate surface, which, depending on the plastic has often only a low tendency to be wetted. Aqueous paint materials on the other hand, because of their high polarity and surface energy, do not wet well. The combination of both aspects represents the situation of waterborne primers and particularly of base coats, directly applied on TPO. Wetting additives help to overcome this problem by reducing the surface tension of the coating.

Clear coats directly applied on plastics, as is already being carried out in the case of polycarbonate-blends, also afford paint material adaptation by additives, since the clear coat formulation is not designed to adhere on thermoplastics. The modification of plastic coating materials with wetting additives might cause failures in other performance characteristics like color shift and interfacial base-coat–clear-coat delamination. For this reason, one often has to correct the formulation by resorting to further additives. Additives, in general, should develop their properties without attacking the substrate. Amine-containing additives like N-alkyl-group containing hindered amine light stabilizer (HALS) can catalyze saponification of polyesters and PCs.

9.1.5.2 Car-Body Color

9.1.5.2.1 Regional Aspects
Plastic-part manufacturers and finishers in North America and Europe often cannot use the same coatings technology, even if they are painting the same plastic substrate for similar parts for the same OEM customer. The difficulties can be categorized in several areas – process infrastructure, performance requirements, environmental concerns, and cost implications. In both regions, the classic plastic offline paint built up in car-body color consists of a three-layer system of primer, base coat, and clear coat. Differences can be noticed in the chemistry of paint materials, bake schedules, and film thicknesses. For example, in North America, thermosetting substrates like reaction injection moldings (RIM) and SMC play a more important role than in Europe, which resulted in the establishment of a high-bake process for many coating lines. Coated TPO bumpers are baked at 120 °C in North America, whereas in Western Europe stoving temperatures are limited to 80–90 °C for primers and clear coats [28] (see Table 9.6). Consequently, different baking temperatures can mean different curing chemistry. TPO in North America, which is typically of different substrate quality with higher filler and EPDM content, dispenses with flaming but uses an adhesion promoter in thin-film builds.

Apart from this standard concept, there exist more or less variations in each area of the paint process. The old two-layer primer–top coat technology in passenger cars has been widely replaced by the three-layer system primer–base coat–clear coat. Special effect coatings for luxury cars or sports cars in Europe can consist of many-layer systems, including two base coats and two clear coats. On the other hand, in Germany, primerless painting of TPO on offline parts has been implemented successfully in serial production. In Western Europe, plastic coatings move strongly toward waterborne materials to fulfill environmental restrictions, whereas in North America, standard solvent-based technologies tend more to

Table 9.6 Differences in painting TPO bumpers in North America and Europe [5]

North America	Europe
Substrate: TPO with high amounts of EPDM and relatively high levels of fillers	Substrate: TPO with moderate amounts of EPDM and fillers
Power wash	Power wash + flame treatment
Adhesion promoter (sb) (CPO), 2–8 µm	Primer (CPO) or other (20 µm)
Bake 15 min, 120 °C	Bake 30 min, 80 °C
Base coat (sb/wb)^a 15–25 µm	Base coat (sb/wb) 15–25 µm
Flash off	Flash off
Clear coat (sb) 40 µm	Clear coat (sb)
Bake 30 min, 120 °C	Bake 30–45 min, 80–90 °C

a sb solvent-based, wb waterborne.

be supported by installing abatement systems like thermal oxidizers, carbon absorption scrubbers, and other methods.

9.1.5.2.2 Paint Systems

Primer The key functions of the primer are as follows:
- smoothing the substrate surface, covering defects;
- promoting adhesion of the colored base coat;
- absorbing stone chipping;
- acting as barrier for top coat solvents to the substrate;
- hiding the substrate color for increasing the hiding power of base coats.

Primers for plastics have light gray to dark gray shades or are even top coat specific colored. Although there is a global trend to move toward the more environmentally waterborne products, many paint shops of job painters for molders are small companies that have not invested in the necessary stainless steel equipment and climate control, to work smoothly with waterborne products. These job painters, in many cases, still use solventborne primers, which present a higher chemical complexity (see Table 9.7).

Solventborne primers for high impact resistant plastics are typically composed of a polyester resin and a polyisocyanate cross-linker as a 2-component (2K) PUR system. Flexibility can be adjusted either by choice of polyester resin or the type of polyisocyanate cross-linker. Hexamethylene-diisocyanate (HDI)-based biuret or allophanate types are more flexible than isocyanurates, especially those of isophorone-diisocyanate (IPDI) [29]. The main resin is often blended with high flexible polyester–urethanes. Such primers are feasible for TPO, PC+ABS, PC+PBT (Polybutylenterephtalate), PA and PUR–RRIM. Apart from this high flexible primer, there are other traditional solventborne primer technologies with lower propagation. Of these, the 1K polyurethane systems are suitable for either

Table 9.7 Typical technical data of solventborne and waterborne 2K primers

Property	Dimension	Solventborne	Waterborne
Color		Light to dark gray	Light to dark gray
Gloss		Low to high	Low to high
Solid Content	%	32–66	45–65
Density	g cm^{-3}	1.1–1.7	1.3–1.7
P/b[a]		0.7–2.0	1.4–2.7
Pot life	h	5	1–2
Viscosity at gun (20°C/ISO cup)	s	17–25	17–25
Curing	min/°C	30–45/80	30–45/80
Film thickness	μm	15–25	25–30

a Pigment to binder ratio.

flexible or semiflexible plastics. Blends like PC+ABS or PC+PBT, of which flexible and semiflexible adjustments are obtainable, can also be primed with a flexible 2K PUR polyester or polyacrylic system. Both primer technologies are also suitable for stiffer substrates like reinforced PA or SMC [30]. Adhesion promoters for TPO contain CPOs, which are supposed to interlock with the near-surface rubber-chains in the weak boundary layer of TPO. CPO primers help to compensate insufficient flaming on complex geometries. They have a very low solid content and are, in the long run, likely to be substituted by more environmental friendly products, like waterborne adhesion promoters or primers.

For a long time, waterborne primers, often called *hydroprimers*, were regarded as being weak in adhesion. Water has a much higher surface tension than paint solvents and therefore, cannot wet well and cannot act as a swelling agent, except with PAs. Despite these disadvantages, there have been developments in a waterborne primer technology that has improved adherence on different substrates, inclusive of flamed TPO. These 1K PUR and 2K PUR hydroprimers, meanwhile, have been successfully commercialized in Western Europe, where they are state of the art [27]. The resin backbone consists of a hard and soft segmented block-polyurethane modified with dimethylolpropionic acid (DMPA) to support emulsifying the resin in water. In the case of 2K PUR primers, this blocked PUR system has additional hydroxyl groups to react with a water-compatible polyisocyanate of low viscosity.

Since the 1980s, SMC parts can be primed by the IMC process, predominantly in use in France, Germany, and the United States. The coating process is transferred to the molding form, which needs to be opened up to a small gap, so that the coating can be injected. After closing the tool, the IMC is cured and fused with the SMC-carrier material.

The primed parts are then coated with conductive primer and assembled at the car body to be integrated into the OEM painting line in an inline-process (see Section 9.1.2). Problems can occur with gassing at the edges of the SMC parts, especially at high baking temperatures. For this reason, low-temperature-curing primer systems, for example, air-drying conductive primers on alkyd-basis are in use. Owing to the prolongation of the paint process by oxidative curing, other fast and room-temperature-curing primers are being sought. UV-curing sealers can be considered as one of the options. As shadow zones of three-dimensional shaped assembly parts also need to be cured, hybrid paint systems with a dual-curing chemistry were the starting point of more exploratory work on UV systems for large plastic body parts [31].

Base Coats For achieving optimum color matching and reducing product complexity, modern base coats are based, in principle, on the OEM base coat formula (see Chapter 6). Optional adaptation of color tint, depending on the color of the undercoat/substrate and individual offline paint process, can be realized by using a tailored concept of color pastes, which are added to the batch. Waterborne base coats, today, are state of the art for plastic parts, and are used by many processors. There is a variety of resin formulas existing for waterborne base coats including physical drying aqueous PURs, acrylics, PUR-acrylic hybrids, and polyesters (see Table 9.8).

Clear Coats Clear coats for plastic parts have to be more flexible compared to OEM clear coats, as it is known that the clear coat significantly affects the impact resistance of the plastic parts.

In North America, owing to higher bake conditions at 120 °C, both 1K and 2K clear coats are in use, whereas in Western Europe, plastic-coating lines work with low-bake 2K polyurethane systems at about 80–90 °C (see Table 9.9). Resins are mostly thermosetting polyacrylics blended with polyesters, both functionalized with hydroxyl-groups. Flexibility and mar resistance can additionally be adjusted by cross-linking with mixtures of HDI–IPDI-polyisocyanates. The standard UV-additive-package for OEM clear coats is used, as well as sag control agent (SCA)-modification for improved surface smoothness on vertical areas of the parts.

The small group of waterborne clear coats for offline painting comprises self-emulsifying polyacrylic resins and low viscous polyisocyanates, typically modified with polyethylene oxide groups. Waterborne primers and base coats for plastics are state of the art, whereas waterborne clear coats for exterior plastic parts have not, so far, been able to establish themselves on the supplier market because of their more limited processing window compared to solventborne clear coats. Therefore,

Table 9.8 Typical technical data of waterborne base coats for application on plastic parts

Property	Dimension	Metallic color	Solid color
Solid content	%	15–20	30–40
Density	g cm^{-3}		
Viscosity at gun (20 °C)	MPa per 1000 s^{-1}	70–100	70–100
Flash off	min	15–20	15–20
Film thickness	μm	12–20	20–30

Table 9.9 Typical data of solventborne 2K clear coats for application on plastics

Property	Dimension	Solventborne 2K clear coat
Solid content	%	38–52
Density	g cm^{-3}	0.95–1.00
Pot life	h	5
Viscosity at gun (20 °C, ISO cup)	s	16–21
Curing	min/°C	30–45/80–90
Film thickness	μm	25–40

it is not expected, generally, that waterborne products will be the environmental friendly solution of choice for clear coats. High-solid-UV or UV-hybrid clear coats could become a more probable option.

9.1.5.3 Contrast Color and Clear Coat on Plastic Systems

For painting in contrast color, mostly black, with low gloss, 1K polyurethanes, 1K acrylic, 2K polyurethanes of polyester and isocyanates are used. Contrast colors have declined in volume owing to change in fashion in favor of coatings in car-body color.

There are some examples for the plain application of a clear coat on plastic assembly parts. Headlamp diffusers made of polymethylmethacrylate (PMMA) are traditionally coated with a UV-cured solventborne or solventfree clear coat.

Bulk-colored plastics, for example, the hang-on parts of the SMART car, are coated with a special 2K clear coat that has been modified toward more flexibility. A special requirement for such a clear coat is to provide direct adhesion on the substrate, in this case, PC–PBT blend, even under humid conditions and after extended weathering tests. Therefore, radical scavengers have to be aminoether-functionalized and to be of low basicity [32].

9.1.6
Technical Demands and Testing

9.1.6.1 Basic Considerations

On the one hand, the demands on plastic coatings in exterior use are aligned widely with the specifications for OEM coatings, but on the other hand, there are some deviations and special requirements. In this chapter, the focus is on these differences and particularities. Some have to do with fact that firstly the substrate is not metal, but plastic and that most of the surface is not flat, but shaped. However, the most important difference in the examination of OEM and plastic coatings is derived from the angle of the individuality of the assembled parts. Depending on the car manufacturer, approvals can cover paint material, or the entire part, or both. The role of the plastic substrate will not be discussed any further, as the characteristics have already been described. However, it is worth emphasizing the fact that different specifications exist for painting parts made of thermoplastic or duromeric plastics.

To begin with, in the application of the wet paint, OEMs, more or less, have an influence on offline paint shops as far as ecological material characteristics and application parameters are concerned. Often, requirements regarding the quality of pretreatment, like power wash, are also part of the system-approval process.

Nevertheless, the intensity of OEM prescriptions for offline paint shops in regard to selection of paint materials and design of the application process differ over a large spectrum, depending on the OEM and the assembly part. Painting of bumpers might afford a different accentuation in specification in comparison with that for door handles or registration plate supports, resulting from nonidentical built-on locations. This flexibility is expressed by differences in demands on optical and functional properties, including safety aspects. On the other hand, tailored fitting of performance profiles to parts represents economic risks of high product diversification.

9.1.6.2 Key Characteristics and Test Methods

9.1.6.2.1 **Film Thickness** One of the first actions in a testing procedure of painted plastic parts is to determine the film thickness. As for all paint materials, properties vary with the film build of the dry coating. In contrast to coatings on metal, electrical and magnetic methods cannot be used. There are a few other options, which can be divided into methods with and those without destruction. Using the paint inspection gauge (PIG), according to DIN EN ISO 2808 : 1990-10, or a microscope, according to the American society for testing and materials (ASTM) B 748 : 1990, will damage the part. Often, it is advisable to determine film builds on different locations of the part. For each measurement, a piece of painted plastic has to be taken from the part with a knife, and a cross section prepared. This is inspected under a light microscope equipped with a setup for linear measures.

Nondestructive determination of film thicknesses is possible using ultrasonic or thermal waves. Both methods are based on the determination of temporal differences of waves reflected at interfaces.

9.1.6.2.2 **Functional Properties** Adhesion is a crucial point in the performance profile of plastic coatings, and delamination is one of the commonly occurring failures. The standard test here is crosshatching, and drawing away a tape attached on the crosshatch according to DIN EN ISO 2409:1994-10, performed as an initial adhesion test, and after humidity impact, extreme temperatures, or weathering. Scratching with a knife is considered a helpful tool for experienced inspectors to get a quick impression about adhesion. This method is, however, not suitable to make an objective judgment about adhesion performance. As a very tough testing method for adhesion, the steam-jet test has been established by various OEMs like DaimlerChrysler DBL 2467. In principle, this is a simulation of a car wash by steam-jet procedure on damaged coating surfaces. The damage is mostly a cross cut down to the plastic surface. Then a hot steam jet is brought to the cross cut, resulting in a strong impact on all interfaces, of which the starting point for attack

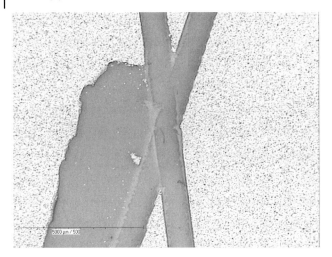

Fig. 9.4 Body-color-painted plastic part showing delamination from the substrate after testing with a steam jet.

is at the boundary of the cut (see Figure 9.4). The result – adhesion or delamination – depends not only on the adhesion forces, but also, to a high extent, on the specific OEM test parameters. These include water, temperature, pressure, duration of impact, geometry of the spray nozzle, distance and angle between the steam jet and the surface.

Much effort has been put into understanding the adhesion of coatings on plastics by using more scientific methods [33–37]. The difficulties that remain to be overcome include the separation of adhesion and cohesion forces in delamination, and application of exact adhesion measurement on weathered plastic parts.

Flexibility has been stressed as one of the key factors that make coatings on plastics successful, so it may be a surprise that the measurement of pure flexibility is not routine in plastic coatings. There are various methods to test the flexibility of coatings, most of them delivering not pure coating flexibility but values composed of film and substrate, overlapped with other physical characteristics. Moreover, typical coating test methods performed on steel, such as deep-drawing and bending steels, are not suitable for plastic substrates. In contrast to these performance tests, tensile experiments of free films give stress–strain curves, of which the elongation at break represents a value for flexibility. In fact, pure information about coating flexibility alone, just for plastic coatings, is not that useful, as the whole system substrate/coating has to be considered and each substrate has a different viscoelastic character. In any case, the viscoelasticities of coating and substrate need to be matched. This counts, in particular, for rubber-containing polyolefine substrates (TPO), which are the dominant substrates for bumpers. What makes coating flexibility really necessary is loss prevention in impact resistance at low temperatures, compared to the nonpainted parts. A standard test method to investigate the influence of the coating on the mechanical-impact performance of the substrate is the determination of puncture-impact behavior by instrumented

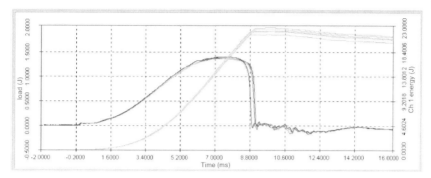

Fig. 9.5 Instrumented high speed impact testing according to ISO 6603-2 (source: Plastics Technology Laboratories, INC).

high speed impact testing, according to ISO 6603-2. The high speed impact test is used to determine toughness load-deflection curves and total energy absorption of impact events. This test can be part of the approval requirements, and is performed as follows: painted plastic coupons are clamped onto the testing platform of the apparatus. The crosshead is raised to the appropriate height and released, so that it impacts at a specified speed. A load-deflection curve is produced (see Figure 9.5). The test is performed from both sides of the painted specimen at different temperatures to get information about ductile/rigid transitions. The temperature point can be reduced by going from room temperature down to minus degrees, when the quantum of absorbed energy as a measure for tenacity drops significantly. The instrumented high speed impact test has been demanded for approval of new plastic coatings.

Mechanical-impact performance will receive more attention in the future, owing to increased safety considerations, and as a consequence, legislation for the safety of not only the passengers, but also the pedestrians. Complete parts are examined in pendulum-head impact tests to simulate vehicle-crash tests. Damaging of the part has to take place without splintering. In general, the outcome of impact tests is not only dependent on the viscoelastic properties of the coating composition, but can also be affected by the release of internal tensions in the substrate, resulting from the effect of aggressive solvents.

Stone chip resistance is routinely tested, as for OEM coatings. Mirror housings, front grills, bumpers, as well as fenders provide extremely chip exposed areas, which need a multilayer coating system with coatings carefully adapted to the substrate and to each other, for excellent adhesion and chip resistance performance.

Scratch resistance is gaining importance for exterior plastic coatings as well as for OEM coatings [38–40]. Body-color-painted large parts are prone to critical examination, owing to frequent customer complaints about scratch-sensitive bumper surfaces. Compared to the OEM clear coat, some of plastic clear coats can take advantage of flexible resin-compositions for a short period of time after application, owing to the self-repair potential of scratches by reflow [41].

Putting so much effort into improving the flexibility of plastic coatings often goes along with some drawbacks, potentially resulting in a lack of hardness. Assembly parts, for example, mirror housing and bumpers, are packed up soon after final inspection. Lack of hardness can result in marks from packaging, because of the pressure of the packaging. Soft clear coats can cause problems during polishing, and primers can suffer from limited sandability. What is called *hardness of coatings* is, in reality, not a definite property, but either penetration, pendulum, or abrasion hardness. Moreover, hardness is a complex mixture of clear physical characteristics such as stiffness and firmness, which can be obtained by physical investigation of the film. Nevertheless, it is common to perform hardness testings to get comparable data for plastic-coatings performance. Standard test methods cover pendulum hardness and diverse penetration methods, of which the universal hardness, according to DIN EN ISO 14577:2003-1-3, can deliver information about stiffness, plasticity, and creep-behavior.

Plastic coatings have to basically fulfill equal demands in durability, that is, weather and chemical resistance as do OEM coatings. Owing to higher resin flexibility and lower bake, glass-transition temperatures of exterior plastic clear coats often are lower compared to those of OEM clear coats. However, this can be taken not as a universal rule, but only as a trend. Clear coats virtually need as many varying glasspoints and resistances as there are parts. For example, in contrast to flexible bumper plastics, door handles made from glass-filled polyamides may be equipped with a more rigid and media-resistant clear coat, owing to increased exposure to sweat from hands.

Taking into account the resistance toward weather, humidity, and extreme temperatures, the reactivity of plastic substrates and their interaction with the coatings under those conditions, have always to be taken into account [42]. Plastics coatings have to show adhesion on the substrate even after impact at frost. They also have to stick on the surface when the substrate undergoes substantial altering, losing softeners, or absorbing water like PAs do.

9.1.6.2.3 Optical Properties

In the early times of plastic coatings, parts were not only of smaller size, but were also usually painted in a contrasting color. Efforts to match the color to that of the car body were less demanding than they are today, because either the contrast color was black, or if not, a little form integration, a high proportion of simple straight shades, and only a few metallic colors helped to keep the demands low, compared to the present situation. Today, colored exterior parts are expected to show the same appearance and color as the car body. Effect colors dominate the range of base coats, many of them showing interference with a strong angle dependence. The number of base-coat colors has also increased and so has the surface portion of exterior plastic, whereas separate black trims between parts and car body have almost disappeared. These factors make color matching between form-integrated assembly part and car body a central issue in testing optical properties of plastic coatings, today.

Examination of color matching is not exclusively a matter of judgment by eye but more and more a matter of physical measurements. Multi-angle-instruments are used in daily approval processes, despite the fact that coloristic theory is still not completely developed for effect coatings. Additionally, the geometric designs of some cars make it hard to achieve color harmony, as its perception also depends on the smoothness of the geometric passage between car body and assembly part. Instrumental colorimetry by portable multiangle apparatus help personnel to approve or disapprove painted parts in built-on trials online. This method works on flat surfaces with large body parts quite well, but has its limitations with shaped smaller parts. Overall, it can be summarized that color matching between assembly part and car body has become an expensive but feasible task. From this point of view, several concepts have been developed to encompass the online painting process and the use of identical base-coat batches for OEM and plastic in offline coating [43].

9.1.7
Trends, Challenges, and Limitations

The overall surface area of painted plastic parts on cars is now expected to grow at a slower rate. Further development of the exterior plastic coatings market will be driven mainly by increasing environmental regulations and cost reduction. Secondly, it is expected to be influenced by car-safety issues and by attendant design trends. The factors mentioned will influence the selection of substrates and composition of coating materials. Finally, they will promote the evolution of new coating techniques, which will use less paint or perhaps even no paint.

9.1.7.1 Substrates and Parts

In the near future, cost pressure on the whole process chain of automotive manufacturing will become an increasing factor on decisions for material concepts. How does this affect the use of painted exterior parts? At first, it is sufficient to state that there are fewer 'low fruit hanging opportunities' for plastic in automotive-exterior design. This is because of the fact that the less design/challenging components of the automobile were converted already to plastic many years ago, in the 1970s, 1980s, and early 1990s [3]. Today, when it comes to decisions for or against plastic for assembly parts, the most probable approach seems to be the evaluation of every individual case on the basis of special, technical, and cost requirements. For the calculation of the latter, a decisive aspect is the number of produced pieces.

Where are the further chances and options for plastics and their coatings in exterior-body design? New legislations for cars include provisions for the protection of pedestrians. One concept is that of giving large exterior parts an airbaglike function by using plastics as a type of passive protection. On the other hand, this is only one of several protection concepts, and the car-passenger has to be protected too; this calls for a stiffer and stronger body construction. As steel has some significant technical advantages, it is expected to remain the dominant exterior material [7]. Nevertheless, the extension of plastic to large parts like hoods and roofs represents an attractive challenge. Fiber-reinforced duromeric plastics are believed to be the feasible material for horizontal parts, whereas thermoplastics seem to be in disfavor for this application owing to limitations in their mechanical properties and thermal expansion. Overall, polyolefine-based materials are expected to replace other plastics, because of their advantageous combination of mechanical properties, cost profile, and recyclability. This kind of cannibalization is expected to decrease the use of more expensive thermoplasts and high density materials like PUR-RRIM. It is driven by two factors – lower raw-material costs and lesser material per component [3].

With respect to the efforts in sustainable development, the preservation and recycling of plastic raw materials will gain in importance. According to European legislations, old vehicles will have to be recycled by material at 80% and by energy by 5%. From 2015 on, these limits may be raised to 85 and 15% respectively. To limit recycling costs, cars in future should be easily disassembled in modules consisting of compatible materials. It seems clear that this trend will ask for an assessment of plastic parts and their related substrates, as well as the substrate spectrum [7].

9.1.7.2 Paint Materials

The options to reduce solvents in plastic paints are mainly waterborne and UV-curing coatings, and these have already been described. Waterborne primers and base coats are state of the art, and are expected to receive extended use. Waterborne clear coats for plastics are under development but have only recently been tried out and introduced into serial applications. Upcoming tighter legislations on solvent emission in the European Union may exert additional pressure on all concerned parties to work more on the existing problems with waterborne clear coats, mainly process robustness, and to lift this technology to a competitive level.

Radiation curing lacquers provide considerable potential for more widespread use with exterior plastic parts than just the headlamp diffusers, which have been its main application for UV-coatings to date [44]. Apart from the prospect of bringing about a major reduction in solvent emissions, UV coatings constitute an attractive alternative for plastics processors, who paint parts offline, for the following reasons:

- immediate processability, since curing takes place within seconds;
- less finishing work, thanks to the lower exposure time for soiling;
- space-saving design of paint shops possible;
- no waste water/air problems in case of solventfree formulations.

Presently, research and development on UV coatings for plastic is concentrated on two areas. First, analogous to OEM paints, scratch and mar resistance in combination with good etch resistance are targets of modern clear coat development for plastic parts also. Second, there is potential for UV-curing primers that have the ability to seal SMC-substrates to prevent gassing and subsequent coating defects. However, in both cases, UV is not the only option and it also has some drawbacks.

Exterior plastic parts are mostly three-dimensional in shape, and need to be coated by spray application. Conventional UV-coating materials contain reactive diluents, which evaporate in a spray dust to a significant level, highly depending on the atomization parameters. With an increased emission of reactive diluents by spray application, and the dependence of cross-linking density on atomization parameters, the call is for waterborne UV-curable systems, which are free from reactive diluents. A prerequisite for UV-paint technology in serial automotive as well as in plastic parts coating is that a way must be found to enable curing in shadow areas. One option is the utilization of dualcure materials that cure thermally in the shadow zones, and in addition, by UV in the light exposed surface areas [45, 46]. Monocure coating systems would need some adaptations in hidden curing zones like programmed UV lamps on robots [47].

The sensitivity of the UV-curing process to inhibition by oxygen requires additional action. A step forward to overcome the oxygen inhibition has been taken recently with the curing of UV-painted three-dimensional objects under CO_2-atmosphere [48]. In the property spectrum of UV coatings, there are different features that have to be balanced out with respect to being feasible for automotive-exterior build-on parts. Radiation-cured coatings can either be constructed hard-brittle or flexible-soft [49].

A flexible character should be in favor for plastic parts, but the protective tasks of clear coats afford certain reactivity and cross-linking density. As the reactivity increases, the resistance to thermal and photochemical yellowing falls. Coatings that cure fast to a dense polymer network tend to shrink significantly, resulting in a weakening of adhesion to the coating underneath or plastic substrate. Therefore, the recipe for UV-curing resins, in particular, has to reflect the special demands in viscoelasticity of the plastic substrate.

There are already exterior plastic parts being coated with outdoor-feasible powder. These include online- and inline-coated SMC and PPE+PA. Powder coating of heat-sensitive substrates like TPO is yet to be achieved. A further expansion of the powder technology into exterior plastic coatings would presume either an increased use of high-bake substrates, or a breakthrough in the development of low-bake powders. Low-bake powder has limitations in storage stability. UV-curable powders have the advantage that flow can be separated from curing, which then would be performed by irradiation. Though already used for medium density fiberboard (MDF) and poly(vinyl chloride) (PVC) floors, UV powders for automotive-exterior plastic parts are still in development. An important prerequisite is electrical conductivity to enable the electrostatical powder application and to reach high transfer efficiencies. The easiest option to achieve this task is the application of electrically conductive primers prior to powder coating.

In serial production, it can be seen that the high demands in color matching of assembly parts up to now, succeed only at high expenses. It seems that a further increase of the plastic with body-colored surface depends largely on the cost structure. Whether this can be accomplished by intelligent logistical and technical painting concepts, or by different techniques for decoration, still has to be shown. Although there are still customer complaints about the mechanical sensitivity of colored car bumpers, it is unlikely to turn the wheel back; rather, it would be necessary for suppliers to work on weaknesses like scratch resistance. Another growing trend is the use of metalized plastic, which brings back a nostalgic touch to the car body; however, this has been resorted to, up to now, only on small surface areas.

9.1.7.3 Processes

9.1.7.3.1 **Paint Layer Reduction** A very simple means of reducing painting costs, and increasing at the same time, the ecological efficiency, is the saving on coating materials by dispensing with individual layers. Positive results have been achieved with primer-free coatings, that is, by using only base coat and clear coat, in the area of bumper painting. The primer will still continue to be in use in many processes, in the future, because it assumes a number of key functions (see Section 9.1.5). It will continue to find use especially in cases where the substrate quality does not allow direct base-coat application without loss in smoothness, adhesion, and color matching. Glass-filled substrates need a primer, in any case, to receive a smooth and even surface, so that the appearance of the clear coat matches that of the OEM car body. Optimum coordination of the color shade of the painted built-on part with the painted body is frequently only possible through the application of a light primer, or even several colored primers, to the often dark gray and untreated part [8]. Color red primer technology has been established, for example, in the plastic coating lines of Audi and Peguform.

9.1.7.3.2 **Molded In Color Plus Clear coat** Clear coating of mass-colored plastics is being performed successfully with Daimler's SMART car. Adhesion of clear coats on PC+PBT seems to be controlled, but has to be carefully tested for every other substrate, in particular, after weathering and humidity impact. A strength of the clear coat-on-plastic technology is the simple realization of straight-shade colors [1]. The spectrum of options for decoration is further extended by tinting the clear coat with color pastes, or by adding effect pigments. On the other hand, effect colors identical to the OEM car body have not been realized by mass coloring with satisfying color matching, up to now, as the mica and aluminum pigments do not receive the desired flat orientation to an acceptable degree. For these reasons, the clear coat on exterior plastic parts has up to now not been extended to large volume car models and has remained a niche product.

9.1.7.3.3 **Foil technology** Studies on the North American automotive plastic-coatings market in 2001 estimated a significant shift from classical painting to in-mold colored or film-laminate technologies called *foil technology*, forecasting a stagnant or even negative growth for traditional plastic coatings [3]. Basically, this forecast could be made for Europe, as well. There is no doubt that with regard to the unification of surfaces, free choice of substrates, and cost savings by centralized manufacturing of body-colored plastic parts, foil application offers large opportunities, especially when fiberglass-reinforced thermoplastics are used [50]. However, the challenges are high in terms of color and painted surface matching, as well as application. Apart from optical and economical demands, foil technologies have to be deep-drawable and adherent to backside molding materials, with no negative impact on the quality of the built-on part [51]. High-gloss foil application is already being performed in serial production for various exterior plastic parts like radiator grilles and wheel caps since the mid-1990s [52]. With the use of an all-plastic foam-laminated roof for the SMART car in 1998, for the first time, a completely one-step produced and surface-finished large body part has been realized with 'paintless film mold' (PFM) technology (PFM-system of BASF). The roof, at that time had a textured surface, so the next step, a class A-surface, was attained in 2002 with a modified PFM technique (Figure 9.6). The high-gloss black SMART roof module is the first large vehicle body part in which glass and plastic with similar stylistic characteristics have been combined to produce a high-quality appearance. It consists of an in-mold laminate of a coextruded PC/acrylnitril–styrene–acrylester (ASA)-blend (Luran S), capped with two layers of PMMA (Plexiglas). Subsequent thermoforming to the shape of the roof is followed by injection of a layer of long-glass-fiber-reinforced PUR foam into the inside surface [53].

Whereas straight shades and simple effect colors can be realized by this kind of compact foil technology, there are limitations with the use of more sophisticated effect colors. Some technical issues cover repair and effect pigment orientation. A solution to these problems could open the door to wider use for new module concepts [1]. A solution, which has the potential of being color-matchable, is painting the coextruded foil with effect-color-base coat and clear coat prior to molding. The carrier foil would be line-coated with base coat and UV clear coat, the wet paint layers predried, then the foil thermoformed, and the clear coat UV-cured. The decorated thermoformed foil is then placed in a form, where it receives injection with fiber-reinforced plastic material from below [51]. Another option is to first thermoform the carrier foil and then to paint the deep-drawn foil

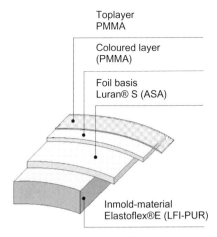

Fig. 9.6 Paintless film molding (PFM-system) (source: BASF AG).

with a conventional two- to three-layer buildup, after it is joined with the injected fiber-reinforced plastic. This is called *offline coatable* foil technique.

Foil-coated modules do offer additional options in assembly. For example, a complete roof, with all interior equipment and supply, should render superfluous overhead operations in assembly, and therefore, increase ergonomy [4]. Progress in assembling the car body, greater flexibility in producing car models, and the potential of reducing process costs are arguments for this kind of further outsourcing in the carmaker's production processes. If the thermal expansion of plastic substrates and the technical performance of foil and coating materials comply with all demands to lower overall costs, then modulized production concepts and foil techniques would be the best suited options, and this would open up new market shares for plastic parts.

Finally, it can be stated that as new and alternative decorative concepts for exterior plastic parts evolve, their acceptance and speed of market penetration will be strongly dependent on cost and quality issues, and therefore, impacted by the existing well-established coating infrastructure and painting capacities already in place [28].

9.2
Interior Plastics

Stefan Jacob

9.2.1
Introduction: the 'Interior' Concept

The decision to buy a given car is not only governed by the preferred body design but also, increasingly, by the appeal of the interior design (see Figure 9.7).

Fig. 9.7 Car interior.

Apart from the fleeting moments when we appreciate the car from the exterior, we actually experience the vehicle only from its interior perspective. The interior design is formed extensively by trends and the current feeling for life of each individual, thus also making it an expression of one's own personality.

The design and quality of a vehicle's interior can be significantly enhanced with a multitude of selected thermoplastics that are finished with appropriate coating materials. This enhancement is carried out under consideration of various functional and economic aspects. Apart from the fulfillment of automobile-specific demands, interior paint systems render additional functions such as high chemical and scratch resistance, antiglare features, and protection from UV radiation. Furthermore, these painted surfaces contribute to the harmonization of the vehicle's interior through individual haptics, a matt gloss, and the possibility of laser inscriptions for controls.

After the much quoted demand for mobility, there has been an ever greater desire for individuality. Consequently, the car is no longer a mere means of mobility, but also an expression of one's self-understanding, image, and joy of life. A multitude of color and effect combinations are available to the designer to support individuality.

Color coding of controls with similar functions reinforce the clarity and structure of the interior's concept. Variable strip elements with special effects accentuate the character of the vehicle. The design is in line with the spirit of the times in unison with long-term trends. The paint systems make it possible to fulfill the necessary flexibility of both development directions, and to realize these in keeping with the philosophy of the given car manufacturer.

The vehicle's interior must challenge and support the sensitivity of the people in an industrially characterized and often 'uninspiring' environment. The use of different shapes, colors, structures and haptics inside the vehicle influences the awareness of the driver.

9.2.2
Surfaces and Effects

Paint concepts covering the described attributes for coatings can be classified into four categories: metallic, haptic or 'softfeel' touch, textile, and laser coatings. All paints are dominantly either solventborne or waterborne systems with the typical surface improvement over unpainted plastics according to Table 9.10 [54].

Table 9.10 Technical requirements and performances of car-interior paint systems compared to bare plastics

- Increased light/aging stability
- Increased mechanical surface strength (scratch and mar resistance)
- Increased chemical resistance
- Harmonizing color and gloss in the interior
- Antiglare properties

Coatings improve not only the design aspects and the appeal of interior parts but also technical performances like durability, scratch resistance, and resistance against chemicals.

The use of 'sparkling' metallic effect paints produces surfaces with clearly discernible metallic structures. By varying the metallic particle size, their concentration, and the chemical after-treatment of the given effects pigments, the finest metal surfaces can be delicately highlighted, or coarse and robust structures can be produced. Yet, the individual metallic effects particles are still clearly discernible to the beholder. The color effect is formed independent of the viewing angle, thereby heightening the optical perception of the painted component shape. In this manner, very specific accents can be set. Together with the paints used for the bodywork, metallic effects with a high color depth can be achieved.

Series tested paint systems with galvanic effects have become available in recent years. With these effects paints, the individual metal-effect particles are no longer discernible, so that thermoplastic elements coated with these effects paints are given an authentic metal look.

Compared with the traditional galvanizing processes, the big advantage resulting from the use of this coating material to achieve and produce a surface that looks just like real metal is the greater ecological compatibility and much simpler process control, thereby boosting ecoefficiency.

Haptic is the term relating to the sense of a human feeling of touch, and is concerned with the physical and psychic influencing magnitudes in the perception of surfaces. The psychic influences in assessing soft coating surfaces for each individual are particularly difficult to define. This gave rise to the demand for a measuring method that would allow objective assessment of haptic senses.

Each time soft coating systems are touched, a signal relating to a high surface quality is consciously or subconsciously transmitted to the brain.

Although a number of different measuring systems have been tested during the last 10 years (microhardness, surface roughness, and slip effect), there is still no measuring instrument that can reliably 'measure' the complex sensations that are triggered in a person that touches a surface.

However, by defining the possible sensations when touching a surface, it has become possible to assess the parameters of haptics so that a more precise description of the desired surface can be given.

Under consideration of the haptics parameters summarized in Figure 9.8, it is possible to create haptic fields in conformity with Figure 9.9 which, in turn, allow the comparison of sample surfaces and reference surfaces.

Nowadays, designers have different decorative and soft coating surfaces at their disposal (Figure 9.10). By applying widely differing haptics, each touch transmits signals to the client that support the perception of quality and enhance the image of the given vehicle.

Consciously or unconsciously perceived, assemblies made touch-friendly by soft coating create the impression of a high-quality surface. The structure and haptics of the self-explanatory functions of controls support the demand for the often under-rated 'sensuous' perception of the environments.

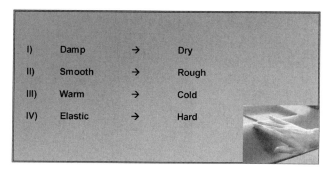

Fig. 9.8 The haptic of a touched surface – sensoric factors.

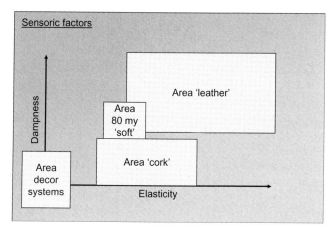

Fig. 9.9 Soft coating paint systems for different sensoric attributes (ALEXIT: Mankiewicz).

As a result of the widely differentiated character of individual vehicle types (roadster, van, limousine), and a culture-formed feeling for high-quality surfaces, a multitude of soft coating systems are available (see Figure 9.10).

As a result of their distinctive texture, textile-effect paints are being increasingly used as an 'optical bridge' between the textile-clad roof and the lower area of the dashboard.

Owing to their excellent cleaning properties, high resistance to scratching, very low gloss levels, and light fastness of such surfaces, these products are being increasingly used to replace textiles, which have taken the back shelf. Simple process control associated with these paint systems, in comparison with the elaborate procedure of textile coating and the associated reduction of costs, definitely supports this trend.

9.2.3
Laser Coatings

Next to the traditional methods of surface inscription, such as pad printing and hot embossing, growing use is being made of so called laser coatings, followed by

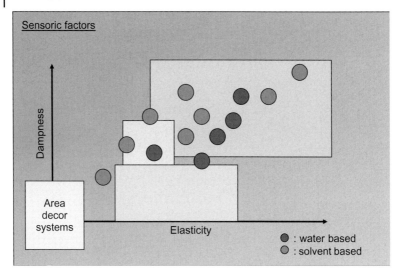

Fig. 9.10 Variety of ALEXIT soft-coating paint systems (Mankiewicz).

Fig. 9.11 Use of laser coatings in the car interiors.

laser inscription [55]. This combines the technical and design-specific advantages of painted surfaces with high-contrast representation of a day–night design. The symbol is characterized by sharp contours and high resistance to abrasion.

With the development of new coating systems, these surfaces are being applied with ever greater frequency for switches, keys (Figure 9.11), and controls for motor vehicles, the home entertainment area, and for general industry.

The high flexibility of the inscription units even makes laser engraving in small series an economical proposition.

The peculiarities of this technology and the framework condition for reliable application will be briefly described.

After cross-linking of the coating material has been completed, the symbol is created by selective burning of the coating material with an inscription laser. Inscription is contactless, even at points that are difficult to access. Various laser types, in different performance classes, are used for the inscription. The inscription laser of the type Nd:YaG (wavelength = 1026 nm) in different output ratings is mostly used for laser engraving of painted plastic parts.

Translucent thermoplastics (e.g. PC) are applied to achieve a day–night design by laser engraving.

Waterborne and solventborne top coats as well as soft coatings are used for application.

Contour-sharp inscription requires a paint system with a high laser-energy-absorption rate. This can be achieved by the development of an appropriate paint formulation specifically for this purpose. In other cases (Figure 9.11) the transmission of the laser energy to the thermo-sensitive plastic substrate can damage the base (foaming, carbonizing, etc.).

An optimal inscription result is dependent on the following:
(a) choosing a substrate that is thermally stable at temperatures of $>130°C$;
(b) the use of a sensitive (highly absorptive) paint system;
(c) the achievement of uniform paint layer thicknesses;
(d) the development of appropriate inscription parameters.

9.2.3.1 Substrate Requirements

Next to the high transparency demands expected of the substrates, thermal stability is also of very special importance. The customary PC types currently on the market differ significantly from each other in this respect.

Special additives, as well as the quality of injection molding, can exert an influence on the inscription results. Efforts should be concentrated on achieving high thermal stability to support paint removal by the laser without damaging the substrate.

9.2.3.2 Requirements to Be Fulfilled by the Paint Systems and Coating

To ensure reliable laser inscription, the employed systems require a high absorptive capacity for the energy input transmitted by the laser.

The representation of soft coating systems, often in a single layer, has been developed to maturity for series production in recent years. Since then it has been very successfully employed in various applications and has proved to be a special technical challenge.

It is obvious that black surfaces will have the highest laser-energy-absorption rate. Therefore, black-pigmented systems were used in the early days of laser-etchable paints, their black surfaces absorbing the thermal energy of the laser, thus burning out completely. Owing to the paint's high absorption rate, the substrate is not influenced by the thermal effects. At the end of the laser-etching process, a sharp-contoured symbol emerges, without any thermal damage done to the substrate.

However, the demand for colorful and brighter tones increases the complexity of the paint formulation, particularly the high thermal stability of some pigment groups, as well as the reflective capacity of bright surfaces. This makes it necessary to increase laser energies, and this can narrow down the processing window of laser inscription.

If a high laser energy density is required to burn out temperature-resistant pigmented paint systems, this will lead to thermal damaging of the thermoplastic substrate. One solution lies in using a special laser-etchible primer, which acts as a buffer.

High opaqueness of the entire assembly is particularly important for the application of a day–night design. The measure of opaqueness of the surfaces – defined in each individual case – prevents light scatter from a light source of a backlit symbol. Special pigment groups may have to be used for specific colors, and these have a lower level of opaqueness than black. In such cases, it is necessary to decide between a more opaque substrate and a primer that can be laser engraved.

The developed laser parameters define the energy input that is necessary to achieve reproducible combustion of a defined paint-coat thickness.

This gives rise to some risks:
- With comparable symbol engraving, the base will be thermally damaged with a diminishing dry film thickness of the coating material – the symbols show signs of burning (foaming, brown discoloration).
- The stipulated laser energy required to engrave the coating material will no longer be sufficient when the required dry film thickness is exceeded. The symbol surface will contain residues of the coating material.
- Component painting, followed by laser engraving, therefore requires the use of an automatic painting system (e.g. automatic surface paint sprayer). This will ensure that the necessary layer thickness tolerance of $\pm 3\,\mu m$ will be maintained to achieve reproducible laser results.

9.2.3.3 Demands Expected by the Inscription Technique

Laser coatings are inscribed with ND:YaG Lasers. The individual models differ by their maximum output, wave length ($\lambda = 1026\,nm$/frequency doubled $\lambda = 513\,nm$), number of inscription heads, and resonator.

The skills and know-how of the inscriber exert a major influence on the inscription result. This know-how includes finding out the optimum energy level characterized by the laser's frequency and amperage. Also, the number of passages and the method by which the symbol is burnt out (line by line or in a circular way) greatly influence the symbol's quality.

Proceeding from the hitherto customary use of hard black top coats, a wide range of inscriptions can be produced with only a few standard laser parameters.

9.2.4
Performances of Interior Coatings

9.2.4.1 Mechanical and Technological Demands

The original purpose of painting was to overcome different gloss and color effects for the interior of motor vehicles, as a result of injection molding and use of master batches.

The 'plastic environment', often regarded to be inferior, is visually harmonized to create the impression of a high-quality ambience. These requirements regarding a refined surface finish are supplemented by technical demands, which are often only assured as a result of the protective effect of a 20–40 µm thick dry coat (Table 9.10) – and this is achieved with colored top coats.

The qualitative demands expected of painted surfaces are defined in the specifications that give a 'time accelerated' description of the stress experienced during the life cycle of a car. These technical specifications differ significantly in many important details between car manufacturers and countries. Weather stability is a primary consideration for exterior paintwork, whereas the interior specifications place emphasis on special tests. Consequently, the mere transfer of the proven exterior paint system to the interior of a vehicle is rarely possible. This makes it necessary to develop paint systems that are specifically adapted to the requirements of the motor vehicle interior.

Twenty years ago, the 'smell' of a new car was connected with a positive association of first-time use, whereas, nowadays, the emissions of plastic components, with regard to quality and quantity, are the object of critical assessment on the test stand. In this context, the emission limit values of defined combinations (partly within the ppm range) or the total carbon emission have been stipulated for the development of interior paint systems.

Figure 9.12 covers the development of requirements expected from interior paint systems, both from the design and technological standpoints, thereby underlining the need for a permanent process of further development of the corresponding formulations.

Changes to the shape and application of some components inside cars can result in more severe stressing of these parts during the car's life. One example is the ergonomic design of the interior door-closing handle over the past years. On account of the user-friendly shape, passengers also rest their hands on the handle while traveling. This is invariably associated with a significantly increased exposure to moisture and perspiration, with the result that specific simulation cycles had to be adapted to the resulting practical situation.

9 Coatings for Plastic Parts

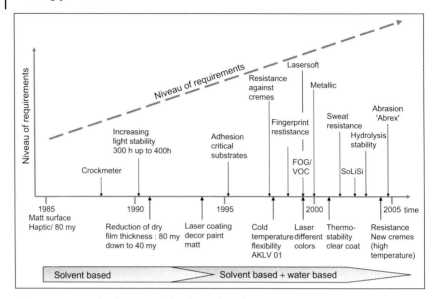

Fig. 9.12 Historic development of the demands and performance requirements of interior paint systems.

As a result of the worldwide distribution of many motor vehicle models, knowledge of the chemicals (also cosmetics) that are commonly used in different countries is required. The hitherto known specification requirements, therefore, have to be extended to take this into account (also climatic peculiarities – temperature, humidity, sun radiation).

The lasting quality, and thus also the brand image of a car manufacturer, is characterized by the assurance of a continuous high-quality level in combination with long-term assured use properties.

Close cooperation of design aspects with technological requirements can only be assured by intensive further development of the formulations.

9.2.4.2 Substrates and Mechanical Adhesion

Table 9.11 and Figure 9.13 show the typical substrates of interior car components. In the simple case, these materials can be a mono-substrate, that is, PA or PP, or in complicated cases, they may involve a blended module. A typical example is the ventilator nozzle unit, which involves the painting of preassembled components made of PA 6.6/GF30 lamellas together with PBT, polypropylen oxide (PPE), and PC enclosures. Painting is performed by an offline process (see Section 9.1.2) with a paintwork composition specifically validated for this purpose.

In addition to the demands expected of standard-conform mechanical adhesion, the homogeneity of the painted surface – relating to color shade and gloss consistency – represents a special challenge. Minor measured gloss differences affect the visual impression created by the surface. Even such minor deviations as

9.2 Interior Plastics

Table 9.11 Typical automotive interior substrates and their applications

Substrate[a]	Application
PP, PP/EPDM, PP/PE	Dashboard, interior door-closing handle, consoles
PBT/PA 6.6 , PA6.12 GF (30–50)	Lamella, door-actuating lever
PC (sometimes PMMA)	Frame/day–night design
ABS, ABS/PC	Consoles, miscellaneous parts
PPE	Side covers

a Abbreviations according to DIN EN ISO 1034-1.

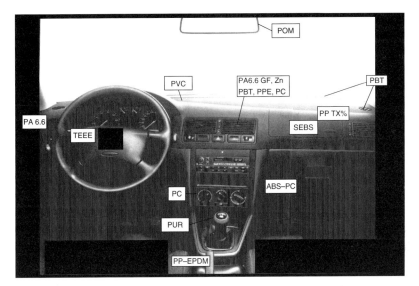

Fig. 9.13 Examples for the variety of plastic substrates in car-interior applications.

≥ 0.6 RW 60° between the individual components can result in the rejection of the parts by the OEM.

Paints for single-layer adhesion are available for a multitude of bases. However, next to the substrate type, many injection-molding parameters are involved in the production of a molded part and, possibly, also the molding auxiliaries contained in the part (light stabilizers, antioxidants, solvents, parting agents, emulsifiers). All these factors can exert a decisive influence on the surface quality and adhesion of the subsequent painting process. Consequently, binding statements have to be made with regarding to the painting processes for a given component.

9.2.4.3 Ecological and Economical Requirements

The demands of the paint processing industry are summarized in Table 9.12. The priorities change individually, according to the given application and assignment. For instance, when a drastic reduction of organic solvents is required, this

Table 9.12 The expectation of the paint user, for applying interior paints

- Process reliability
- Simple process control (cost, time)
- Large 'processing window'
- Lowest possible emission of organic solvents
- Single-layer application on, for example, PP, PBT, PA, and so on
- Universal application of the paint components (complete modules)
- Human safety

will result in the use of waterborne materials. This may result in a significant narrowing of the processing window. In addition to the special requirements expected of the painting installations (paint-conveying elements made of stainless steel, air-stream volumes), waterborne coating materials are far more prone to environment-induced failure, for example, fluctuations of the humidity and temperature during processing.

Paint chemists have a large variety of organic solvents at their disposal, which they can use in connection with conventional, solventborne systems to compensate such variations. With the waterborne coating materials, however, all that is available is water as a solvent, the physical properties of which cannot be changed!

The additional effort and expenses for temperature and humidity control in the painting installation must be taken into account, and compared with the savings potential resulting from simpler fire and explosion protection and thermal incineration.

Table 9.13 gives a greatly simplified survey of the basic property profiles of the material classes – 1-/2-component (1K/2K) and conventional/waterborne coatings. Nowadays, waterborne alternatives with volatile organic compounds (VOC) $<250\,g\,l^{-1}$ are available for the substitution of conventional systems having more than $450\,g\,l^{-1}$. Compared with conventional systems, reliable paint lines for waterborne coating systems require a more complex technology, including distinct air-conditioning, power wash for the pretreatment and paint-conveying pipes, pumps, and walls made of stainless steel.

9.2.4.4 Equipment for the Application of Interior Paint Systems

For the application of automotive interior paint systems, highly automatic equipment based on pneumatic atomization is used. To obtain extremely matt surfaces, the interior paint formulation includes high quantities of matting agents causing thixotropic effects in the rheological behavior of the liquid paint systems. To achieve a homogeneous surface, the atomized paint droplets must be very small and narrowly distributed, which is why for pneumatic application, a nozzle diameter of 1.2–1.4 mm and atomization pressure of 2.5–4.0 bar is advisable.

Over the last five years, tests have been carried out with the view of reducing overspray, and increasing coating efficiency by use of new equipment like the HVLP (high volume low pressure). Depending on the geometries of both the spray gun and the object to be coated, several methods were discovered to reduce the overspray.

Table 9.13 Basic Properties of Different Classes of Interior Paints

Paint system	1K system		2K PUR system	
	Conventional	Waterborne	Conventional	Waterborne
Film properties				
– Chemical resistance	0/+	0/+	++	++
– Mechanical resistance	0/+	0/+	++	++
– Light fastness	+	+	++	++
– Adhesion	++	+	++	++
Economics				
– Material costs (in % relative)	70	80	100	110
– Efforts for the cleaning process (%)	100	110	100	110
– Air-conditioning of the painting installation (%)	100	110–130	100	110–130
– Two pack -installation	–	–	Recommended	Recommended
Process security				
– Tolerance against residual dirt	++	0	++	0
– Stability against variance of relative humidity in the spraybooth	+	0	+	0
– Pot life	–	–	1–4 h	1–4 h
Ecology				
– Emission VOC	–	++	+/0	++

++ = very good, + = good, 0 = neutral, – = poor, – – = bad.

However, the actual benefit during series application, requiring adaptation of both the equipment and the rheology of the paint system, still needs to be proved.

The overspray may further be reduced by use of ESTA. As several institutes discovered, the overspray may be reduced by more than 40%, resulting in a solid-content increase of 60–85%. To practically implement this, the whole application process including frames, paint formulation (according to electrical resistances), voltages, and additional electrodes must be adjusted accordingly. As is known, not every part geometry is suitable for this kind of application technology. Parts creating a Faraday cage are not suited to application of ESTA technology.

To comply with the high demands of the automotive manufacturers' surface specifications, 2-component, polyurethane-based paint systems are most commonly used. Owing to the high content of matting agent, these interior paint systems have a pot life of only 30–60 minutes. For a paint shop to get approved in the automotive sector, the use of a 2-component paint unit to ensure a safe application process for the coating of several thousands of parts per day is indispensable.

Nowadays, various 2-component paint units are available in the market. The efficiency of the mixing process is important, and depends on the technology behind the measuring cells, the mixing geometry and the volume rates (velocity and mixing energy) within the mixing process.

About 10 years ago, when waterborne paint systems were first applied in the car interior, mixing problems occurred while using the waterborne paint in combination with a solventborne hardener. As a result, with the increasing use of waterborne paint systems in the European automotive interior sector, both the paint and the machinery manufacturers managed to adapt to these technologies, so that nowadays safe series application is a matter of course.

Owing to the huge variety of substrates, part geometries, equipment, and, last but not least, paint systems, individual adjustment of the application parameters is required at the start of a new project.

9.2.5
Raw-Material Basis of Interior Paints

The multitude of paint systems used for car interiors makes it impossible to give a detailed coverage of all the raw materials employed. Moreover, there is also the know-how of the individual paint manufacturers that is incorporated in special paint formulations, and this includes polymers (resins), matting agents, cross-linking partners, and additives.

Until six years ago, a single-component top coat system with low solids content was primarily used – especially in the French and Japanese markets. These systems, based on non-cross-linking polyacrylates (solvent level up to 85% >600 g l^{-1} VOC) were applied to cover typical molding faults (flow lines) with a dry film that was only 15–20 µm thick. The mechanical and chemical resistance properties were sufficient for the demands expected of surfaces at that time. However, the demands expected of beauty and haptics could no longer be satisfied with this very hard decorative paint surface. This resulted in the extensive replacement of these single-component systems by 2-component PUR systems.

As already indicated (see Section 9.2.3.1), the requirements of the car's interior from the standpoint of design and technology were subject to a process of constant adaptation that was simultaneously associated with rising expectations to beauty and lasting quality of the surfaces.

These demands could only be fulfilled by the use of soft coatings and top coats based on 2-component PUR systems. Nowadays, a large variety of, in some cases, highly dulled, polyester- and polyacrylate-polyurethane systems are available (see Table 9.14).

A specialized product in this area is the soft coating system. Most of the polyester–polyurethane based, touch-friendly surfaces are produced with relatively low-functional, and therefore highly flexible, basic polymers. Compared with the conventional exterior clear coats, the lower cross-linking density required for soft coating systems resulted in the exclusive use of these systems for the interior. In the meantime, there are a multitude of soft coating systems on the market.

However, it is not possible to generalize the properties with regard to haptics, nor long-term stability vis-à-vis exposure to chemicals and mechanical stress.

Table 9.14 Typical Technical Data of Soft Feel Coatings and Top Coats for Interior Plastic Application

Type of paint	Solid content (1 h/110 °C)	VOC (g l^{-1})	Baking Conditions (min T^{-1})	Film Thickness (μm)
1K solventborne (only top coat)	30%	>600	20/>50°C	20–25
2K solventborne	50%	>450	30/80°C	>35
2K waterborne	40–45%	<250	30/80°C	>35

The stability of soft-coating PUR networks can differ greatly. As a result of the formulation know-how of the paint developer, soft coating systems with a gentle touch have been created with a relatively long lasting chemical resistance of the cured networks.

Nowadays, waterborne 2-component PUR systems are being used for car interiors in Germany with a share of about 65%, throughout Europe approximately 50%, in the United States approximately 10%, and in Asia approximately 3%. The relatively high share of the waterborne systems with the lowest VOC in Europe is primarily a result of the implementation of the 'European VOC Directive' [56] throughout the European Union in the last five years.

The reliable and safe application of waterborne coating materials requires technological framework conditions summarized in Table 9.13. These are being fulfilled as a result of the development work in recent years leading to an expectation of further expansion of this technology.

The continued reduction of emissions from organic solvents has become a permanent research assignment, and is being realized through the consistent further development of high-solid and waterborne paint systems. Waterborne 2K PUR systems have a disadvantage compared with conventional systems owing to required greater technological complexity. This will have to be further optimized by persistent raw-material research and development. The day-to-day expansion of the practical experience gained by the printing workshops and the paint manufacturers in handling waterborne 2-component PUR systems will invariably contribute to the continued stabilization of the processes for safe serial application.

9.2.6
Summary/Outlook

In an environment of increasing expectations of the customer with regard to individuality, the demand for special coating materials will continue to grow. Together with the use of new high-performance substrates as well as the use of low-price plastics, early integration of the paint supplier will produce the best result. Early discovery of special technological performance characteristics of the coating materials have increased efficiency through the use of more cost-effective standard plastics and the use single-application coating materials (process cost reduction).

For the continued reduction of emissions from organic solvents, conventional coating materials are being increasingly replaced, partly or entirely, in Europe by waterborne 2-component PUR paint systems. To increase process reliability in the processing of waterborne paint systems, further optimization of existing raw materials, and concentrated development of new ones, will be necessary in close dialogue between the developers of raw materials and paints.

References

1 Rothbart, F. (**2001**) *Kunststoffe Plast Europe*, **8**, 30.
2 The ChemQuest Group, Inc. (**2003**) *An Overview of the North American Plastic Coatings Market.*
3 The ChemQuest Group, Inc. (**2001**) *Proceedings of SME Finishing Conference*, Chicago.
4 Stracke, K.F., Demant, H.H. (**2002**) *Kunststoffe im Automobilbau – Innovationen und Trends in Kunststoffe im Automobilbau*, VDI-Verlag, Düsseldorf.
5 Slinckx, M. (**2000**) *Proceedings of XXV Fatipec-Congress*, Turino.
6 Klein, G. (**2001**) *Kunststoffe*, **91**(8), 72.
7 Steuer, U., Seufert, M. (**2002**) *Erweiterter Kunststoffeinsatz in der Rohkarosserie – Chancen und Risiken in Kunststoffe im Automobilbau*, VDI-Verlag, Düsseldorf.
8 Wilke, G. (**1999**) *Kunststoffe Plast Europe*, **89**(3), 29.
9 Kettemann, B. (**2002**) *Proceedings of DFO-Tagung "Kunststofflackierung"*, Dresden.
10 Saechtling, H. (**2004**) *Kunststoff-Taschenbuch*, 29 Auflage, Carl Hanser, München Wien.
11 (a) Maß, M. (**2000**) *Metalloberflache*, **54**, 42. (b) Harper, C.A., Petrie, E.M. (**2003**) *Plastic Materials and Processes*, Wiley, Chichester, New York.
12 Hellerich, W., Harsch, G., Haenle, S. (**2004**) *Werkstoff-Führer Kunststoffe*, 9 Auflage, Hanser, München Wien.
13 Meichsner, G. (**2003**) *Lackeigenschaften Messen und Steuern*, Vincentz, Hannover.
14 Goldschmidt, A., Streitberger, H.-J. (**2003**) *BASF-Handbook on Basics of Coating Technology*, Hannover.
15 Kouleshova, I. (**2002**) *Farbe + Lack*, **108**(6), 41.
16 Goldschmidt, A., Streitberger, H.-J. (**2003**) *BASF Handbook on Basics of Coating Technology*, Vincentz, Hannover, p. 290.
17 Blunk, R.H.J., Wilkes, J.O. (**2001**) *Journal of Coatings Technology*, **73**(918), 63.
18 Ryntz, R.R., Ramamurthy, A.C. (**1995**) *Journal of Coatings Technology*, **67**(840), 35.
19 Ma, Y., Farinha, J.P.S., Winnik, A., Yaneff, P.V., Ryntz, R.A. (**2004**) *Macromolecules*, **37**(17), 6544.
20 Ma, Y., Farinha, J.P.S., Winnik, A., Yaneff, P.V., Ryntz, R.A. (**2005**) *JCT Research*, **2**(5), 407.
21 Ryntz, R.R. (**2005**) *JCT Research*, **2**(5), 351.
22 (a) Hanser, W. (**2002**) *Journal für Oberflächentechnik*, **42**(12), 38. (b) Tacke, P. (**2006**) *Journal für Oberflächentechnik*, 24.
23 Osterhold, M., Armbruster, K., Breucker, M. (**1990**) *Farbe + Lack*, **96**, 503.
24 Jänichen, K. (**2003**) *Journal Of Science and Technology*, **17**(12), 1635.
25 Möller, B. (**1999**) *Kunststoffe*, **89**(12), 40.
26 Pfuch, A., Heft, A., Ertel, M., Schimanski, A. (**2006**) *Kunststoffe*, **96**(3), 147.
27 (a) Glausch, R., Kieser, M., Maisch, R., Pfaff, G., Weitzel, J. (**1998**) *Special Effect Pigments*, Vincentz, Hannover. (b) Patent Appl., DE 42 12950 A1.

28 Carlisle, J. (**2002**) *Proceedings of DFO-Tagung "Kunststofflackierung"*, Dresden.
29 Bock, M. (**1999**) *Polyurethane*, Vincentz, Hannover.
30 *Lacksysteme für Kunststoffe*, Company Brochure by Akzo Coatings.
31 Joesel, K. (**2002**) *Coatings World*, **7**(4), 32.
32 US-Patent 5 145 893.
33 Lawniczak, J.E., Williams, K.A., Germinario, L.T. (**2005**) *JCT Research*, **2**(5), 399.
34 Ryntz, R.A., Britz, D., Mihora, D.M., Pierce, R. (**2001**) *Journal of Coatings Technology*, **73**(921), 107.
35 Asbeck, W.A. (**2005**) *Journal of Coatings Technology*, 48.
36 Moore, D.R. (**2001**) *Surface Coatings International Part B: Coatings Transactions*, **84**(B4), 243.
37 Kinloch, A.J., Lau, C.C., Williams, J.G. (**1994**) *International Journal of Fracture*, **66**, 45.
38 Britz, D., Ryntz, R.A., Ardret, V., Yaneff, P.V. (**2006**) *Journal of Coatings Technology*, 3.
39 Ryntz, R.A., Britz, D. (**2002**) *Journal of Coatings Technology*, **74**(925), 77.
40 Jardret, V., Ryntz, R.A. (**2005**) *JCT Research*, **2**(8), 591.
41 Gräwe, R., Schlesing, W., Osterhold, M., Flosbach, C., Adler, H.-J. (**1999**) *European Coatings Journal*, 80.
42 Ryntz, R. (**2005**) *Journal of Coatings Technology*, **2**(18), 30.
43 (a) Wegener, M. (**2001**) *Metalloberfläche*, **55**(7), 25. (b) Kernaghan, S., Stegen, H. (**2002**) *Proceedings of 6. Automobilkreis Special-Fachtagung*, Bad Nauheim.
44 Wilke, G., Meichsner, G. (**2006**) *Kunststoffe*, **96**(3), 142.
45 Fischer, W., Weikard, J.W. (**2001**) *Farbe + Lack*, **107**(3), 120.
46 Fey, T. (**2003**) *Proceedings of Radtech Conference*, Berlin.
47 Döttinger, S. (**2004**) *Proceedings of Api-Tagung*, Ulm.
48 Beck, E. (**2002**) *European Coatings Journal*, **3**(2), 24.
49 Schwalm, R. (**2000**) *Farbe + Lack*, **106**(4), 58.
50 Zukarnain, D., Schlotterbeck, U., Gleich, H. (**2005**) *Kunststoffe*, **95**(3), 123.
51 Warta, H. (**2004**) *Proceedings of DFO-Tagung "Kunststofflackierung"*, Aachen.
52 Grevenstein, A., Kaymak, K. (**2003**) *Kunststoff Plast Europe*, **11**, 39.
53 BASF News Release, Oct. 11 (**2002**).
54 *Modern Paint and Coatings* (**2001**), 27.
55 Jacob, S., Bernau, C. (**2004**) *Journal für Oberflächentechnik*, **44**(3), 30.
56 Drexler, H.J., Snell, J. (**2002**) *European Coatings Journal*, 24.

Further Reading

57 DE-Patent 196 44615 C1.

# 10

Adhesive Bonding – a Universal Joining Technology

Peter W. Merz, Bernd Burchardt and Dobrivoje Jovanovic

10.1
Introduction

Adhesive bonding has become an important joining technology in automotive industry. Since adhesives have many similarities with paints and coatings, like process and adhesion issues, and furthermore are often applied on coating layers, this topic has been included in this book and principles for bonding and sealing in automotive industry are dealt with. The basic application can be allocated to the body shop and trim shop (i.e. final assembly line).

10.2
Fundamentals

10.2.1
Basic Principles of Bonding and Material Performances

The understanding of the basic principles of bonding – where adhesives bond two parts together often on coated surfaces – is a major topic. The key issues are as follows:
 1. types of adhesives
 2. adhesives are process materials
 3. advantages of bonding
 4. application.

10.2.1.1 Types of Adhesives
Adhesives systems are classified into high-structural, flexible-structural, and elastic bonding types. The sealants are presented as the fourth group. Each of these various adhesive and sealant systems has a unique, reversible elastic deformation and shear modulus, as in Figure 10.1.

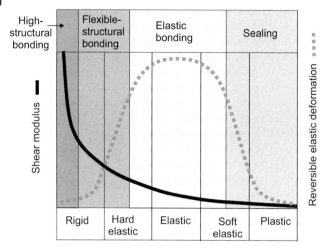

Fig. 10.1 Different adhesive and sealant systems: reversible elastic deformation versus shear modulus (schematic).

10.2.1.1.1 **Definition of Reversible Elastic Deformation** After deformation of an adhesive or a sealant system to a certain degree, the deformation back to the original position is completely reversible, when the system behaves as perfectly elastic. A very high deformation potential with reduced elastic recovery is a typical sealant property.

Figure 10.2 shows the shear modulus and elongation at break of the four adhesive and sealant systems. The shear modulus defines the degree of stiffness or load transfer depending on deformation forces. Also it is the base for FEM (finite-element method) calculations of a bonded structure.

Even stress distribution in the bond-line is a fundamental principle of bonding technology. Figure 10.3 is an example where the load-bearing capabilities of high-structural, flexible-structural, and elastic bonding are compared with respect to the bond overlap. The load-bearing capability is the area below the curve of the stress level.

High-structural bonding adhesives are widely used for structural applications. The adhesives are based on epoxy, acrylic, and polyurethane (PUR) resins and can be formulated as room-temperature two-part or heat-curing one-part adhesive systems. When testing lap shear samples bonded with such high-structural adhesives, high peak stress at the edge of the bond and an uneven stress distribution are observed, leading to a reduced load-bearing capability in the joint.

When the overlap is increased from 12.5 to 25 mm, the load-bearing capability does not increase to the same degree, since practically the inner area of the joint does not contribute to the load transfer (see Figure 10.3). The peak load at the edge limits the strength of bonded joint. This behavior of high peak stress combined with a low 2% elongation results in poor impact resistance.

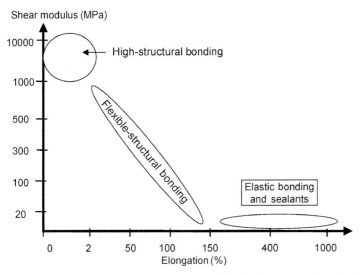

Fig. 10.2 Shear modulus and elongation at break for bonding materials and sealants.

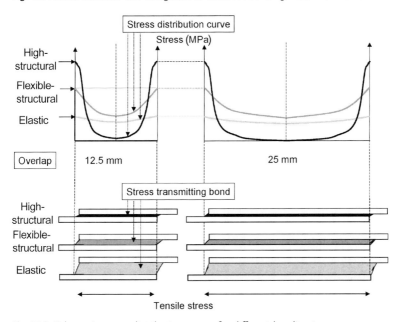

Fig. 10.3 Schematic stress distribution curves for different bonding types.

Elastic bonding exhibits a much lower stress level, but in contrast to high-structural bonding, the stress distribution is rather even. By enlarging the overlap, the load-bearing capability increases almost proportionally. Reactive adhesives for elastic bonding are available as one-part and two-part systems, based on PUR,

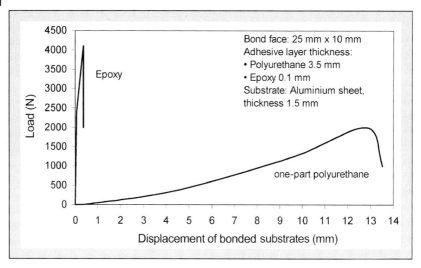

Fig. 10.4 Tensile lap shear strain.

polysulphide, silicone, or silane terminated polymers, such as PUR hybrids and modified silicones. These adhesive systems can be applied cold or warm.

The ability of elastic adhesives to undergo deformation and then recover makes them very forgiving when subjected to sudden stresses or brief periods of overload. In moving vehicles, such stresses may result from vibrations or from sudden impact with an obstacle. Whether an adhesive bond can withstand overloading without damage depends on its strength and, above all, on the fracture energy. This is the energy required to deform the adhesive layer before failure occurs. It is proportional to the area beneath the curve of a graph of the tensile lap shear strain (see Figure 10.4). In this diagram, the thin, rigid bond made with a high-strength epoxy adhesive exhibits very little deformation under high loads. By comparison, the fracture energy required for the elastic PUR adhesive bond is much higher. The result is a significant gain in safety.

The flexible-structural bonding is the combination of high-structural and elastic bonding and provides optimized mechanical properties and maximum possible elongation in combination with structural strength and stiffness. The load transfer of the flexible-structural adhesives is lower than that of high-structural adhesives and higher than that of the elastic adhesives. The elasticity in the flexible-structural adhesives overcomes the disadvantages of high-structural adhesives and they perform better with respect to impact, crash, and durability. Even with an increase of overlap to 25 mm, the load is still transferred to a significant amount even within the joint. The flexible-structural bonding adhesives are in principal formulated as high-structural adhesives; however, they are toughened with a flexibilisator to achieve a higher elongation at break together with a high level of mechanics (see Figure 10.2 and Section 10.3.1.4).

10.2.1.1.2 **Definition of Structural Bonding** There have been many attempts at defining structural bonding, but till date there has been no general accepted definition. The following definition reflects our present knowledge and understanding of structural bonding: structural bonding means the joining of substrates, which can transfer loads over the expected lifetime and under all expected conditions. This can be calculated and the durability can be predicted. As per our experience, shear modulus is required for this calculation instead of lap shear strength. Lap shear strength is dependent on the geometry of the samples and the properties of the substrate, whereas shear modulus is an intrinsic property of the adhesive.

10.2.1.2 **Adhesives are Process Materials**

Like paints and coatings, adhesives are process materials. This means that the final product is created during the production process. The parameters of this process are the key elements of the final product. The specific product design, the right process, and correct application are the preconditions for the quality of the final product.

Correct joint design is crucial and needs to take into account the specific mechanical properties of the substrates and the adhesives used. Simply substituting adhesives for rigid fastening – such as a riveted joint – will not provide the desired result. To find the best joint design, several disciplines must work together. Cooperation between chemists, mechanical engineers, and process engineers is fundamental. It is important to discuss the requirements, design, and other relevant issues such as aging behavior in the design phase.

The chemist needs to have knowledge about the basic chemistry of reactive systems and must design and formulate the adhesive according to identified needs. Also, surface preparation is usually a decisive factor.

Mechanical engineers will design the joint, calculate the loads to be transferred, and predict durability and strength through calculation.

The application and process engineers have to pretreat the surface of substrates in the right way, apply the adhesive to the right place and in the right amount, correctly cure the adhesive, and therefore control almost the whole process (see Figure 10.5).

All system elements are of equal importance to create the best bonding solution. This comprehensive approach with process materials enables the successful use of bonding and brings all the benefits of a bonded structure.

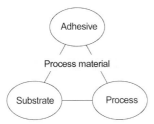

Fig. 10.5 Elements of a process material.

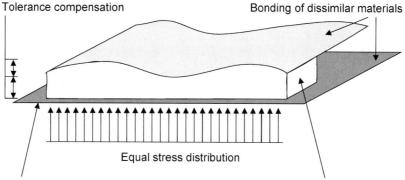

Fig. 10.6 Key performances of bonding.

10.2.1.3 Advantages of Bonding

Bonding is a modern and highly effective joining technique with a number of innovative performance characteristics, which is a welcome addition to the standard repertoire of mechanical assembling technologies. Bonding provides economical benefits. The key advantages of bonding are as follows (see Figure 10.6):

- bonding and sealing in one operation
- tolerance compensation
- bonding of dissimilar materials
- even stress distribution
- no thermal distortion
- corrosion protection
- improved acoustics
- freedom of design.

Bonding and Sealing in One Operation
  Because elastic adhesives also act as sealants, they offer a relatively simple but effective way to protect a joint against the ingress of gas or water.
Tolerance Compensation
  Manufacturing tolerances for components in the transport industries are usually measured in millimeters and can exceed 1 cm. Bonding technology allows manufacturers to bridge gaps of this order without loss of strength. The elastic bond-line, that is the interfacial layer of permanent elastic adhesive, should be, in general, minimally 3 mm thick. For example, at the direct glazing the tolerance is up to 2 mm.
Bonding of Dissimilar Materials (see Figure 10.9)
  The need to join different materials is often associated with the lightweight designs of cars, trucks, trains, and their components. Synthetic materials and plastics, including fiber-reinforced or composites, are also being increasingly used in lightweight construction. These adhesive-bonded body panels contribute directly to the structural strength of the vehicle, reduce the combined weight of a structural

frame and a nonload-bearing sheet metal skin, increase torsional stiffness, and allow weight saving in substrates, which can be thinner or lighter.

Even Stress Distribution

To ensure a durable connection and maximize the service life of the materials, an even distribution of stresses throughout the assembly is essential. Conventional joining methods such as bonding with rigid adhesives, welding, riveting, screwing, or bolting cause localized peak stresses at the joint itself (see Figures 10.10 and 10.24).

The distribution of stresses can be revealed in photo-optical models of joint assemblies made from a transparent material that becomes doubly refractive when subjected to stress.

Figure 10.7 shows a thin- and a thick-layer rigid adhesive bond. The expansion and deflection of the bonded substrates causes peak stress at the ends of the overlaps, which is the reason why the adhesive layer begins to break at this point. The central area of the bond layer, on the other hand, contributes very little to the load-bearing capacity of the joint.

Figure 10.8 shows a thick-layer elastic adhesive bond. Here, the stresses in the bonded substrates are uniformly distributed along the bond-line, indicating that the whole area of the bond-line contributes to the strength of the joint. Elastic-bonded joints can transmit relatively large forces simply by increasing the area of the bond face (length of overlap).

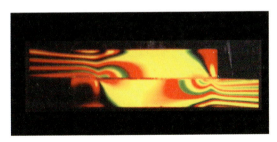

(a)

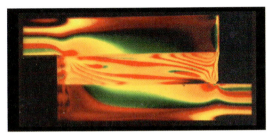

(b)

Fig. 10.7 Photographies of stress pattern of a thin-layer (a) and a thick-layer (b) rigid adhesive bond.

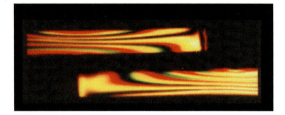

Fig. 10.8 Photograph of stress pattern of a thick-layer elastic adhesive bond.

No thermal distortion

Joining methods, like welding or soldering, involve the application of heat to the components. This can cause thermal distortion and lead to deformation and change of the material's internal structure. Correcting such damage is usually very costly and labor-intensive. With adhesives this corrective work is generally not needed.

Corrosion Protection

The causes for corrosion are complex. In case of chemical resistance, adhesive bonding has been shown to offer better protection against service corrosion and galvanic corrosion than many other mechanical assembling techniques. However, this presupposes the use of adhesives that are themselves effective electrical insulators when bonding dissimilar materials like steel and zinc to avoid the generation of galvanic elements (see Figure 10.9). When metals are bonded with adhesive, water has to be kept away from the adhesively bonded parts, thus preventing corrosion.

Improved Acoustics

Personal comfort can be improved when using bonding techniques. Highly effective noise and vibration damping can be engineered into the products. In addition, owing to higher structural stiffness, the resonance frequency is shifted toward higher wavelengths.

Freedom of Design

Contrary to mechanical fastening, bonded parts do not generally deform the substrates and enable aesthetically improved surfaces. Bonding with adhesives introduces little or no local stresses. This is an advantage for thin metal sheets, for lightweight metals like aluminum, plastics, and composites (see Figure 10.10).

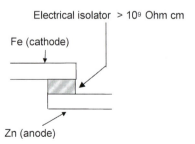

Fig. 10.9 Bonding of dissimilar metals to avoid the generation of galvanic elements.

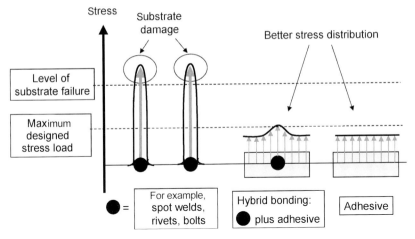

Fig. 10.10 Equal stress distribution with flexible-structural adhesives in contrast to mechanical fastenings.

The maximum design stress load is much higher when a flexible-structural adhesive is used. Owing to the equal stress distribution realized by the adhesives, a better fatigue resistance and a longer lifetime are achieved in comparison to mechanical fastening.

10.2.1.4 Application
Adhesives in automotive production are applied in various forms depending on the design and process parameters:
- bead, triangular, or round bead (Figure 10.11a);
- multidots (Figure 10.11a);
- swirl spray, ultrathin streams of adhesive: as an example, in the body shop, hem-flange adhesives are applied automatically by use of a robot (Figure 10.11b).

10.2.2
Surface Preparation

One of the most important factors in adhesive bonding is the surface condition of the substrate, that is, the surfaces of the materials to be joined. Since adhesion takes place only at the interface between the component and the adhesive, it is evident that surface preparation is of crucial importance for the quality of the adhesive bond.

Surface preparation includes various types of pretreatments:
- simple cleaning of the surface to be bonded by use of a liquid system;
- mechanical abrasion;
- thermal oxidation processes such as flame treatment;

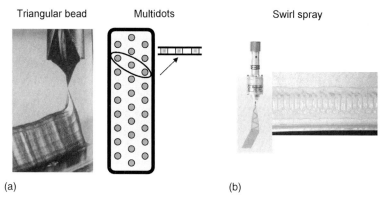

Fig. 10.11 (a) Extrusion application and (b) swirl application.

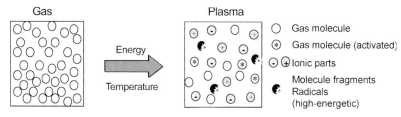

Fig. 10.12 Plasma consists of different species of gas (source: PlasmaTreat GmbH).

- specialized physical–chemical techniques, such as corona, low-pressure plasma treatment, and open-air plasma treatment (see Figure 10.12);
- etching, for example, with phosphorus acids;
- adhesion promoters, activators;
- primers.

The targets of pretreatments are as follows:
- to clean the surface;
- to flatten the topography;
- to introduce interacting groups, for example, polar groups, adhesion promoters;
- to harden and reinforce the surface;
- to provide UV protection, for example, black primers;
- to improve corrosion protection.

10.2.2.0.1 **Example, Open-Air Plasma Treatment** The PlasmaTreat process introduces active carboxylate, ether, ketone, or other oxygen-containing polar groups on the plastic surface without the use of a primer. For example, treating of polypropylene with plasma increases the surface tension from 30 to 60 mN m^{-1}. The substrate to be bonded is briefly passed under a potential-free plasma flame at atmospheric pressure, thus providing good wettability and allowing application of the adhesive immediately. No electric arc is created, which would damage parts as the ions generated by the plasma are filtered off at the plasma jet nozzle. Thus, the potential-free plasma jet can be used on or around metals.

10.2.2.1 Substrates

With adhesive bonding, a large variety of substrates can be joined, such as glass, ceramics, plastics, fiber-reinforced plastics, metals, wood, and paints. But for each substrate, it is necessary to know the real status of the surface. Release agents and specific lubricants may destroy the adhesion.

10.2.2.2 Adhesion

For durable bonding, the substrate must be free from surface contaminations, like oil and dirt. However, adhesion can also be achieved on an oily surface.

The automotive industry is known for bonding on oily or even dirty surfaces. They have realized that a steel surface with no oil protection usually corrodes very quickly – thus producing a loose iron oxide layer on the surface, which is extremely prone to further corrosion and poor adhesion. In the automotive industry paint process, one-part systems are used, which cure at temperatures of about 180°C. At this temperature, the viscosity of the adhesive drops to a lower level, thus leading to better wetting and the absorption of a definite amount of oil. Thus, excellent adhesion can be obtained. Of course, the compatibility of adhesive and oil must be checked initially. This process eliminates the need for pretreatment of the metallic sheets and reduces the process costs significantly; see Figure 10.13.

Contrary to high-temperature (high-bake) curing adhesives, cold-curing adhesives need pretreated metals. When using a cold-curing two-part system without curing at elevated temperature, the adhesive's oil absorption ability is limited because the diffusion at room temperature is very slow compared to heat-curing systems. Additionally, the fast curing speed of two-part adhesives reduces the time available for oil absorption and wetting may not be sufficient, resulting in poor adhesion.

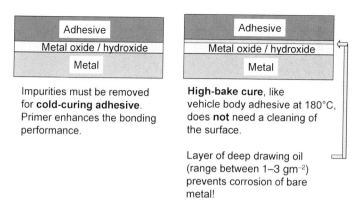

Fig. 10.13 Difference between cold- and high-bake curing adhesives.

10.2.2.3 Durability and Aging of Bonded System

Aging performance is determined using severe accelerated testing procedures, where a combination of temperature, humidity, and salt influence are tested. The test conditions must be set up in such a way that the real aging process is simulated in a short-term test. In these accelerated aging tests, there should be no conditions that damage the adhesive in an unrealistic manner. Similar to the tests of paint and coatings, every car manufacturer has its own preferred test cycle based on its tradition and its experience.

According to the experience of vehicle manufacturers and also that of the authors, the adhesion of elastic bonding adhesives and sealants is generally best proved by the peeling test of a bead on which aging is carried out.

In practice, a bead is applied to the substrate and a sequence of aging storage conditions are used, where after each storage test the bead is peeled off from the substrates (see Figure 10.14). With a gripper the bead is clamped and peeled vertical to the horizontal lying substrate by cutting the peeled bead near to the substrate with a knife.

Typical aging tests follow the sequence of testing conditions outlined below:
- 7 days storage at RT;
- the test specimens are placed in water for 7 days;
- the test specimens are stored at 70°C and 100% relative humidity, which is called *cataplasma testing*;
- the adhesive is exposed for 1 day to heat at 80°C and up to 5000 hours of UV radiation.

The requirements are that the peeling tests should result in more than 75% cohesion failure.

Even if no values for peel strength are measured, the peeling test shows very sensitive adhesion or cohesion failure, which is sometimes the only criterion to be

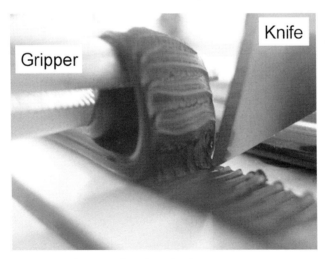

Fig. 10.14 Peeling test of an elastic bead.

checked. However, the mechanical values of the adhesive can be determined by other methods. If cohesion failure occurs, the result is generally good. If there is no influence of humidity and temperature on the adhesion, this peeling test will usually show the same.

Numerous cyclic aging tests are defined for testing, in particular, adhesion on metallic surfaces. In general, structural and flexible-structural adhesives are tested with these cyclic aging tests. A selection of cyclic aging tests are illustrated below. At present, some European vehicle manufacturers use the VDA(Verband der Deutschen Automobilindustrie) cyclic aging test (see Figure 10.15) for testing adhesion, for example, on steel and galvanized steel.

Some car manufacturers, like the Volkswagen Group, apply a similar, but more severe, test to the above-described VDA cyclic aging test, the P1210 cyclic aging test (see Figure 10.16).

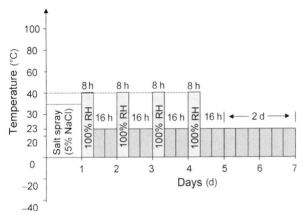

Fig. 10.15 VDA cyclic aging test (10 weeks).

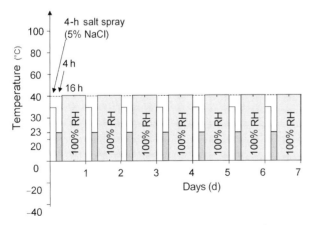

Fig. 10.16 P1210 cyclic aging test of Volkswagen (3 months).

364 | *10 Adhesive Bonding – a Universal Joining Technology*

An even more demanding cyclic aging test is a combination where P1210 cyclic aging test with a duration of 3 weeks is followed by 1 week of P1200 cyclic aging test, as in Figure 10.17. This combination is repeated three times.

Adhesion performance of aluminum is tested for example with the SCAB cyclic aging test GM SCAB-test, a test creating a scabby surface on aluminum; see Figure 10.18).

The performance established by these aging tests is normally rated by the type/quality of the adhesion failure and the percentage retention of strength. The

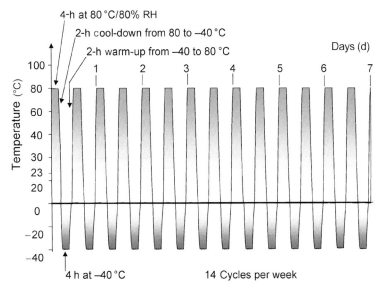

Fig. 10.17 P1200 cyclic aging test (1 week).

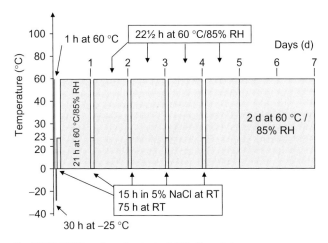

Fig. 10.18 SCAB cyclic aging test of GM (8 weeks).

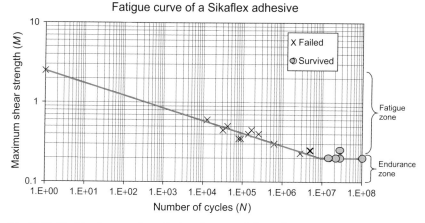

Fig. 10.19 Woehler diagram of an elastic adhesive (Sikaflex, Sika).

requirements are generally that less than 5% of adhesion failure and more than 80% of retention of strength are required.

To establish the expected service life under a certain load, additionally the fatigue strength is tested. This is depicted as a cyclic load (*Woehler*) diagram where the number of load cycles is plotted against the load level on the ordinate or Y axis.

The fatigue behavior of an elastic adhesive is shown in Figure 10.19. The adhesive endurance is optimal below a shear stress of 0.2 MPa.

Figure 10.20 compares the various types of adhesives with respect to fatigue. Single lap shear specimens were prepared and aged in a salt spray over 480 hours. After this, the fatigue strength of these aged specimens was determined. SikaPower, a vehicle body adhesive, retains most of its maximum load carrying

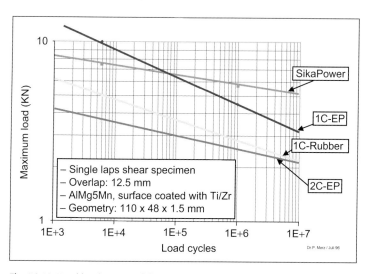

Fig. 10.20 Woehler diagram of flexible-structural adhesive.

Table 10.1 Factors Affecting the Quality of an Adhesive Bond

Influencing factor	Quality measures
Adhesive	Selected to suit the production process and the service requirements to which the finished assembly will be subjected
Substrate	Consistency of composition and surface condition
Surface preparation	Selected to suit the production process and the service requirements to which the finished assembly will be subjected
Application parameters	Working within the process parameters (temperature, cycle time, humidity, etc.)
Joint design	Adhesive-specific joint design, dimensioning of joints to suit functional requirements of finished assembly
Staff training	External or internal courses organized in conjunction with adhesive suppliers

ability and therefore has good durability even after loading cycles at a high percentage of maximum load. This examples show that a flexible-structural adhesive with a lower initial load-bearing capacity has a much better long term durability (logarithmic scale) in comparison with rigid adhesives such as one-component epoxy, one-component rubber, and two-component epoxy adhesives.

10.2.2.3.1 **Principles of Joint Design** In all the aging tests, water generally has the most negative influence on the adhesive properties. At elevated temperatures, hydrolysis reactions can take place, leading to a chemical deterioration of the adhesive matrix structure.

Since water has great potential to damage the adhesive, it is imperative to design and construct joints in such a way as to avoid permanent standing water on the adhesive, for example, with drain holes or perfect sealing of cavities.

10.2.2.3.2 **Factors Contributing to the Final Quality of the Bond** A typical quality management program for adhesive applications is set out in Table 10.1, pointing out, that the task of assuring the quality of the bonded assembly begins at the project stage and does not end until production ceases. This model has been adopted with highly satisfactory results in many areas of the manufacturing industry.

10.3
Bonding in Car Production

The automotive industry produces cars in three major steps:
1. body shop
2. paint shop
3. trim shop (= final assembly line).

10.3.1
Body Shop Bonding

The types of adhesives used here are heat-curable and used in the automotive body shop, which is, in vehicle production, the first assembly step for metal structure of a car. The adhesive contains a hardener, which is activated upon thermal curing and reacts then with the epoxy resin to give the final product.

Such body shop adhesives are applied on oily metallic surface. Compared to paint process, it is surprising that the preparation for a good adhesion of paints needs such a costly process, whereas adhesives afford only a temperature curing.

In preparing the surfaces for the paint process, the assembled car body with hang-on parts is pretreated in a bath where a washing process degreases the surfaces, followed by a phosphate treatment and an electro-deposition coating. To exclude wash-out and soiling of the pretreatment baths, the adhesives and sealants have to be precured by heat (oven), induction, UV radiation, moisture, or use of a two-component system. Depending on the pretreatment conditions, some adhesives can even be used without any precuring.

After pretreatment and electro-deposition process, the curing of the adhesive is completed in the electrocoat oven at a temperature of 160–180°C.

SikaPower, a family of epoxy/PUR-hybrid systems, is a typical product used. A complete family of adhesives is available, ranging from seam sealant and antiflutter adhesives to structural and crash-resistant car body adhesives (see Figure 10.21).

In Figure 10.22, the correlation between strength, modulus, and elongation is shown and it demonstrates that in automotive body shop a wide range of mechanical properties are needed to fulfill the high demands of automotive industry.

10.3.1.1 Antiflutter Adhesives

When the outer skin of the structure is assembled with an inner part, the adhesive bonds the two parts together and adds to the stiffness of the system. But in any case, there should be no read-through, that is, deformation on the outer skin at the area of adhesive after curing, which sometimes limits the strength of the adhesive. The tendency of marking or read-through increases with reduced metal sheet thickness and increased adhesive modulus.

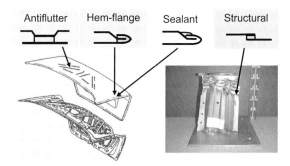

Fig. 10.21 Application of SikaPower adhesives.

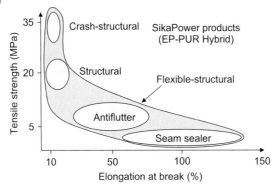

Fig. 10.22 Overview and positioning of SikaPower adhesives in the tensile strength versus elongation at break.

Adhesives for antiflutter applications must be nonsag up to a bead height of 8–15 mm in order to bridge broad gap.

Antiflutter bonding contributes significantly to the stiffness of the structure. Calculations using FEM models confirm that the shear modulus of the adhesive has an influence on the stiffness. A study revealed that a shear modulus of about 100 MPa was necessary to realize nearly the maximum increase of bending strength. The consequence is that even lower modulus adhesives, such as those used for antiflutter bonding, contribute significantly to the stiffness of such structures.

Owing to this low modulus, other important effects like read-through can be avoided and good optical properties are achieved.

10.3.1.2 Hem-Flange Bonding

In the recent years, there has been a felt need for additional mechanical properties of hem-flange adhesives. Some automotive manufacturers are using adhesives with crash-resistance properties because doors have an important function in side impact safety. Therefore, the inner and outer parts of the door must hold together and must undergo deformation without disbonding to play its part in case of a side impact. The process of hem-flange bonding is very tricky. On the one side, all the area in the flange should be bonded to have maximum strength but no corrosion but during production, the hem-flange tools should not be contaminated with the adhesive. Figure 10.23 shows an incompletely filled joint. Therefore, the hem-flange seal encloses air bubbles, resulting in poor appearance and reduced corrosion protection.

Another important issue is air bubbles in parts that are precured and then shipped to other assembly plants. This is caused by water absorption from air humidity or from the pretreatment steps. During the final baking, the water evaporates, forms undesired bubbles and damages the desired smooth surface. One way to minimize air enclosures is to heat up the hem-flange to more than 100°C. This elevated temperature of the hem-flange reduces the expansion of air, thereby avoiding penetration into the sealant and damage of adhesive surface.

Hem-flange sealing is usually carried out in the paint shop (see Section 10.3.2).

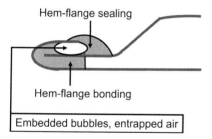

Fig. 10.23 Incomplete filling of hem-flange bonding.

10.3.1.3 Spot-Weld Bonding

In the production of vehicles, parts, and equipment, high performance adhesives are used in combination with the traditional joining techniques, like spot welding, self-piercing, riveting, or clinching. This leads to stiffer and more durable car structures, which allows additional weight reduction.

Figure 10.24 describes the application of structural and crash-resistant bonding systems.

10.3.1.4 Crash-Resistant Adhesives/Bonding

Traditional epoxy adhesives have a high mechanical strength, in particular, high tensile and lap shear strengths. But in case of impact, the bonding with a structural epoxy adhesive appears, in general, to be very brittle and exhibits low toughness. Therefore, it cannot fulfill the requirement of impact resistance in a car crash where high tensile, peel, and lap shear stresses occur at high speed. To improve the toughness of the epoxy adhesive for high impact loads, a different approach with specific molecular designs was necessary.

Figure 10.25 shows the differences between specific adhesive formulations upon rupture in a lap shear specimen. The crash-resistant formula breaks in a corrugated way, whereby the tear propagation stops and starts again many times before final rupture. The unproved product shows a more linear cohesive failure.

The pictures of engine support beams damaged in crash tests in Figure 10.26 point out important differences in crash behavior. The joint opens and is therefore not crash-resistant with SikaPower-490. For SikaPower-496, the engine support beam is folded without the joint opening.

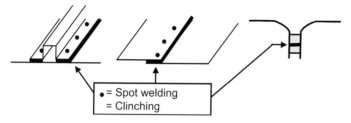

Fig. 10.24 Applications of spot welding in the car body.

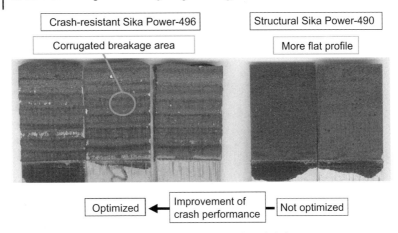

Fig. 10.25 Impact peel test representing high speed crash behavior.

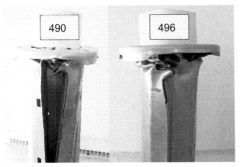

Thickness of metal sheets is about 1mm

Fig. 10.26 Crash behavior tested in engine support beams.

10.3.2
Paint Shop

The automotive industry wants to move the sealing operation from the paint shop to the body shop. This will facilitate automation and help keep the paint shop clean. Polyvinyl chloride (PVC) plastisol is widely used in sealing and is applied manually for sealing the joints of, for example, hem-flanges (see Figure 10.23). The thickness of PVC plastisol is in the range of 2–4 mm, whereas body shop sealants are required to be of maximum thickness of 2 mm. Corrosion protection properties are considerably improved compared to PVC plastisol. This reduction in bond-line thickness partly compensates for higher cost of the epoxy-containing sealants compared to PVC plastisol, which is much less expensive. For under body protection, PVC plastisol is sprayed.

Since the sealant application is being moved to the body shop, the applied sealants must withstand the pretreatment steps and should not take up moisture in the uncured state, which would lead to formation of air bubbles upon high-bake curing in the electrocoat oven. All these aspects can be addressed if the sealant is partially precured, that is, it is not necessary to have it completely cured.

10.3.3
Trim Shop

10.3.3.1 Special Aspects of Structural Bonding in the Trim Shop

At this step of car manufacturing process, the metal surfaces are painted. But the clear coat, which is the last layer of a typical top coat (see Chapter 6), cannot sustain high-structural forces and providing a strong bonded joint. A predetermined breakage zone between the clear coat and the base coat results in a reduction of bonding strength. The reason behind is that, in case of gravels hitting the painted surface, the clear coat is damaged, but the primer surfacer or the base coat below the clear coat remains intact and the color is therefore not changed (see a in Figure 10.27).

The modulus of windshield adhesives must be below the adhesion strength of the paint system. When using adhesives of higher strength with a shear modulus of more than 2.5 MPa according to DIN 54451, the paint will become the limiting factor and other measures have to be taken to improve the strength level and adhesion.

10.3.3.2 'Direct Glazing'

In 1964, the US automotive industry had to fulfill safety standards that required the retention of the windshield in the event of a car crash. The Federal Motor Vehicle Safety Standards (FMVSS) 212/208 regulations specifies that, when a vehicle driving at a speed of 48 km per hour crashes against a barrier, 75% or more of the bond-line must be retained and not more than 50% of the adhesion on one side should be lost.

The windshield bonding technology involved has since been called 'Direct Glazing'. At that time, this was achieved by the use of a material that was very soft like a sealant, but cured to an elastomeric adhesive and exhibited good adhesion on

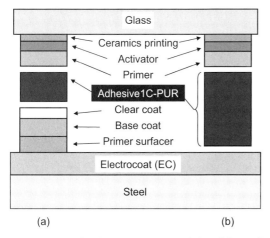

Fig. 10.27 Two bonding versions (a) and (b) of direct glazing.

the glass and steel substrates after appropriate surface preparation. Thus, the first elastic bonding adhesive was used. But in the opinion of many experts, this was a sealant and not really an adhesive.

There are two methods of bonding the windshield into the frame: it can be done either on the clear coat (see Figure 10.27a) or on the electro-coated (EC) steel (see Figure 10.27b).

Version (a) in general needs no pretreatment or in case adhesion is not proper, a pretreatment with a paint primer is required. For version (b) a sprayable masking material or tapes, which are removed before the direct glazing, protects the EC steel against the following paint applications and in general no pretreatment is required.

Since every year about 5% of car windshields are damaged, efficient and reliable replacement methods are of increasing importance. During repair, the windshield is taken out by cutting the adhesive using a vibration-cutting tool or a cutting edge wire. On this precut adhesive layer, the new windshield with the uncured adhesive along the side is bonded.

To achieve a fast 'fit-for-use' of vehicle after windshield repair, a short safe-drive-away time (SDAT) is requested. This needs specific formulations, since a moisture-curing system alone cannot build up enough strength within 1 hour to meet the extremely strict requirements mandated by the above-described FMVSS regulation.

Direct glazing adhesives are multifunctional and can cope with several demanding specifications, such as nonsagging, cutoff string, strength buildup, crash-resistant at SDAT, corrosion resistance, mechanical performance, antenna suitable, and so forth. In a standard climate (23°C/50% RH), the buildup of the adhesive strength varies between 1 and 4 hours.

10.3.3.3 Modular Design

Today, the cars at the assembly line are manufactured not with thousands of parts, but increasingly with modules as shown in Figure 10.28.

Figure 10.29 shows a roof module, where a glass panel is bonded to an aluminum frame. In the trim shop this prefabricated roof module is bonded to the cross beam of the basic module. Preferably elastic adhesives can be used for these two applications.

Another example of modular design using bonding techniques is the construction of modern trains, subways, and trams (see Figures 10.30 and 10.31).

Modern trains owe their stylish looks to a combination of glass, plastics, and lightweight metals bonded together with adhesives. The front cabin of a regional tram is bonded to the train structure and the screens are glazed directly into the front cabin.

10.3.3.4 Other Trim Part Bondings

Another application in the final assembly line of cars and trucks is the bonding of trim parts to the vehicle, for example, to the door of the vehicle. These parts are normally plastics, which must be generally cleaned with a solvent or pretreated with an activator to improve adhesion. The door does not require pretreatment,

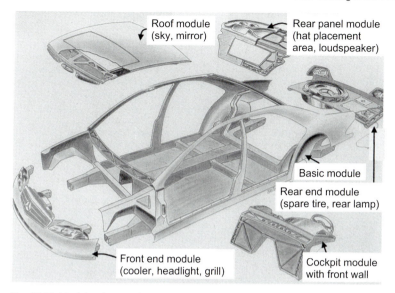

Fig. 10.28 Modular design of a car (source: DaimlerChrysler).

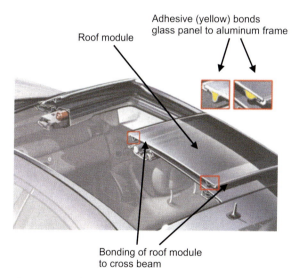

Fig. 10.29 Roof module of a car using adhesive (source: DaimlerChrysler, Sika).

since the paint earlier applied in the paint shop provides a suitable surface for further bonding in the trim shop.

Since adhesives must cure before they build up their strength, a fixation during the curing time is required to hold the parts in place. This can either be done by mechanical fastening, with the help of a double-sided adhesive tape, or with warm or hot-melt type of adhesives, which cool down immediately after application and are able to fix the parts almost immediately.

Fig. 10.30 Bonding of a front cabin to a train.

Fig. 10.31 Glazing of front cabin of a tram.

10.4
Summary

Today, bonding, alone and in combination with mechanical fastening technologies, is fully accepted by the automotive OEM (original equipment manufacturer). The benefits of bonding technology can be reaped best when the adhesive manufacturer gets the opportunity to fine-tune the adhesive to the application technique used as well as to the optimal joint design.

Also careful definition of the durability requirements, substrate selection, and process control will help to ensure that the final assembly fulfills the requirements. A number of products from one-part to two-part systems, cold-curing, and high-bake based on different resins are available.

The mechanical behaviors of such adhesives can be predicted by means of FEM calculations, for which shear modulus, sustainable load, creep behavior, elasticity, and so on, are the basic parameters.

Whereas bonding directly on metal surfaces provides the full strength and load transfer capability of a structural adhesive, bonding on paint surfaces is more critical, since the adhesion and mechanical properties of paint are the limiting factors of such a bond design.

In general, it can be said that the automotive industry is the driving force for the development of new adhesive systems and other industries are followers.

11
In-plant Repairs
Karl-Friedrich Dössel

More than 80% of all new cars shipped to the customer have at least one repair done in one of the layers of the total coating. The inability of the original equipment manufacturer (OEM) paint shop to coat the $10\,m^2$ outer surface of a car with four layers of paint without any defect leads to a situation where most new cars need to be repaired even during the manufacturing process. In this respect, repairing involves sanding, cleaning, polishing, and, in severe cases, total or local paint application. Repairs may be carried out after the main steps of the coating process in four stages:

- sanding and cleaning after electrocoating
- sanding and cleaning after primer surfacer
- sanding, polishing, and paint repair after top coat
- repair after the assembly line (end of line repair).

11.1
Repair After Pretreatment and Electrocoat Application

Most paint shops have a sanding area where the car body is brought after electrocoating. With the first coating layer, some of the defects from the body shop, for example, rough sanding marks or dents in the chassis, become visible. Defects may also be caused during the pretreatment or e-coat process, for example, rough crystallite structure, dirt, or paint sags. In the sanding area, tinsmiths remove the dents and sanding is used to remove the other defects. When sanding goes down to the metal, an epoxy-based flash primer is used to restore corrosion protection in these areas.

The best paint shops have eliminated all repairs in this area (no touch), mainly by eliminating body defects in the body shop and by improved application processes during pretreatment and electrocoating (see Chapters 2 and 3).

11.2
Repair After the Primer Surfacer Process

Most paint shops have a sanding process after the primer surfacer process. Mainly dirt and sometimes sanding marks are sanded to give a defect-free surface before the top-coat process. In cases where sanding goes down to the metal, the epoxy-based flash primer is used. Before the car body enters the top-coat section, great care has to be taken to remove particles deposited in the sanding process. Some car makers have installed power wash to remove all particles, while others use blow off and emu brushes.

With the introduction of wet-on-wet-on-wet or primerless processes, sanding at this stage in the coating process has been eliminated. Top-coat preparation is then moved to the e-coat sanding area.

11.3
Top-Coat Repairs

After leaving the paint shop and before entering the assembly line, the surface of each car is carefully inspected. About 6 to 12 specialists minutely examine the surface using special light sources (see Figure 11.1). Many of the defects visible in the top coat (base coat–clear coat) layer can be removed by sanding the clear coat and polishing it, in which case, no additional coating step is required. This is normally done in about 3 to 5 minutes. To remove the remaining defects, car makers have adopted different strategies. The most common method used today is spot repair [1]. The car is moved into a repair cabin, where the defects are removed by sanding the defective spot of 1–10 cm diameter, masking around the spot, and then applying the base coat–clear coat to that small area. This is done by small spray equipment like airbrush guns and needs some skill on the part of the workers [2]. Most car makers apply the same base coat as used in the main line for this type of repair. 2K PUR clear coat is used in most cases, though some paintshops use catalyzed versions of their 1K line material. Curing is normally done using IR lamps at 80–120 °C for 20 minutes. The overall process takes about 1 hour (see Figure 11.2).

Larger defects as well as too many defects are handled by repairing the particular panel, that is, panel repair. In this case, a larger segment of the car, for example, the hood or the fender, is sanded to remove the defects, and the rest of the car is kept masked as before. The whole car is then put back into the normal top-coat line. These 'reruns' reduce the capacity of the total paint shop and typically account for a maximum of 5% of the total production. Many avoid the manual masking–unmasking operation and would rather have the whole car body resprayed.

Instead of panel repair, some opt for the concept of 'exchange parts'. Parts with coating defects, for example, doors, hoods, or fenders, are removed from the car body and replaced with nondefective parts. The defective parts are sanded to

Fig. 11.1 Inspection zone of an automotive coating line (source: BASF).

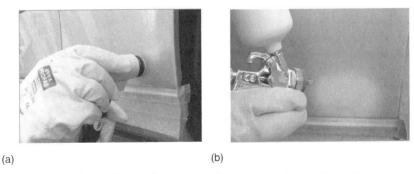

(a)   (b)

Fig. 11.2 (a) Sanding and (b) application step of a spot repair (source: Dupont).

remove the defects, placed on a spare part skid, and rerun through the main line. This can easily be done in the case of the space frame concept (see Chapter 1).

The best car makers achieve a 'no touch' rate of 75%, that is, 75% of the car bodies are handed over to the assembly line without any polishing or repairs in the top coat, 20% need spot or panel repair, and 5% would require total repair.

11.4
End-of-Line Repairs

Defects or damages created or detected after assembly of the car have to be repaired with a low bake process. As these conditions are very similar to field (collision) repairs, many car makers use the standard field repair materials. In some cases, the base coats are taken from the OEM line. Cabins similar to those designed for the spot repair process after the painting line are available for this task.

References

1 *Fahrzeug+Karosserie* (**2004**), **57**(1), 23.

2 *Besser Lackieren* (**2005**), **7**(3), 9.

12
Specifications and Testing

12.1
Color and Appearance
Gabi Kiegle-Böckler

The exterior paint finish and the interior of a car play a major role in how customers perceive its quality. The two main requirements it is supposed to fulfill are to protect the surface underneath and to enhance the value of the overall product. Eye-catching finishes should have a beautiful and rich color, and look like a mirror 'high gloss and perfectly smooth' – and of course, not diminish over time [1].

There are limitations coming from the paint properties, process capability, and most importantly the amount of money that can be spent to improve the finish [2]. As a result, each automotive company defines its own color and appearance standards, which are to meet or exceed the level shown by competitors and customer expectations. Particularly important is the uniformity or also called *harmony among all components*. In case of the exterior finish, it is critical that add-on parts such as bumpers, spoilers, mirror housings, or other decorative trim parts match the adjacent body panels (see Figure 12.1). Differences in color and appearance are especially obvious on models with panels having very tight fits to each other.

12.1.1
Visual Evaluation of Appearance

To visually describe the appearance of a class A automotive finish, a variety of terms are used, for example, glossy, brilliant, dull, wet look, snap, orange peel, or smooth. Like color, appearance also needs to be objectively measurable as it is dependent on the viewing conditions like light source and focus, the surface condition like material and surface structure, and the observer himself. Depending on the viewing conditions, focus of our eye is on the surface of the material or the image of the object reflected by the material (see Figure 12.2).

By focusing on the reflected image of an object, an observer obtains information on the image forming capabilities of the surface. A reflected light source may appear

Automotive Paints and Coatings. Edited by H.-J. Streitberger and K.-F. Dössel
Copyright © 2008 WILEY-VCH Verlag GmbH & Co. KGaA, Weinheim
ISBN: 978-3-527-30971-9

Fig. 12.1 Harmony among different body panels is an essential quality criterion.

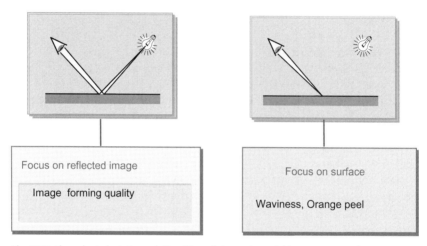

Fig. 12.2 Flow chart depicting relationships of the main variable appearance characteristics.

brilliant or diffuse and its outlines may appear distinct or blurred. Focusing on the surface of an object provides information about the size, depth, and shape of surface structures, which contribute to things such as waviness or directionality of brush marks. However, experience has shown that no single objective measurement of gloss will provide perfect correlation with the integrated subjective appraisal of glossiness that the eye so quickly renders.

12.1.1.1 Specular Gloss Measurement

Specular gloss is defined in ASTM E 284 [3] as the 'ratio of flux reflected in specular direction to incident flux for a specified angle of incidence and source and receptor angular apertures' (Figure 12.3).

This aspect of gloss has been measured most frequently because it is the one for which an instrument is most easily constructed (see Figure 12.3b). The light source, receptor sensitivity, and measurement angles with tolerances are precisely

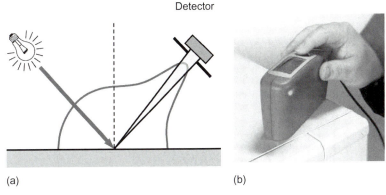

Fig. 12.3 Schematic diagram of a glossmeter (a) and a glossmeter, 'micro-TRI-gloss' (b) (source: Byk–Gardner).

specified in international test methods for specular gloss (e.g. ASTM D 523, ISO 2813) [4, 5].

Experience has shown that single measurement geometry, such as 60°, may not provide instrument readings of gloss that correlate well with the visual observations when comparing different gloss levels. Therefore, three different angles of incidence, namely, 20, 60, and 85° (see Figure 12.4) are being used.

The choice of geometry depends on the gloss level. The 20° geometry is mainly used for high gloss finishes – for example, for top coat evaluation when the 60° gloss values are higher than 70. The 60° geometry is used for semigloss finishes, for example primers when the 60° gloss values are among 10–70. The 85° geometry is used for comparing specimens for sheen or near-grazing shininess. It is most frequently applied when specimens have 60° gloss values lower than 10 (see Figure 12.5).

In case of automotive top coat finishes, typically the 20° gloss should be more than 80 gloss units.

As the measurement of specular gloss is an absolute measurement, it is dependent on curvature and the refractive index of the surface. The international test

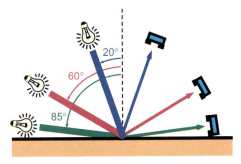

Fig. 12.4 The most common gloss geometries for measuring high, medium, and low gloss coatings.

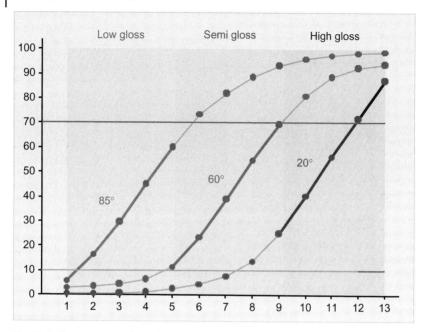

Fig. 12.5 Gloss geometry dependent on gloss range of the coatings.

methods for specular gloss list the gloss values in dependence of the refractive index. A polished glass with a refractive index of $n = 1.550$ has a maximum $20°$ gloss value of 95.4. If the refractive index changes to $n = 1.510$, the $20°$ gloss value will decrease to 84.7. Therefore, gloss readings of materials with different refractive indices, such as one component (1K) and two component (2K) clear coat systems will not correlate with the visual perception. 2K–clear coat system looks as brilliant or even more brilliant as 1K–clear coat system, but due to differences in the refractive index, the 2K–clear coat will result in specular gloss values lower by $20°$. As long as the curvature and refractive index of the material do not change, a glossmeter will give objective measurement results.

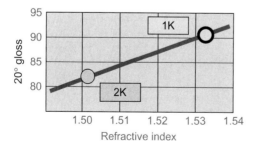

Fig. 12.6 Specular gloss is dependent on the refractive index of the material.

12.1.1.2 Visual Evaluation of Distinctness-of-Image (DOI)

This aspect of appearance is also referred to as *'image clarity'*. ASTM E 284 [3] defines distinctness-of-image (DOI) gloss as 'the aspect of gloss characterized by the sharpness of images of objects produced by reflection at a surface'. During visual observation, the sharpness of the light–dark edge of a reflected object can be observed. Landolt rings have been used by ophthalmologists to evaluate visual acuity for nearly a hundred years [6]. The test consists of locating the gaps in a graduated series of sizes of incomplete rings whose radial thickness and gap are equal to one-fifth the diameter of the ring. For gloss evaluation of transilluminated rings, reflections are viewed on Mylar. Rings have different sizes and different gap orientations. An image-gloss scale is associated with the different sizes of rings. An image-gloss scale ranging from 10 to 100 in steps of 10 was established for 11 sizes of rings from the largest to the smallest. The development of the scale is not documented, but its development took place in the General Motors automotive division sometime around January 1977. Visual observers select the smallest size of ring for which they can call the gap orientation correctly. The visual judgment is influenced by the loss of contrast and sharpness of the outlines. The numerical size of the rings is used as an inverse index of DOI gloss.

12.1.1.3 Measurement of Distinctness-of-Image

A variety of different technologies have been used to measure DOI [7–11]. The two most often used principles are as follows:

1. evaluation of the steepness of the reflection indicatrix, or
2. measurement of contrast in dependence of the structure elements.

ASTM E 430 [12] describes the design of an instrument based on the evaluation of the reflection indicatrix. The instrument illuminates the specimen at a 30° angle and measures the light reflected at 0.3° from the specular angle with an aperture of 0.3° width. This instrument is no longer available commercially.

On the basis of Deutsche Forschungsgesellschaft für Oberflächentechnik (DFO) research project, [13] the visual perception of DOI is influenced by loss of contrast and microstructures distorting outlines. Therefore, in 1999 a new measurement technology [14] was developed to measure light scattering caused by structure sizes smaller than 0.1 mm. This new measurement parameter was named *dullness*. A light emitting diode (LED) light source and a charge-coupled device (CCD) chip similar to that present in digital cameras are used to detect the reflected image of the source aperture (see Figure 12.7).

The dullness measurement determines the amount of light scattered within and outside the aperture in a defined range. The dullness value is a ratio of these two values. Therefore, this parameter of measurement is independent of the refractive index. In addition, an adaptive filter is used to separate between the inner and outer image, which minimizes the influence of curvature (see Figure 12.8).

The DOI value used by various carmakers like General Motors, Daimler Chrysler, and Volvo is calculated on the basis of 'dullness'-measurement and microstructures

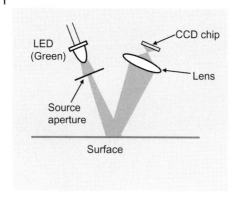

Fig. 12.7 Schematic diagram of dullness measurement.

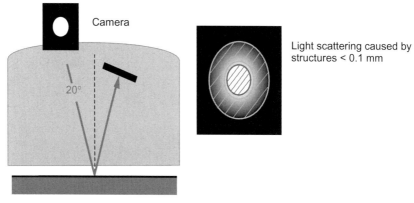

Fig. 12.8 Dullness measures the amount of light scattering within and outside the aperture in a defined range.

(Wa, Wb) as described in the following sections. The DOI value is dependent on the color family being dark/light metallic or dark/light solid color. Light metallic finishes are expected to have a DOI above 70 and dark solid colors should have it above 85.

12.1.1.4 Visual Evaluation of 'Orange Peel'

One obvious type of waviness is designated 'orange peel'. ASTM E 284 [3] defines orange peel as 'the appearance of irregularity of a surface resembling the skin of an orange'. A surface may be described as exhibiting orange peel when it has many small indentations that are perceived as a pattern of both highlighted and non-highlighted areas [15]. An observer interprets this pattern as a three-dimensional structure of hills and valleys.

The automotive industry in the United States established a physical standard for orange peel consisting of 10 high-gloss panels with various degrees of orange peel structure in the 1970s (Set of orange peel panels can be obtained from Advanced Coatings Technology, 273 Industrial Dr., Hillsdale, MI 49242). The panels are

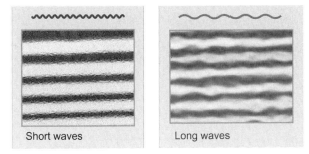

Fig. 12.9 Appearance changes in relation to structure size.

visually ranked from 1 to 10 with panel number 1 depicting very pronounced orange peel and panel number 10 denoting no orange peel. The visual observer can use these panels as a supportive tool to evaluate the degree of orange peel.

To understand this visual ranking, it is important to realize that structure size and the observing distance influence our visual impression (see Figure 12.9).

The structured surface of a coating appears to our eye as a wavy pattern of light and dark areas. The real height differences cannot be resolved with our eye. The maximum detectable height difference is approximately 20 μm. The mechanical amplitude of an automotive coating is typically in the range of 1 μm. Thus, our physiological impression is dependent on the contrast sensitivity and resolution capability of our eyes. The contrast sensitivity of the human eye has its maximum resolution at three periods per degree (see Figure 12.10).

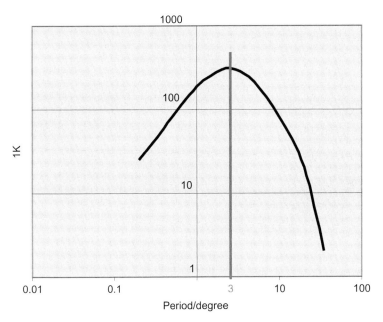

Fig. 12.10 Contrast sensitivity of our eyes is highest at three periods per degree [16].

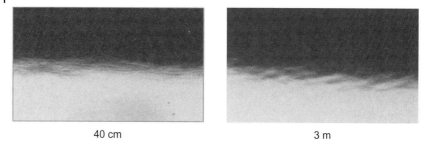

Fig. 12.11 Same panel viewed at different distances.

Consequently, the detection of different texture sizes is dependent on the observing distance (see Figure 12.11).

At a 40-cm distance, we can see structure sizes between 0.3 and 10 mm with a maximum at approximately 1–3 mm. While at a 3-m distance, we can only see larger structures of 3–30 mm (see Figure 12.12). Structures smaller than 0.3 mm are mainly responsible for the image forming quality of a surface, also referred to as *DOI*.

12.1.1.5 Instrumental Measurement of Waviness (Orange Peel)

Waviness has been evaluated by visual means and by use of a profilometer. The correlation between profilometer measurements and visual perception is satisfactory for surfaces with similar optical properties. The operation of a profilometer, however, is very time-consuming and limited to laboratory use. When the eye of an observer is focused on a painted surface, various types of waviness can be

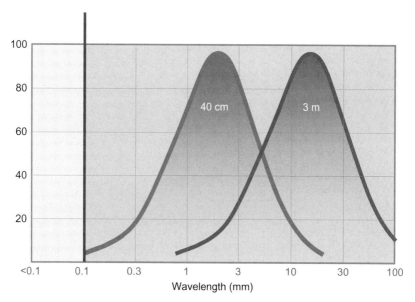

Fig. 12.12 Sensitivity of the human eye to structure sizes at a distance of 40 cm and 3 m.

identified that involve size, structure, and shape. Variations in process or material parameters can cause differences in surface structure. For example, poor flow or leveling properties of a coating will usually cause a long-wave structure often called *orange peel*. Changes in substrate roughness, on the other hand, will exhibit a short-wave structure of higher frequency. Because waviness is often caused on the production line, it is important to control it there.

After considerable research, an instrument was introduced in 1992 to provide an objective evaluation of waviness with structure sizes between 0.3–12 mm. In 1999, the measurement principle of this instrument was further developed to increase the resolution and expand the measurement range from 0.1–30 mm (see Figure 12.13). The new instrument 'wave-scan DOI' is now capable of measuring waviness and DOI in a single instrument (see Figure 12.14).

A diode laser source is used to illuminate the specimen at 60°. The reflected light intensity is evaluated at the specular angle. During the measurement, the

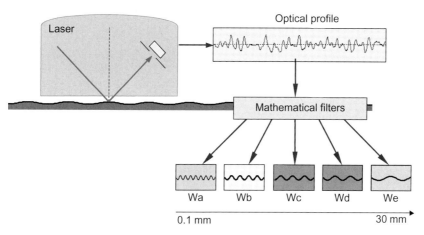

Fig. 12.13 Measurement principle 'waviness' of wave-scan DOI.

Fig. 12.14 The 'wave-scan DOI' instrument for orange peel and DOI measurement (source: Byk–Gardner).

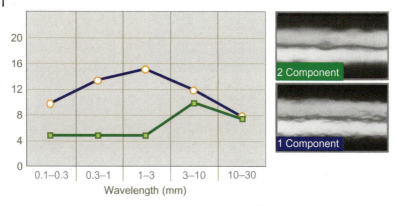

Fig. 12.15 Spectrum of wavelength structures of two different coatings surfaces.

instrument is moved along the surface for distance of about 10 cm. The intensity of the reflected light is maximum when coming from a concave structure element. The detector receives less light from a convex structure element. The human eye cannot resolve the actual heights of the structural elements of a painted surface, but the contrast between light and dark areas provides an impression of depth. The contrast of a surface structure can be expressed by use of the statistical parameter 'variance'.

The final measurement results are divided into five wavelength ranges (Wa–We) using electronic filtering procedures. As a final result, a structure spectrum is obtained for structure sizes between 0.1–30 mm (see Figure 12.15).

The combination of two measurement principles in one instrument simulates the visual perception at different distances and permits categorization of structure sizes with their causes.

On the basis of above information, many of the major automotive makers and their suppliers are using the wave-scan DOI as their standard tool to objectively evaluate the appearance of a car body. Because of the need of having a uniform appearance on the entire vehicle, small and highly curved parts need to be measured objectively. Therefore, the wave-scan DOI was further developed to allow measurement of automotive add-on parts, such as bumpers, gas tank doors, mirror housings, door handles, decorative trim, or motorcycle parts. This instrument, as seen in Figure 12.16 , uses the same measurement principle as the wave-scan DOI with miniaturized optics and redesigned electronics. This guarantees a good correlation to the wave-scan DOI and allows measurement of parts with curvatures as small as 300 mm and a sample size of 25 × 40 mm.

12.1.1.6 The Structure Spectrum and its Visual Impressions

As already pointed out, the visual perception of a surface is strongly influenced by changes in its spectrum of structures. For better understanding of how various structure sizes influence the visual perception, changes in the structure spectrum can be compared with those in the color measurement. The visible spectrum of color measurement is between 400–700 nm (blue to red). For simplicity, three

Fig. 12.16 The instrument 'microwave-scan' for orange peel and DOI measurement of small and curved parts (source: Byk–Gardner).

colors can be allocated to three structure sizes: long waves correspond to red, short waves correspond to green, and microstructures (DOI) correspond to blue. The ideal surface would be perfectly smooth, that is no waves. In colorimetric terms, we would be dealing with the 'black' (no waves = no color). Coating with severe waviness is to be improved, whereas the main target is to reduce the longer waves in the example depicted in Figure 12.17a. The original surface has a high amount of long and short waves corresponding to the color gray.

A reduction of the longer waves can be achieved by optimizing the base coat–clear coat application, for example by higher clear coat film thickness. As a result, the new look will be dominated by the shorter waves (see Figure 12.18), which would be equivalent to a blue/green color. In practice, a dominance of shorter waves can be caused by a poor-quality primer.

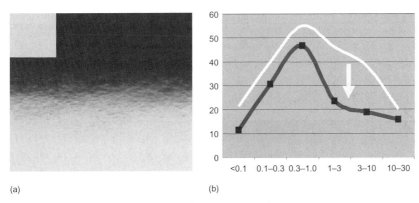

Fig. 12.17 Top coat with high amount of waviness (a) and the respective structure spectrum measured with wave-scan DOI (b).

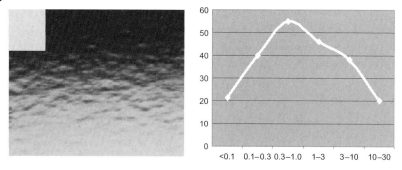

Fig. 12.18 Visual perception is dominated by short-wave values.

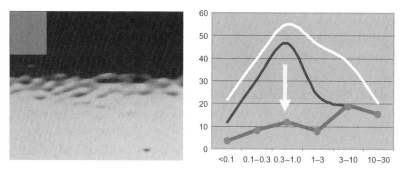

Fig. 12.19 Owing to low short-wave values, the longer waves dominate the visual perception.

This high amount of shorter waves can be reduced by, for example sanding the primer. Figure 12.19 demonstrates the reduction of the shortwaves and microstructures as compared with the amount of longer waves.

In color measurement, red would be dominating the visual impression. The same is true in the instrumental appearance measurement: the very low levels of short waves make the surface very brilliant with a high DOI, and consequently, the small amounts of long waves are very obvious.

As shown in the previous example, substrate roughness of the primer or the steel substrate can telegraph through the clear coat and can cause a 'fuzzy' appearance. As electrocoat and primer coatings are often semigloss finishes, so far the only way to measure was by applying a clear tape – highly user dependent, and meaningful only for long-wave values. A new development was started to measure orange peel of lower gloss surfaces [17]: the new instrument wave-scan dual as shown in Figure 12.20 allows measuring of the structure spectrum on high gloss and medium gloss surfaces. Thus, the surface quality can be optimized and controlled after each coating layer.

12.1.1.7 Outlook of Appearance Measurement Techniques

The actual appearance level can be maintained on the basis of typical specification sets of the automotive industry according to Table 12.1. But as mentioned,

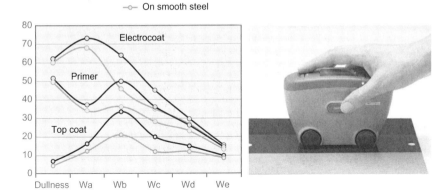

Fig. 12.20 'Wave-scan dual' for objective surface quality control of high and medium gloss surfaces (source: Byk–Gardner).

Table 12.1 Automotive specifications for exterior appearance of coatings

	Horizontal	Vertical
20° gloss	>80	>80
DOI (ASTM E430)	>80	
Orange peel		
Ranking[a]	>7	>5
Long wave(LW)	<10	<20
Short wave(SW)	<35	<35

a Analog ACT (advanced coatings technology, MI).

it is important to realize that the impression of gloss is a multidimensional phenomenon like color. By changing the ratio of different structure sizes, the visual appearance can be dramatically changed. The structure spectrum is like the fingerprint of a surface and can be compared with the spectral curve of colorimetry. Currently, the correlation of the different wavelength ranges to the visual perception is under investigation. The goal is to obtain a measurement system for appearance similar to the L^*, a^*, b^* system of color measurement [17].

12.1.2
Visual Evaluation of Color

12.1.2.1 Solid Colors
Uniform and consistent color is essential to achieve the impression of a high-quality finish and to avoid customer complaints. This is important not only at the time of purchase, but also throughout the lifetime of the vehicle. According to Figure 12.22,

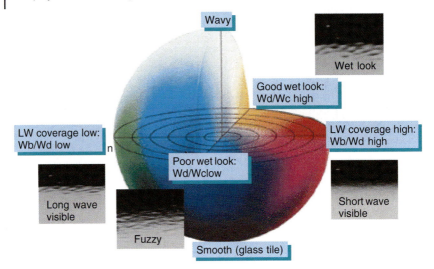

Fig. 12.21 New three-dimensional appearance system to correlate to the multidimensional perception of appearance.

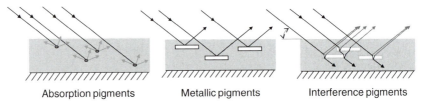

Fig. 12.22 Main types of pigments for automotive colors.

three main types of pigments are used in the automotive industry to create a range of solid and metallic colors: absorption pigments, metallic pigments, and interference pigments.

The special effect coatings play a dominant role in automotive applications as they make an object distinctively appealing. In contrast to conventional solid colors, the metallic finishes change their appearance with viewing and lighting conditions. Interference colors and special effect colors show not only a lightness change with viewing angle, but also a chroma and hue change. And in the latest development of special effect pigments, additional sparkling effects are created with changes in the lighting conditions from sunlight to cloudy sky. Thus, the color of a car comes alive that makes the car look not only pretty but also stylish as depicted by the cars in Figure 12.23.

Visual perception of color is very subjective and has shortcomings. One of those is that with increasing age, the eye gets fatigued and the lens turns yellowish with the consequence that an elder person judges colors redder and more yellow. Second, our mood influences our color perception. It is also proven that color vision is dependent on gender. The inherited types of color vision defects, which are caused by a sex-linked, recessive gene are far more common in men than in women. Men

Fig. 12.23 Metallic and special effect finishes accentuate the contours of a car.

who inherit a single such gene will show some form of color deficiency. Women, on the other hand, must inherit the gene from both parents to be affected. Only about 0.5% of women have defective color vision, while approximately 8% of men will show some form of deficiency. The most common defect is the red/green blindness. One should take into consideration that the phrase 'color blind' is misleading. Humans who see only shades of gray are very rare. Thus, the first step for an objective color-control procedure is to standardize the observer.

Reflected light from a colored object enters the human eye through the lens and strikes the retina. The retina is populated with light-sensitive receptors, the cones and the rods. The cones are responsible for daylight color vision, whereas the rods for light/dark detection at night. The retina has varying degrees of sensitivity. In the foveal pit, which is within 2° of the visual axis, all of the receptors are cones. Moving outwards from the foveal pit, the number of rods increases. Each eye has about 7 million cones and 120 million rods. There are three different types of cones each containing a different photosensitive pigment giving us sensitivity in the red, green, and blue regions of the visible spectrum.

In 1931, Guild and Wright performed an experiment to standardize the human observer (see Figure 12.24) [18].

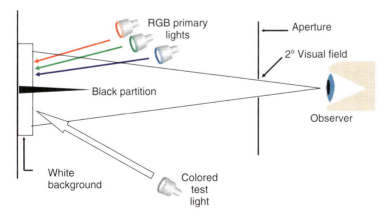

Fig. 12.24 The experimental set up to standardize the human observer.

Seventeen human subjects were used to make observations. Light from a test lamp was projected on a white background through a 2° field of view opening. Three monochromatic light sources for red, 700 ; green, 546 ; and blue, 435 nm were projected on a second white background. Each observer was to adjust the three light sources until the color appeared to exactly match the color of the test lamp. The 2° angle was used for strict foveal vision using only the cones. On the basis of these results, the Commission International de l'Eclairage (CIE) 2° supplementary standard observer was defined [19].

As in practice, the achieved numbers did not correlate with human color perception, the experiment was repeated in 1964 with a larger field of view of 10°. This took into consideration that usually the observed sample size is larger, stimulating both the cones and rods and resulted in the CIE 10° supplementary standard observer. The field of view is similar to the size of a tennis ball, whereas the 2° field of view is similar to viewing your thumbnail at arm length. The experimental work resulted in the establishment of the three color matching functions for red, green, and blue (see Figure 12.25) describing the sensitivity of the human eye.

The 10° observer allows for greater spectral sensitivity in the blue and red areas of the visible spectrum and gives better correlation to the visual assessment of larger samples [20–22].

Color also changes with the lighting conditions. There are many different light sources that cause different appearance: daylight represents a sunny day without clouds, incandescent light simulates a warm atmosphere, like at home, and a cool white light simulates the atmosphere of a departmental store .

Therefore, standardized illuminants have to be used to guarantee objective color evaluation (see Figure 12.26).

The CIE standardized light sources by the amount of emitted energy at each wavelength (= relative spectral power distribution). The most common illuminants include D65 to simulate natural daylight, illuminant A to represent incandescent or tungsten light, and F2 or F11 representing a fluorescent departmental store light

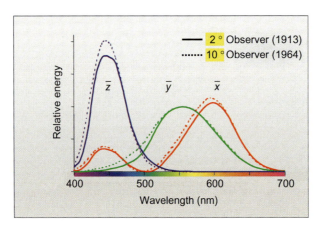

Fig. 12.25 Color matching functions for 2 and 10° supplementary standard observer.

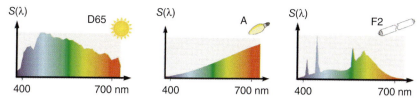

Fig. 12.26 Spectral power distribution for standard illuminants D65, A, and F2.

source. In industry, there is an appearance condition where two objects appear to visually match when viewed under one light source, but do no longer match when viewed under a second source. This is called *illuminant metamerism*. Metamerism is caused by a difference in pigments used to color the two objects. When viewed under the matching light source, the dissimilar pigments interact with the light source to give low visual impact. When viewed under the second light source, the differences in the pigments for each object are amplified, and hence appear to be mismatched. The example in Figure 12.27 shows that the metameric pair has spectral curves crossing at least three times. The number of crossings determines the degree and potential for metamerism. Today's color instruments are able to calculate a metamerism index to evaluate color changes under different light sources.

12.1.2.2 Metallic and Interference Colors

Visual color perception of metallic finishes is not only dependent on the lighting condition, but also on the viewing angle. Metallic finishes show a lightness change with changing viewing angle. By tilting the sample panel backwards and forwards, this effect is created which is also referred to as 'light-dark flop' [23].

The more rapidly the lightness decreases with angle of view, the better the contours of a car are accentuated. In case of interference coatings, the color hue changes with viewing and illumination angle, which causes an even more spectacular color

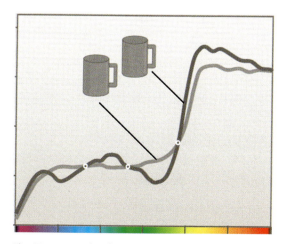

Fig. 12.27 Example of spectral curves of a metameric sample pair.

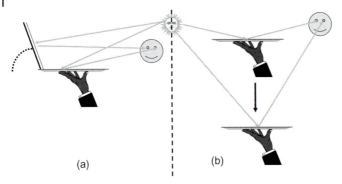

Fig. 12.28 Visual evaluation of effect colors: (a) shows metallic finishes with light–dark flop and (b) shows new effect finishes with color flop.

effect than metallic finishes. To observe this type of interference shift, the panel must be moved to allow increasing or decreasing the angle to the light source [24].

12.1.2.3 Color Measurement of Solid Colors

Light source and observer are defined by the CIE [19] and their spectral functions are stored within color instruments. Optical properties of an object are the only variables that need to be measured. In order to obtain them, a color instrument is used that contains a lamp, which illuminates the sample. The light that is reflected by a colored object is resolved into its spectral components and the amount is measured by a photoelectric detector. This is done at each wavelength and is called *the spectral data*. For example, a black object reflects no light across the complete spectrum (0% reflection), whereas an ideal white specimen reflects nearly all light (100% reflection). All other colors reflect light only in selected parts of the spectrum. Therefore, they have specific curve shapes or fingerprints, which are their spectral curves. A red object reflects light only in the range of long wavelengths (between 600–700 nm).

In contrast, a blue product reflects light in the range of short wavelength between 400–500 nm, while a green object reflects light in the medium range between 500–600 nm (see Figure 12.29).

For communication and documentation of color and color differences, color systems were developed. The basic system is the XYZ system. It is the product of

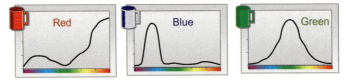

Fig. 12.29 Typical reflectance curves for blue, green, and red samples. X represents the amount of red, Y the amount of green, and Z the amount of blue. On the basis of XYZ values all other indices and color systems are calculated.

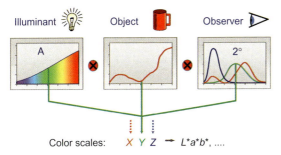

Color scales: X Y Z → L*a*b*, ...

Fig. 12.30 Interaction of illuminant–object–observer determining color perception. If X represents the amount of red, Y the amount of green, and Z the amount of blue, then on the basis of XYZ values, all other indices and color systems can be calculated.

mathematical description for the illuminant (= spectral power distribution), the object (= spectral reflectance) and the human eye (CIE color matching functions) (see Figure 12.30).

The system recommended by the CIE in 1976 [20] and mostly used today is the CIE $L^*a^*b^*$- system. It consists of two axes a^* and b^*, which are at right angles and represent the hue dimension of color. The third axis is the lightness L^*. It is perpendicular to the a^*b^* plane. Within this system any color can be specified with the coordinates $L^*a^*b^*$. Alternatively, L^*, C^*, $h°$ are commonly used (see Figure 12.31), particularly for chromatic colors. C^* (= Chroma) represents the intensity or saturation of the color, whereas the angle $h°$ is another term to express the actual hue.

To keep a color on target, a standard needs to be established and the production run is compared with the standard. Therefore, color communication is done in terms of differences rather than absolute values. The total change of color, dE^*,

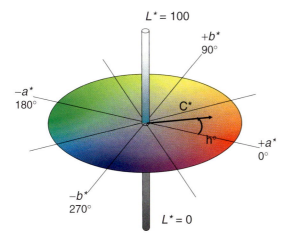

Fig. 12.31 The CIE $L^*a^*b^*$– color system of 1976.

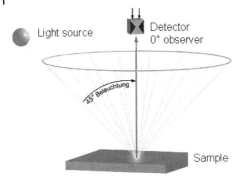

Fig. 12.32 Optical design of 45/0 geometry for agreement with visual assessment.

is commonly used to represent a total color difference. To determine the actual change in color, the individual colorimetric components, namely dL^*, da^*, db^*, or dL^*, dC^*, dH^* need to be used. The color differences that can be accepted must be agreed upon between customer and supplier. These tolerances are dependent both on demands and technical capabilities.

In the automotive industry, there are two classes of instruments that are used to measure color: 45/0 and sphere geometry. The 45/0 geometry uses 45° circumferential illumination and 0° viewing perpendicular to the sample plane (see Figure 12.32). This simulates the normal condition used for color evaluation. A high gloss sample with the same pigmentation is visually judged darker when compared with a matte or structured sample. This is exactly what a 45/0 instrument measures. Applications, where it is necessary to have the agreement with the visual assessment, are batch-to-batch comparisons and assembly of multicomponent products.

As shown in Figure 12.33, a spherical instrument illuminates the sample diffusely by means of a white color coated integrating sphere. Baffles prevent the light from directly illuminating the sample surface. Measurement is done using an 8° viewing angle. For sphere instruments, the most common measurement condition is the specular included mode. The total reflected light including diffuse and direct

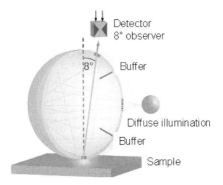

Fig. 12.33 Optical design of sphere geometry to measure the color hue.

reflection is measured there. Color is measured independent of the sample's gloss or surface structure. Typical applications are color strength evaluation, weathering, temperature influence, and color matching.

12.1.2.4 Color Measurement of Metallic and Interference Coatings

To fully characterize special effect colors, they need to be measured at a variety of different angles [24], which is not practical for industrial color control. Therefore, research projects were conducted and it was determined that a minimum of three viewing angles and best five viewing angles are needed to provide sufficient information on the goniophotometric characteristics of a metallic finish [25–28].

On the basis of these studies, ASTM and Deutsches Institut für Normung eV (DIN) were able to standardize the illumination and viewing conditions for multiangle color measurement of metallic flake pigmented materials [29, 30]. The measurement geometry for multiangle measurements is specified by aspecular angels (see Figure 12.34). The aspecular angle is the viewing angle measured from the specular direction in the illuminator plane. The angle is positive when measured from the specular direction toward the normal direction. The reverse geometry is considered equivalent, if all other components of the instrument design are the same.

To characterize a metallic finish, the near-specular angle (15°), the face angle (45°), and the flop angle (110°) are of major importance. The near-specular angle is mainly influenced by spray application differences, which can cause flake orientation differences. The 45° angle duplicates the angle commonly used on many conventional color instruments. The 110° angle delivers the information needed to determine whether the desired lightness-darkness change occurs. For automotive exterior applications, directional illumination is used versus circumferential illumination. A circumferential illumination minimizes the contribution from directional effects such as the venetian blind effect and surface irregularities.

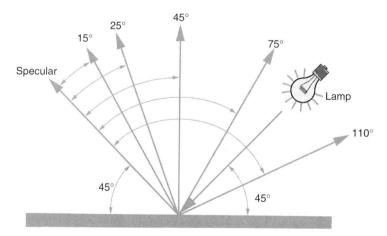

Fig. 12.34 Schematic of illumination and sensing geometry for a multiangle spectrophotometer.

Thus, the averaging of the circumferential illumination would cause the measured color values of two specimens to be the same, while visually the two specimens would not match.

Metallic colors have considerably higher reflectance factors, for example coarse silver can have a reflectance factor of almost 300% compared with the perfect diffuser, which is used for calibration. This causes problems in colorant formulation programs as the traditional Kubelka–Munk theory no longer applies. For color QC, the colorimetric data L^*, a^*, b^* (or L^*, C^*, $h°$) can be used; one cannot use the same tolerances for the different viewing angles. As a rule of thumb, the 45° tolerance is similar to the tolerances used with conventional spectrophotometers in the same geometry. The flop angle (110°) tolerances are about twice those at 45°, and the near-specular tolerances are again about a factor of two above the flop angle.

Another useful index is the so-called flop index, a measure of the change in reflectance of a metallic color as it is tilted through the entire range of viewing angles. DuPont researchers developed an index using the 15, 45, and 110° angle based on visual correlation study [30].

12.1.2.5 Typical Applications of Color Control in the Automotive Industry

In order to have a unique tolerance parameter independent of the color, weighted factors have to be used. Therefore, the automotive companies often have set specifications on delta E_{CMC} based on the British Standard (BS) BS 6923 or the delta E' based on DIN 6175-2 using 3- or 5-angle instrumentation (DIN 6175-2, *Farbtoleranzen für Automobillackierungen*, Teil 2 Effektlackierungen). Delta E' is introducing variable weighting factors through the CIE lab space. Thus, the user can work with a uniform delta E' for all colors. The weighting factors are determined by the car manufacturers [31]. Audi was a pioneer in the development of the equations for these weighting factors. Usually, the weighting factors for batch approval are more stringent than those for production QC of car bodies or plastic add-on parts. The delta E' allows setting up of pass/fail criteria on the basis of a 'green', 'yellow', or 'red' zone as follows:

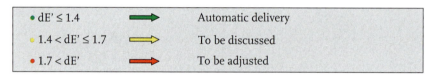

• dE' ≤ 1.4	⟹	Automatic delivery
• 1.4 < dE' ≤ 1.7	⟹	To be discussed
• 1.7 < dE'	⟹	To be adjusted

The CMC [32] tolerant system describes a 3D ellipsoid with axes corresponding to hue, chroma, and lightness. Their weighting functions are dependent on the color value of the standard. They are default parameters and already included in the computation of dEcmc. Therefore, they cannot be individually varied. It allows to modify the shape of the tolerance ellipsoid by varying the ratio $l : c$, where the parameter '*l*' stands for lightness changes and the parameter '*c*' summarizes changes in chromaticity (= chroma and hue). To create a volume of acceptance, CMC uses the commercial factor *cf*. The commercial factor adjusts all axes to the same extent and determines the overall size of the ellipse. For metallic finishes, usually a 'cf' = 1 is used and the ratio $l:c = 2:1$, while solid colors often are controlled with a 'cf' = 0.5 and a ratio $l:c = 1:1$. Delta E_{CMC} allows setting up of pass/fail criteria based on a 'green' or 'red' zone:

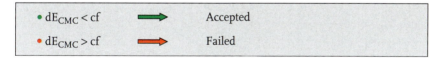

- $dE_{CMC} < cf$ → Accepted
- $dE_{CMC} > cf$ → Failed

As the data interpretation gets more and more complex with the number of measurement data, the individual color component data by angle are usually saved in databases. The specifications are based on the total color difference data. The individual delta color components are reviewed when production data are no longer within agreed upon tolerances.

In Figure 12.35, fenders and fascias of actual production were compared with the master standard panel. The dE^*, dL^*, da^*, and db^* of all 5 angles for each part are displayed. The Fender on the left side shows the highest difference to the master at 15 and 25°, which is mainly caused by a difference in lightness.

12.1.2.6 Color Measurement Outlook

In the last 2 years, a new generation of special effect pigments has become more and more popular. They are based on innovative substrates like silica or aluminum flakes, very thin multilayer interference flakes, or liquid crystals. For some of these new pigments, for example Colorstream from Merck, Variochrom from BASF, or Chroma Flair from Flex Products, the color travels over a wide range – that is the color change can be dependent on illumination and viewing angle. Therefore, ASTM has formed a new task force to establish a standard practice for multiangle color measurement of interference pigments [33]. In order to fully capture the color travel of these interference pigments, it might be necessary to add viewing (particularly behind the gloss at 15°) and illumination angles (see Figure 12.36).

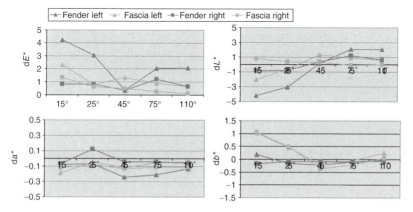

Fig. 12.35 Trend graphic for dE^*, dL^*, da^*, and db^*.

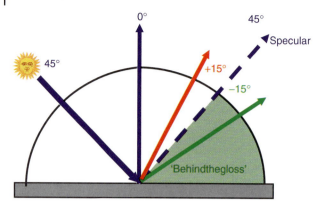

Fig. 12.36 Measurement of color travel 'behind the gloss' at 15° angle.

Another new family of effect pigments like Xirallic [34], which changes the color only in one quadrant, but exhibits an intense sparkle effect when viewed under direct sun light has opened a new way to evaluate the overall color impression. This sparkle effect demonstrated in Figure 12.37a can neither be captured nor sufficiently characterized with traditional 5-angle color measurement. Interestingly, this sparkling effect can also be observed in other effect coatings just on a smaller scale and are usually dependent on the flake size, density, or distribution within the coating. Apart from the sparkle effect under direct sun light, another effect can be observed under cloudy conditions, which is described as an optical texture, coarseness, or salt and pepper appearance as can be seen in Figure 12.37b. This visual coarseness can be influenced by the flake diameter or the orientation of the flakes resulting in a nonuniform and irregular pattern. Thus, several pigment manufacturers and automotive paint manufacturers have started to research the visual perception of effect coatings under different lighting conditions to simulate how a car looks like on a sunny and a cloudy day [35–37].

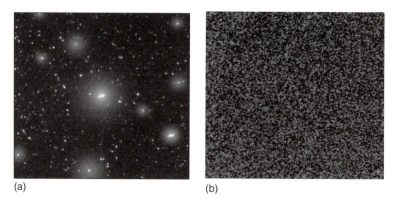

Fig. 12.37 (a) Sparkle observed under direct sun light; (b) visual coarseness observed under diffused light.

For a better performance in terms of robustness of paint in application, new test methods are gaining importance that try to characterize a test panel on a statistical basis under various aspects including color and surface properties [38].

12.2
Weathering Resistance of Automotive Coatings

Gerhard Pausch and Jörg Schwarz

12.2.1
Introduction

The weathering resistance of paints and the light fastness – that is aging in light of coatings for car metals and plastics, and materials used in car interiors are important properties for vehicle manufacturers and should be determined in the shortest possible time. Above all, high temperatures, rain, and the UV part of sunlight cause aging in polymer materials, even behind windows, [39] resulting in color change, gloss reduction, cracking, and blistering [40].

Improved corrosion protective coatings, light, and weathering– resistant pigments, special fillers, and additives as well as UV absorbers in clear coatings have led to a noticeably longer guarantee period and service life for vehicles in recent years [41]. This furthermore represents the challenge to testing technology for new equipment to reduce the time for relevant product development results. It is only through previous practical knowledge about the effects of weathering and its causes, as well as a well-planned combination of laboratory weathering, outdoor weathering, and complementary analytical studies for interpreting test results that it is possible to come up with well-founded forecasts about the life span of new paints within a short time. The following overview is intended to help users to select the most appropriate weathering methods for their needs from the wide choice that is available.

New cars are expected to last at least 15 years or longer. During this period, on one hand the visual appearance, which includes color and gloss, and on the other the performance and service life of the exterior paint work has to remain fully effective. This requires that no cracks or blisters occur in the paint structure and with regard to the corrosion properties of the phosphating (see Chapter 3) and priming by cathodic electrodeposition (see Chapter 4) along with galvanization wherever applicable, that primers and top coats continue to remain impenetrable to light, water, acids, corrosive gases, and salts. Furthermore, it is expected that the initial glossy appearance preserved as much as possible – that is the clear coating is expected to retain a high level of gloss and the base coat a good level of color fastness – that is no significant color fading.

Stains formed by the effects of synthetic and natural chemicals, particularly the effects of bird droppings and tree sap cannot be completely avoided over the long term, but this should only minimally detract from the appearance of the car's external shell. In addition, 'acid rain' can lead to irreversible damage to the top coat–clear coat.

Materials used for car interiors are expected to have a good level of color and light fastness even under the effects of intense sunlight and temperature, for example in the extreme climate of Arizona.

As mentioned previously, modern automotive paint work is a multiple layer system made up of different individual layers of coatings with properties that are specifically adapted to each other. Now in order to also ensure the mentioned high-quality requirements of the overall paint work structure on new developments using water-based primers, water-based base coat, and clear coating, paint developers resort to a whole range of experience and knowledge (see Chapters 5, 6, and 9). Because of the very short development times demanded these days for new automotive colors and paints, often not all the desired results are available until they are introduced, especially the long-term weathering results on original bodywork. This can be helped by carefully selecting previously tested new materials, which have been subject to preliminary work in screening tests, from manufacturers of paint and raw materials with regard to weathering resistance. However, when introducing a new paint formulation for mass production, extensive short-term and Florida outdoor tests are still necessary for evaluating long-term weathering resistance.

The usual requirement of base coat systems in the automotive industry includes good results after 12 or 24 months of outdoor weathering in Florida (Miami) and passing short-term weathering tests of at least 3000 hours. Furthermore, automotive clear coatings are expected to demonstrate acceptable resistance after at least 3 years Florida exposure. They are sometimes also tested for up to 10 years with appropriate intermediary assessments.

To shorten development time and ensure a higher degree of safety when evaluating the weathering resistance of automotive paint work systems, mainly of the base coat–clear coat system, light-stabilizers and quencher (see Chapter 6) have been examined and thresholds specified where UV transmission is concerned.

12.2.2
Environmental Impact on Coatings

Coatings are subject to a number of influencing factors during service life, which, depending on the coating formula, the intensity, and duration of environmental impact can lead to degradation of the polymeric network and the described defects. UV light, heat, and humidity have an effect on materials in places all-around the world with different intensity depending on the climate and weather conditions. These three factors are simulated in all forms of laboratory weathering in a controlled manner. Beside this, materials in the open are often subject to corrosive gases and microorganisms. These factors are only recreated in laboratory tests in special cases. The simulation of acid rain in assessing the life span of automotive coatings has gained in significance over recent years [42–44].

Exposure to Radiation by Sunlight

Exposure to radiation exposure can be from direct and indirect sunlight. Overall radiation is also called *global radiation*. As shown in Figure 12.38, the

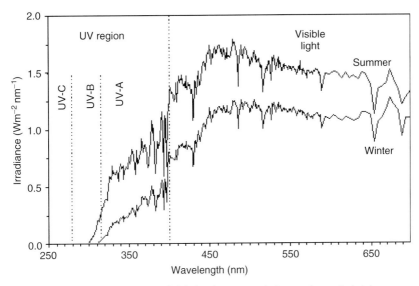

Fig. 12.38 Spectral radiation level of global radiation in relation to the sun's height.

electromagnetic waves emitted from the sun start at a wavelength of about 300 nm, pass through the visible light band at a maximum, and continue into the infrared band. The spectral radiation level shows the amount of energy falling on a surface unit at a defined wavelength per wavelength unit. The radiation level results from adding all relevant wavelengths and also by integrating the time.

Heat Exposure

Due to the dependence of the speed of chemical reactions to increasing temperature, weathering impact on paints is accelerated considerably by an increase in temperature. Table 12.2 shows the typical temperatures of the most commonly used commercial outdoor weathering locations.

It should be noted that dark surfaces exposed to direct sunlight heat up by up to 30 °C above the temperature of the ambient air. Therefore, when describing weathering conditions, besides the ambient temperature the surface temperature of comparable standard metal sheets is also given. The so-called black panel temperature is the one most used. This is measured by means of a black panel thermometer using a temperature sensor on a black painted sheet of metal.

Alongside this, so-called black standard thermometers are used in connection with the heat conductivity properties of plastics. In this case, a 0.5-mm-thick black sheet of metal is placed on a 5-mm–thick sheet of plastic. Owing to the lower conductivity of the black standard thermometer, black standard measurements are taken at 60 °C, which is about 5 °C higher than the black panel temperature.

Light-colored surfaces heat up less than dark ones in the sun. The difference can be shown by using a white standard thermometer. The difference between the black and white standard temperatures in the midday Florida and Arizona sun is in the region of 15–20 °C.

Table 12.2 Climate data for outdoor weathering in Florida and Arizona

	Florida	Arizona
Air temperature		
Average peak temperature in summer (°C)	32	40
Annual average (°C)	24	21
Humidity		
Average top value in summer (% RH)	93	28
Annual average (% RH)	70	35
Rainfall		
Monthly maximum (mm)	237	28
Annual total (mm)	1420	186
Typical irradiation per year		
Overall radiation (300–3000 nm)[a] (MJ m^{-2})	6588	8004
UV radiation (295–385 nm)[a] (MJ m^{-2})	280	334

[a] Angle 25° south in Florida and 34° south in Arizona.

Exposure to Humidity

Paint and plastics absorb water in the surface region when rainfall, humidity, and dew occur. This water can contribute toward mechanical degradation as a result of the following mechanisms amongst others:

- Water can contribute toward the hydrolytic decomposition of chemical bonds.
- Water acts as an oxygen carrier and thus accelerates oxidation processes.
- Mechanical forces are associated with swelling and contraction whose repeated action can lead to the formation of cracks.

Exposure to humidity from dew can be particularly intensive because in the right climate and weather conditions, it lasts through the night into early morning. The major significance of exposure to humidity can also be seen from the fact that the damaging effect of weathering on paints is much more evident in the damp heat of Florida than in Arizona which is even hotter but drier.

12.2.2.1 Natural Weathering

The resistance of a coated surface on a vehicle can vary widely in different locations due to climatic differences. Therefore materials are subject to extreme reference climate conditions of humid/warm and dry/hot. Materials age quicker under weathering in these areas than in those located further north in the world. Subtropical South Florida for humid/warm conditions and Arizona for dry/hot conditions have been established as standards. There are service providers in these regions that take care of exposing samples and regularly checking and evaluating them. There are also weathering centers at other locations, for example in the south of France, on the European Atlantic coast, in Australia and South Africa. The latter two centers are sometimes used for exposing specimens during our winter months due to their location in the southern hemisphere. Weathering centers also used to be operated in the Alps. Because of the low temperatures, however, materials got slightly damaged resulting in these centers being closed.

To record climate variations over the weathering period and to compare weathering tests spread over different times, weather data such as sunshine, temperature, humidity, and precipitation are continually recorded and charted at the weathering sites using sensors.

The test specimens can take almost any form and be made of almost any material, for example painted metal test panels, textiles, plastic panels, molded plastic parts, and complex products, which are made of various materials (composites) (Figure 12.39). Resistance studies are conducted on all products that are intended for outside use.

International norms and standards dictated by vehicle manufacturers and industrial associations specify the place of exposure and method, according to which the alignment and angle to the sun are determined. These standards make the comparison of weathering and corrosion tests easier.

On every set of weathering tests it is important to correctly evaluate and quantify the results. The recommended procedure is to always simultaneously subject a reference material of known resistance to weathering, which is of similar composition to the test specimen. In primary studies, at least three specimens of each test material should be exposed to compensate for variations in the test conditions.

Some changes to materials, such as color, gloss, and transparency can be measured with instruments; other changes such as crack formation, flaking, chalking, blistering, and rust have to be judged visually and classified by standardized analysis criteria. Normally, the properties that are of interest are measured or evaluated

Fig. 12.39 Example of specimen for weatherability tests.

before weathering. Assessments then follow at monthly or quarterly intervals to determine the progress of any degradation.

As part of a comprehensive set of tests, other special services or treatment may be required, such as washing, polishing, scratching, and weighing. The test report contains all the relevant information including the measurements of the selected material properties in the specimen's starting condition and over the course of weathering.

An outline of the most important outdoor weathering locations as well as selected versions of natural and accelerated outdoor weathering (see also Table 12.2) are given in the following chapters.

The Most Important Outdoor Weathering Locations:

>Florida: (Subtropical – high radiation levels, high temperatures, periods of heavy short rainfall, and high relative humidity)

Southern Florida is the only truly subtropical region on the US mainland. This area is characterized by high radiation levels with a large amount of UV, average of high annual temperatures, extremely high amount of rainfall, very high humidity, and low levels of dust. This combination is characteristic of the extreme climate that makes Miami the ideal place to test the weather resistance of products. The wet, damp climate with intensive dew overnight has proven to be extremely damaging to many materials like coatings and plastics.

This climate provides exposure to humidity particularly for the automotive multiple layer systems, while Florida exposure offers early recognition benefits compared with laboratory weathering when testing for delamination. It is thought that because of the so-called rest phase in the test overnight, degradation processes are accelerated in the polymer by more intensive diffusion of water vapor.

>Arizona: (Desert climate – dry, high radiation levels, and extremely high temperatures)

Because of high irradiation levels and all year-round high temperatures, Arizona is a reference location for outdoor weathering tests in desert climates. Arizona has about 20% more radiation from the sun than Florida over the year and summer temperatures are extremely high. The maximum air temperature is usually 8 °C higher than what it is in Miami. Black panel temperature can even reach 11 °C higher. In summer, temperatures in Arizona often reach 46 °C. An exposed black panel would reach those conditions at the surface temperature of more than 71 °C.

This extreme climate has proven to be particularly critical to certain materials. Above all, the following are affected:

- color and gloss of coatings
- color fastness
- aging from heat and the physical properties of plastics
- coatings on metal and plastic
- the light fastness
- tensile strength of textiles.

12.2.2.1.1 Procedures of Outdoor Weathering

Direct Weathering For most materials, direct weathering is the most commonly used process. Test specimens are laid on an open frame and normally directed toward the south. In case of weathering, 'open backed', test specimens are subject to the effects of weathering on both sides. Weathering tests that are 'backed' have a plywood underlay, which supports flexible test specimens and 3D objects, and leads to higher surface temperatures.

The weathering rack can be set at any desired angle. Mostly an angle of 45° is chosen for weathering. An angle corresponding to the latitude is also commonly used – that is 25° in Miami and 33° in Phoenix to maximize the irradiation level. Also, an angle of 5° is common, which equates to a flat roof or another horizontal position. On flat panels, the frame provides a covered test area, which can be used for comparison purposes. Final evaluation should be made against a retain.

Biological Attack Degradation from microorganism and fungi is sometimes detected on vehicle coatings. The humid/warm climate of subtropical Florida is especially well-suited for this problem [48]. The sides of the panels to be tested are pointed northwards to reduce the sun radiation and the surface temperature and to increase the humidity.

Exposure Behind Glass In order to simulate the natural weathering exposure for indoor materials, ventilated or sealed glass covers are used (refer Figure 12.40). While the ventilated version is often used on household textiles and floor coverings, the unventilated option is predominantly applied to automotive interior parts, such as the dashboard.

Accelerated Versions of Outdoor Weathering For conducting these procedure, the samples are subject to the natural environmental conditions. For acceleration purposes, the temperature or the exposure to the irradiation is increased in different ways . Some procedures also feature regular spraying with water, salt, or acidic solutions.

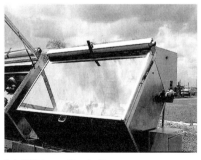

(a) AIM box with shutter   (b) Wood backing box, sealed or ventilated

Fig. 12.40 Exposure behind glass for interior car parts.

Fig. 12.41 Black box for increased temperature of the panel during exposure.

Black Box To simulate a higher temperatures on car applications, the test panels are mounted in such a way that they form the surface of a black box (shown in Figure 12.41), which thermally isolates the back of test specimens and increases the temperature of the test panels. In Florida, the black box is normally set at an angle of 5° to the horizon.

Concentrated Radiation Using Fresnel Mirrors (Q-Track or EMMAQUA) This weathering procedure uses the sun's natural radiation as a source of radiation. An array of 10 mirrors is used to reflect and concentrate the sunshine onto the test specimen.

At the same time, the device automatically follows the sun's position from the morning to the evening, thus maximizing the irradiation from the sun. A programmable water spray device can wet the samples and create temperature shocks. The increased exposure to heat is kept as low as possible by high performance cooling fans but can still be above the maximum practical exposure.

When using this method, the maximum allowable temperature of the material to be tested should therefore be known.

The average total annual UV radiation from 295 to 385 nm is 1400 MJ m^{-2} in Florida.

This range is monitored because it is the most damaging part of the sun's radiation. The annual dose of radiation in the Q-track is equivalent to about 5 years of sunshine in Florida or 4.25 years in Arizona [46].

Depending on the dose of radiation, tests in an accelerated weathering unit using Fresnel mirrors, according to DIN EN ISO 877, ASTM G 90, D 4364, and SAE J1961, cost considerably less than similar tests with xenon lamp equipment in the laboratory. Weathering with Fresnel mirror systems has therefore found a justified application.

Fig. 12.42 Example of a Q-track for increased irradiation of test panels.

(a)   (b)

Fig. 12.43 Salt-spray-test equipment and outdoor corrosion testing.

Accelerated Outdoor Corrosion Test Using Salt Spray Painted test panels or objects are mounted on a usual weathering rack (depicted in Figure 12.43). In addition, the test specimens are sprayed with a 5% salt solution twice a week. This method provides for a quick and realistic form of exposure to corrosion for car materials as compared with others. It can be conducted either in Florida or in Arizona. A special version of this is the so-called 'acid rain test', in which a special test solution is sprayed to simulate the industrial atmosphere in Jacksonville in the northern part of Florida [47].

In the case of accelerated outdoor corrosion, test specimens are sprayed twice a week with a salt solution.

12.2.2.2 Artificial Weathering

Coatings are exposed using artificial weathering or artificial radiation to duplicate aging processes in the laboratory similar to natural or the other procedures described above. As opposed to outdoor weathering, artificial weathering is done under limited numbers of parameters, which can be adjusted to accelerate aging. This subjects the materials to varying exposure from radiation and/or wetting at increased temperatures.

Clear relationships between the aging processes on artificial weathering and outdoor weathering can only be expected if the important parameters, namely, distribution of the radiation level in the photochemically effective range of the spectrum; test temperature; and type and intervals of wetting cycles and relative humidity are brought in line with a specified outdoor weathering location. Without details of the above-mentioned test conditions, results from artificial weathering can therefore be compared with outdoor weathering results only with a large degree of reservation. Even a comparison of laboratory tests at different places is not possible without this information. Thus, the designation 'good resistance in QUV (trade name of Q-Lab Inc.) apparatus', for example, only gives a very inaccurate assessment of a material's weathering resistance.

Although the automotive industry demands short-term results from accelerated weathering tests, it also expects a high degree of correlation between weathering exposures in the lab and outdoors. Unfortunately, these two objectives cannot be realized in a single laboratory weathering test as the most important routes to fast results like extraordinarily high test temperature and irradiation with unnaturally short wavelengths have a negative impact on the correlation.

A simple rule for good correlation can be given in one aspect: the results from outdoor weathering show the best correlation in those accelerated weathering devices whose source of radiation comes closest to the spectral distribution of sunlight in the photochemically sensitive area of the test. Xenon arc lamps simulate the largest part of the sun's spectrum very well, especially visible light and long-wave UV, while fluorescent UV lamps are better at representing the short-wave area of the sun's spectrum.

In test procedures, a distinction is made between exposure procedures through 'irradiation' and 'weathering'. In the case of irradiation, radiation is simulated 'behind window glass', that is radiation without UVB and reduced UVA part of light and without periods of water spray.

Historical Development of Testing Equipment and the Latest State-of-Art The light sources on the first weathering devices (1918) were carbon arc lamps. Later, mainly the light sources were the focus of improvements as well as standardization. The xenon lamp systems developed in the early 1950s were gradually improved over the years and have now fully replaced carbon arc systems.
- 1918 – carbon arc lamp
- 1933 – Sunshine carbon arc lamp with Corex D filter
- 1954 – Tester with xenon arc lamp
- 1970 – QUV tester with fluorescent FS-40 (UVB) lamp

- 1977 – QUV tester standardized in ASTM G53
- 1983 – Global UV tester, type BAM, with different fluorescent lamps
- 1987 – UVA–340 type fluorescent lamp for QUV
- 1992 – SolarEye irradiation control system for QUV testers.

Xenon Lamp Systems Xenon arc lamps are equipped with filter systems as shown in Figure 12.44, which adapt the spectral distribution to different conditions required by the automotive industry, particularly in the ultraviolet and visible spectrum. Among others, this enables good simulation of sunlight or sunlight behind a 3-mm-thick window pane in the visible spectrum too.

As regard to the degradation of the tested materials, this is only required for testing a few pigments.

The time of the tests is further reduced by a quartz-boron filter combination. This allows a smaller but more energetic portion of short-wave UVB irradiation below the solar cutoff of 290 nm to the samples. Respective norms are SAE J2527, formerly J1960. The very short, but common humidity phases of 3 minutes were extended to 18 minutes as part of the 1990 ISO harmonization in ISO 11341. Certain automotive specifications, such as SAE J 2527, increase the humidity phases to 60 minutes. The 4-hour dew cycle in accordance with ISO 11507 offers a more realistic simulation to natural nighttime humidity from dew.

Because of the aging of the xenon arc lamps and optical filters, the intensity and the spectrum of emitted radiation changes. Even if the intensity of some types retains a certain energy level at 340 nm, the rest of the remaining spectrum is altered. This happens particularly in the UV spectrum, which is photochemically important for macromolecular substances. Therefore, not only is the duration of

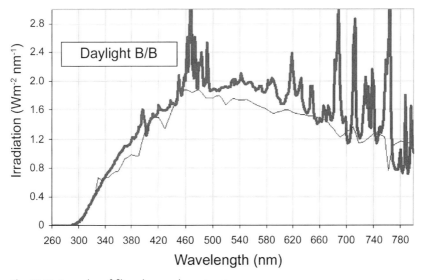

Fig. 12.44 Examples of filtered xenon lamp spectra.

12 Specifications and Testing

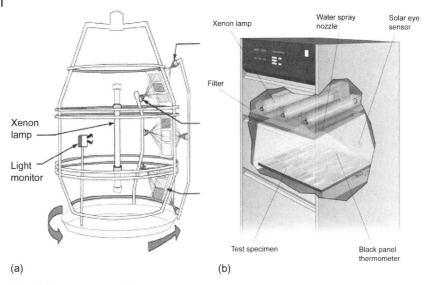

Fig. 12.45 Arrangement of specimens in a xenon system.

radiation measured but also the radiation in the wavelength range below 400 nm or at a certain wavelength, for example 340 nm is used as a reference value for the aging on coatings. The results of tests with new lamps are therefore only comparable to a limited extent with those done within the service life of older lamps.

With short-term weathering apparatus with xenon lamps, the costs of buying, maintaining, and operating the equipment are considerably higher than on the later described QUV systems. When it comes to larger systems, the choice is between two different models, which are outlined in Figures 12.45 and 12.46. Besides that, there are also small, simplified tabletop systems, which are more than adequate for screening purposes.

Both arrangements have proven to be equally effective in terms of the results achieved and the test options available. The specimens form an interior wall of a rotating test carousel on the drum type xenon testers.

In both types of apparatus, the radiation level is continuously recorded and used to control the radiation output. In newer apparatus, a radiation level higher than the maximum radiation from the sun can be set to further accelerate the test. Furthermore, the specimens in both systems can be sprayed with water. To check the temperature, black panel or black standard thermometers are installed at sample level in each case. These record the surface temperature of a black reference sample. Both types of apparatus mentioned are available in versions with additional control of the humidity and test chamber temperature.

The rotation type systems come in several sizes and are also suitable particularly for large sample numbers. The static version offers cost benefits with limited sample numbers due to its simpler design.

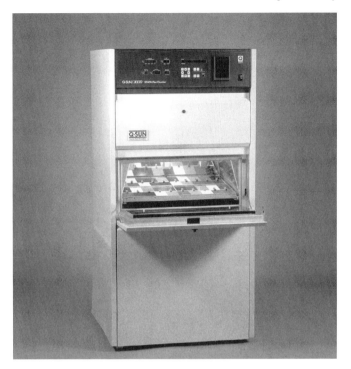

Fig. 12.46 Example of a Xenon system with static sample arrangement.

Figure 12.46 shows an example of a standard system with static sample arrangement, which is outlined on the right of Figure 12.45. The specimen tray measuring about 45 × 72 cm can take 39 samples of size 7 × 10 cm.

QUV Systems Only a fraction of the sun's radiation, which reaches the earth's surface through the so-called optical window in the atmosphere with wavelengths between 290 and 1400 nm, causes nearly all aging degradation by destroying the polymers. We are talking about the short-wave section in the UV region. If meaningful statements are to be produced using short-term weathering apparatus, the radiation level has to be largely recreated in the range up to 400 nm of the solar radiation.

The QUV tester was introduced in 1970 using UV fluorescent lamps which, unlike xenon lamps, emit radiation only in the photochemically effective range under 400 nm. As this type of apparatus hardly produces any visible and infrared radiation, there is no need to cool the specimens. Wetting the surface of the samples is normally done by a dewing process lasting several hours, as is also the case outdoors at night .

The principle of QUV apparatus is based on alternation between UV radiation and wetting through dew or spraying in freely selectable cycles. The specimens themselves form two sides of the test chamber according to Figure 12.47. The

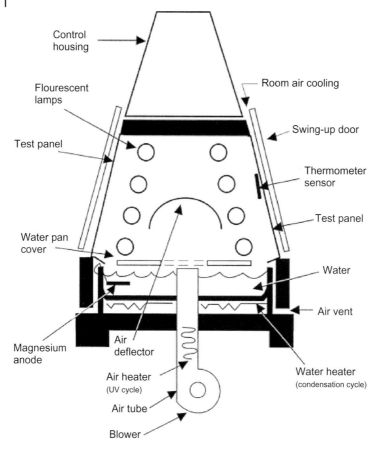

Fig. 12.47 Interior of a QUV tester.

ambient air cools the back of the specimens, thus enabling drops of water to condense on their surface giving sufficient thermal conductivity to the specimens.

The UV lamps are arranged in groups of four fluorescent lamps on top of each other, which irradiate the specimens diffusely with UV light. Just like with the newer Xenon testers, the lamps' drop in the performance caused by aging can be compensated by an irradiation measuring and control system. The typical service life of lamps in the QUV with irradiation control is 5000 hours. Originally, the unavoidable aging of lamps was compensated for by manually changing them over in turn every 400 hours.

The specially designed test chamber, on which the specimen holders can be directly accessed after opening the door, leads to the design of the apparatus shown in Figure 12.48.

UVA lamps having the wavelength range of $315 \leq \lambda \leq 385$ nm reproduce the maximum global radiation up to about 350 nm almost exactly, meaning that the photochemical disintegration process from the effect of solar radiation is perfectly

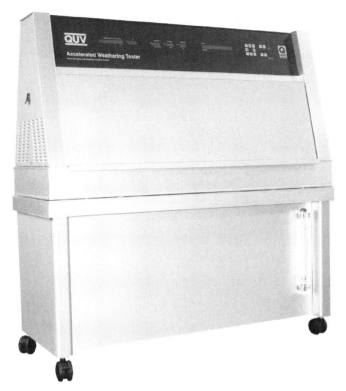

Fig. 12.48 Exterior view of a typical QUV tester.

simulated for many materials. UVB lamps with the range of $280 \leq \lambda \leq 315$ nm preferably emit 'unnatural' UV radiation as shown in Figure 12.44.

Figures 12.49 and 12.50 show the radiation levels of the QUV lamps most commonly used.

Although in many cases, despite the 'unnatural' UV radiation, the correlation to outdoor weathering is thoroughly satisfactory, the use of these lamps is recommended only for comparative tests. In order to forecast outdoor weathering behavior, parallel tests using UVA lamps should be carried out to ensure that the results of materials examined are not significantly skewed on UVB weathering in terms of the degradation relevant to practical use.

Even in QUV lamps, the radiation level in the apparatus with built-in radiation control can be increased by a factor of about 1.75. Besides this, an increased radiation level can be realized by choosing the UVB-313 lamp as shown in the example in Figure 12.50.

How to Use Laboratory Weathering Best? As automotive coatings have to survive different climates and weather conditions, a single laboratory trial can only determine a part of their practical behavior. No generally valid assertions about the life span of a coating can therefore be deduced from a single laboratory weathering trial.

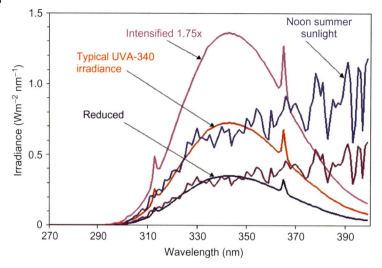

Fig. 12.49 Spectral radiation level of a UVA–340 lamp.

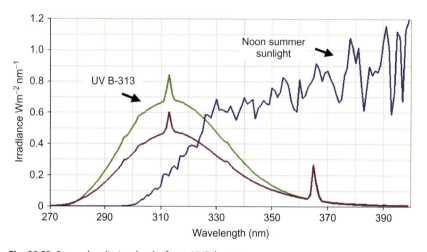

Fig. 12.50 Spectral radiation level of two UVB lamps.

Thus, it is not recommended to drop individual variations on new developments just on the basis of one or two negative test results, particularly if extremely high black panel temperatures and UV ranges were used to shorten test times.

The following procedure has proven to be effective for an initial comparison of new paint variations:

Step 1
- start with harsh conditions to have a great acceleration factor
- establish a ranking for 1–2 key properties
- compare with a reference paint tested in parallel whose weathering behavior is known.

Step 2
- use test conditions without increasing temperature and UV compared to conditions in Florida or your preferred reference climate
- vary the conditions of humidity and temperature without increasing UV compared to conditions in Florida or your preferred reference climate.

This results in indications, amongst other things, as to how paint variations affect the degradation mechanisms [48].

Light Fastness Test (Aging from Hot Light) Coated and self-colored parts in car interiors are damaged by solar radiation and the heat associated with it. To simulate this degradation, coated components are tested in special boxes in Florida and Arizona with regard to their color resistance (light fastness).

In laboratory apparatus, the exposure is simulated by radiation 'behind window glass' (achieved by an appropriate radiation filter in xenon systems or by using UVA-351 lamps in QUV testers). Here it should be noted that temperatures inside cars can reach over 100 °C.

The black panel temperatures recommended in ISO 11341 and 11507 are well below 100 °C. Therefore, higher temperatures should be selected while testing coatings for car interiors. Alternatively, the corresponding American automotive standard SAE J2412 (formerly J1885) specifying a black panel temperature of 70 °C can be applied. Furthermore, it must be noted that with regard to radiation transmission, laminated automotive glass differs considerably from 3 mm window glass. Special factory standards exist at each of the automotive groups on this subject.

12.2.2.3 New Developments

In recent years, two developments to conventional weathering testing have gained significance. Firstly, climate alternating tests are increasingly combined with UV radiation also for automotive coatings. Second, attempts, both in outdoor and laboratory weathering, are being carried out to reproduce degradation to coatings through acid rain.

Climate Alternating Test With UV Radiation It has been shown that corrosion is also affected by degradation in the top coat and primer, particularly if these layers have previously been damaged, for example from being hit by stones. Degradation to the top layers of coatings increases the permeability for water, salt, and corrosive gases, thus aiding corrosion.

In addition to standard cyclic corrosion tests (CCT), experiments have also been underway using the usual cycles with additional weathering by using UV radiation and dew. This method is described in ISO 11997-2 'Resistance to cyclic corrosion conditions – wet (salt fog)/dry/humidity/UV light'. The specimens are subject to a CCT with salt spray and a QUV test with UVA–340 lamps in each case for a week in accordance with the following conditions: 1 week QUV test (4 hours UVA–340 at 60 °C, 4 hours condensation at 50 °C), 1 week CCT (1 hours salt fog at 23 °C, 1 hour dry at 36 °C).

The correlation to natural exposure is very good on many industrial paints. However, this combined test cannot be conducted in one set of apparatus, and the weekly manual moving of specimens scares off many users.

Simulation of Acid Rain Paint degradation from acid rain considerably detracts from the appearance of a car body, especially on new cars with dark colors. For this situation, an annual 14–week exposure period in Jacksonville's harbor area in northern Florida has been established as a standard test. These tests can only be modestly reproduced due to the annual weather variations and the limitation of only conducting the test once a year. Therefore, corresponding lab tests have been developed by several groups. In these lab tests, the specimens are regularly sprayed with acid mixtures and are subjected to a weathering cycle that matches the climate in Jacksonville.

As an alternative to the extensive tests by Trubiroha and Schulz [42, 43], a xenon tester was specially modified by Boisseau and others [49] to simulate acid rain. The starting point for this is a natural reproduction of 'events' in Jacksonville according to Table 12.3.

On the basis of these observations, the test cycle shown in Table 12.4 was developed. Figure 12.51 shows digital evaluations of etched surfaces by enhancement.

The first comprehensive tests have shown that the proposed test correlates well with the results from exposure in Jacksonville. This is illustrated by comparing the Jacksonville results on four typical coatings with the results from the lab test in Figure 12.52 and in Figure 12.53.

After 200 hours in the Q-sun, the relative ranking order was well established and remained unchanged throughout the exposure. After only 200 hours, the lab test gave the same ranking as the Jacksonville exposures.

Table 12.3 Etching test (source: BASF)

Factor	Field	Q-sun technique
Temperature (°C)	70–80	80
Rainfall	10–15 real events, lowest pH 3.5	13 acid sprays, pH 3.4
Dew	Evening dew	Dark step sprays
Humidity (RH)	Typically 80% or greater	Maintain 80% throughout test
Orientation	Zero to five degrees	Zero degrees
UV light spectrum	Variable spectrum	Noon summer sunlight
UV light intensity (W m^{-2} at 340 nm)	0–0.68 (changing)	0.55 (constant)

12.2 Weathering Resistance of Automotive Coatings

Table 12.4 Cycles of laboratory etching test (source: BASF)

Step 1	1 min	Dark exposure with acid rain spray
Step 2	3 h 50 min	Dark exposure, 38 °C panel and air temp, 80% RH
Step 3	12 h	Light exposure 0.55 W/m2/nm at 340 nm, daylight filters, 80 °C black panel temp., 55 °C chamber air temp., 80% RH
Step 4	27 min	Dark exposure, 38 °C panel, and air temp, 80% RH
Step 5	1 min	Dark exposure with pure DI water spray
Step 6	3 h 50 min	Dark exposure, 38 °C panel and air temp, 80% RH
Step 7	1 min	Dark exposure with pure DI water spray
Step 8	3 h 50 min	Dark exposure, 38 °C panel and air temp, 80% RH
Step 9	Repeat	Step 1

Fig. 12.51 Digital enhancements of etching test results.

After 400 hours, it produced the correct Spearman rank order (rho = 1,0) and approximately the same level of etching as seen after 14 weeks of a Jacksonville exposure in 2001.

12.2.3
Standards for Conducting and Evaluating Weathering Tests

For a clear description of weathering results, it is necessary to show both, details of the weathering conditions and the method of evaluation. Both items are described not only in numerous national and international test standards, but also in automotive groups factory standards partly, for example German Verband der Deutschen Automobilindustrie (VDA) and American Society of American Engineers (SAE).

Apart from one exception, the following illustration only deals with the most important ISO standards. These are also often used as national and European

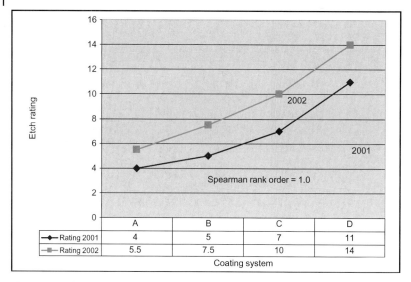

Fig. 12.52 Etching outdoor test results in Jacksonville, Florida.

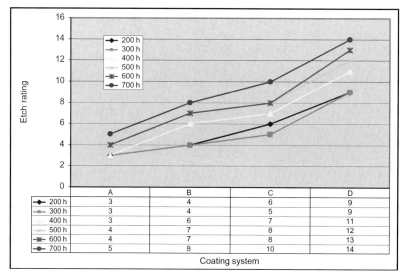

Fig. 12.53 Laboratory test results of Q-sun/BASF.

standards, and as a result, they are available in several languages. For example, ISO standards, which are valid both as a German and European norm, are identified with the prefix DIN EN ISO.

Outdoor weathering is normally conducted by commercial service providers or paint manufacturers according to specifications. Irrespective of this, the outdoor weathering of coatings is described in ISO 2810 and also in accordance with VDA 621-403.

As set out in the section 'laboratory weathering', users have a choice between QUV and xenon light systems when it comes to artificial accelerated weathering.

The test parameters that are recommended in the most important test standards have been put together in Table 12.5. All the standards mentioned, contain amongst other things, detailed advice about operating the apparatus and conducting tests, as well as references to numerous standards for evaluating the tests. Besides the parameters listed in the table, each standard also describes alternative weathering conditions. In particular, ISO 11341 advises that a black panel temperature of 50 °C is preferred for tests on colorfastness.

Standards ISO 11507 and 11341 also contain different test methods for radiation behind window glass.

The fluorescent lamp UVA–351 is used for this in QUV apparatus, while solar radiation behind 3–mm window glass is simulated in xenon testers using appropriate filters. However, this does not cover all the test requirements of the automotive industry for coatings in vehicle interiors. In-house standards partly require extremely high black panel and test chamber temperatures as well as radiation filters that match car windows.

In view of the extremely high resistance of modern automotive clear coats, a very high black panel temperature of 70 °C is required in the SAE standards mentioned,

Table 12.5 The most important standards for laboratory weathering on coatings

Standard radiation source	Radiation	Cycle	Temperatures Irradiation phase		Dark phase
			Black panel temperature	Test chamber temperature	
ISO 11507 Fluorescent lamp UVA–340	Adapted to CIE 85 in the UV, about 0.9 W m^{-2} at 340 nm	4 h radiation 4 h condensation	60 °C		50 °C
ISO 11341 xenon arc lamps	Closest adaptation to CIE 85, about 0.5 W m^{-2} at 340 nm	102 min radiation without rain 18 min radiation with rain	63 °C	38 °C	
SAE 2527/1960 xenon arc lamps	Adapted to CIE 85 with increased UV 0.55 W m^{-2} at 340 nm	40 min radiation without rain 20 min radiation with rain 60 min radiation without rain 60 min dark phase with rain	70 °C 70 °C	47 °C 38 °C 47 °C	38 °C

along with higher UV proportions under 320 nm compared to sunlight. These 'unnatural' increases compared with ISO 11341 help toward reducing the duration of test. The German automotive industry has also adopted harsher conditions of the SAE standards as a VDA specified recommendation.

For plastics, there are ISO standards that had been harmonized with the above-mentioned ISO 11505 and ISO 11341. The corresponding standards are ISO 4892 Part 1–3 under the title 'Plastics – Methods of exposures to laboratory light sources, Part 1: General Guidance, Part 2: Xenon-arc-sources and Part 3: Fluorescent UV lamps'.

Apart from this, every car manufacturer has its own corrosion alternating tests that are applied with varying parameters. Data and experiences gathered over the years are available on this subject, which also correlates to practical tests on vehicles.

When evaluating weathering degradation, a comparison is made to the initial condition of the paint. For some of the properties that are of interest in this regard, such as the gloss and color, there are widely accepted measuring procedures available. As for other properties, such as the formation of blisters and cracks, visual classifications are predominantly undertaken. Both types of assessment are largely described in standards. The ISO 4628-1 to 10 series of standards has the general title 'Paints and coatings. Evaluation of degradation of coatings. Designation of quantity and size of defects, and of intensity of uniform changes in appearance'. Part 1, 'General introduction and designation system', describes the system of classifications. Visual impressions are classified by means of pictorial standards on a scale of zero (unchanged) to five (very marked change). Part 2, 'Assessment of blistering', helps to classify blisters regarding their quantity and size. Similarly, Parts 3–5 enable rusting, cracking, and flaking to be assessed. Parts 6 and 7 deal with the assessment of chalking, and parts 8 and 10 involve the evaluation of corrosion. In addition to the properties already mentioned, wherever necessary, especially in the case of outdoor weathering, soiling, fungus buildup, and staining are also evaluated.

All visual classifications are often documented using photographs that are suitably enlarged, if necessary, with the aid of a microscope. For years, a so-called image analysis has been carried out on the basis of standardized photos for the purposes of classification conforming to standards. This image analysis is commercially available in various forms, and the first applications used to assess degradation on coatings are already written into standards, for example ISO 21227.

12.2.4
Correlation Between Artificial and Natural Weathering Results

The degradation process in automotive coatings of base coat–clear coat depends on climatic factors and the chemical structure of the paints. Thus, in the best scenario, a factor for testing time reduction can only be specified for a laboratory weathering test that is adapted to a certain climate. In accordance with this, the SAE standard J2527 is regarded as the accelerated simulation of Florida weathering [42, 50]. As the time reduction in the laboratory weathering is not uniform in the two works that are quoted, the authors cannot give a conversion factor.

According to a study by Ford [41], positive results are required from an outdoor weathering trial in Florida and at other exposure locations lasting at least 5 years before a new paint system can be introduced. During this time, the gloss of a clear coat is clearly reduced, but no cracks occur, and more precisely, the base paint and its adhesion to the clear coating remains intact. In order to deduce reliable forecasts about durability from laboratory weathering tests, besides the evaluation of visually noticeable changes, the degradation mechanisms are examined analytically [45, 46, 51, 52] and models relating to the degradation process are partly derived from the analytical data [39, 53–55].

By means of IR spectra, for example, analytical studies can determine which chemical bonds are broken down by photo–oxidation and to what extent chemical changes during laboratory weathering match those from outdoor weathering. Moreover, analytical studies on microtome sections [41, 45] can demonstrate changes in the layer between clear coat and base paint early on, meaning that it is possible to explain whether or not such changes bring about subsequent delamination.

The works by Wernstahl [42, 46] have already shown that acid rain makes a significant contribution to taking the gloss of clear coatings. In a recently published work by Schulz [43], the current status of these type of tests is compiled using a comparison of the 14–week test in Jacksonville with an artificial exposure to acid from regular spraying during conventional Florida weathering. The authors came to the conclusion that in view of the effects of annual weather variations on the ability to reproduce the tests in Florida, a laboratory weathering test using acid solution sprays [44, 49] shows definite benefits in forecasting real environmental performance.

12.3
Corrosion Protection

Hans-Joachim Streitberger

12.3.1
Introduction

The corrosion protection of cars was a problem of great economic importance. The cost connected with the corrosion of passenger cars was reported to be about 16 billion US$ in US in 1975 [56]. Since the 80s of last century, efforts have been put on this problem by a number of car manufacturers and the coating industry. Today, the level has been down very much. Many car manufacturers guarantee corrosion protection of their cars up to 12 years.

It has been learned that the different ways of building a car as well as the many different materials being used today (see Chapter 2) needs to be judged for potential corrosion problems starting at the design phase of new automobiles. Some of the materials are used for improved corrosion protection like galvanized steel, hot dip galvanized steel, and coil coated steel. But the car body construction is depending on special stiffness characteristics and weight reduction so that many different materials come together. Under those circumstances, the ability of the electrocoat process to cover inner space sections as well as the different pretreated materials (see Chapters 2 and 3) needs to be tested. Even calculation programs exist which may cease those tests carried out with total bodies [57]. Before these so-called 'proving ground' tests are done, a lot of test panels and different type of specimen are tested in specially designed equipment for salt spray and climate- changing treatment. Considering the different coatings on a car, these tests are carried out with electrocoated panels, totally topcoated panels, panels with sealing materials over electrocoat, and waxes over electrocoat just to name the most important conditions. This results in a lot of small test series to assure that the final tests on total cars will not fail. In case of not passing the proving ground test, the total development of the car is in jeopardy because of the fact that corrections in construction and material will cost a lot of time.

These considerations are valid for development of a new car and even more to introduce a new electrocoat with improved performance. This is the reason that all the important paint suppliers for the electrocoat have pilot tanks somewhere in the world where test bodies can be coated, evaluated, and tested (refer Figure 12.54).

In general, today's surface corrosion is negligible because of the mostly used galvanized steel for the outside surface. Nevertheless, each new electrocoat is tested extensively for this behavior over different substrates.

Corrosion problems can also occur if the total cleaning, pretreatment, and electrocoating process is not running in the specified range.

The main types of corrosion on a body, which can become visible to the car owner after some years are as follows:
- edge corrosion
- contact corrosion
- inner space corrosion

Fig. 12.54 Pilot electrocoat tank (source: BASF).

- flange corrosion
- welding spot corrosion
- stone chip.

All aspects will be tested in the laboratory either by standardized test methods or by methods that are mutually agreed upon by the supplier and the car manufacturer.

During development and approvals, not only the electrocoated panels but also the total automotive coating system including primer surface and top coat are tested. Experience so far shows that these tests are only securing tests especially when it comes to the evaluation of surface protection.

12.3.2
General Tests for Surface Protection

Specified test panels of different sizes mostly between 10×20 and 10×30 cm are used for general corrosion testing of electrocoatings. The specification of these panels focuses on type of steel, that is cold roll steel, galvanized steel, or aluminum, type of pretreatment, and storage conditions. They are manufactured mostly from service companies or the pretreatment suppliers according to different specifications of the car manufacturers.

These panels will be electrocoated with standard film build, baked under specified temperature/time curve, and either tested as single coat or tested with the total top coat system including primer surfacer. Before entering the test cabinet, either a scribe vertical to the panel position or a cross is made through the coating down to the steel surface.

The most common test for steel is according to ASTM B117 or DIN ISO 50021, the so-called salt spray test. Duration of testing for automotive primers is 1000 hours. After this time the rust at the scribe must be less than 2 mm wide. Often the scribe will be scratched or treated with adhesive tape for better finding the areas of adhesion loss and better reading. Each type of evaluation is standardized in ASTM D1654 (numbers are from zero to five) and includes the panel preparation in DIN EN ISO 7253.

A first impression of edge protection is monitored even though the edges of the standard panels are not carefully specified. These are the numbers according to DIN on the left side of panels in Figure 12.55.

The salt spray test is only useful for cold roll steel to compare the corrosion protection potential of primers like the common cathodic electrocoating systems (see Chapter 4).

Important corrosion tests became the climate-changing test for example, according to VDA 621-415 or SAE J2334. Those tests simulate the impact of climates to the test panels in cycles and are very much accepted by the car manufacturers due to the fact that these tests duplicate the conditions cars will find around the world during their lifetime. Those cycles often include chipping, water soak, dry periods, heat, and cold treatment as well as salt spray or salt soak periods. Cycles can be 1 week for example at the VDA-test or a day for the SAE test. The test

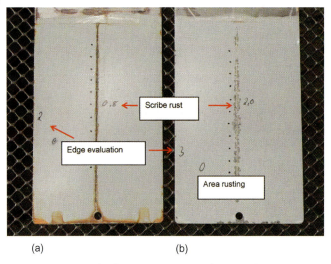

Fig. 12.55 Test panels after corrosion testing by 10 weeks (= cycles) climate change test according to VDA 621-415 ((a) steel, (b) galvanized steel).

panels are prepared and evaluated in the same way as for the salt spray test (see Figure 12.55).

The climate change tests according to VDA 621-415 is run 10 cycles (= 10 weeks), the SAE test is run mostly 60 cycles (= 60 days), and they are suitable for cold roll steel as well as galvanized or hot dip galvanized steel. The depth and type of the scratch is very crucial for the test result. Normally, the Zn layer should only be

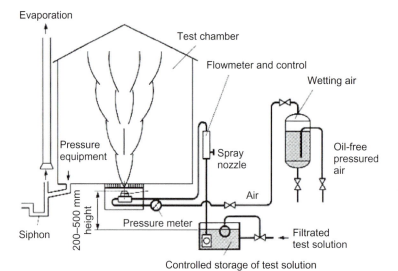

Fig. 12.56 Scheme of a corrosion test chamber (source: BASF Handbook).

Table 12.6 Typical data of corrosion test results according to VDA 621-415 after 10 cycles

Evaluation	Steel		Galvanized steel	
	Electrocoat only	Top coat	Electrocoat only	Top coat
Rust at scribe (mm)	<2	<2	<2	0
Surface rust [0–5]1	0	0	0	0
Edge rust [0–5]1	<1	0	0	0

scratched. In case of uncovered iron surface in the scratch, the result will become worse and less reproducible. (see Table 12.6).

All car manufacturers have their specifications for these tests and their own description of test procedures, but are very similar in terms of the requirement.

The condensed water constant climate tests according to DIN 50017 and DIN 50018, known in Germany as the 'Kesternich' test, are not frequently used anymore.

For aluminium, the copper accelerated acetic acid spray (CASS) test according to ASTM B368 is commonly used as a basic information. Sometimes, filiform-corrosion occurs on aluminium panels, which generates after the test under various conditions. DIN EN ISO 4623 describes a test, which is able to test this specifically.

For those tests on the described specimen, special cabinets are available in the market that are able to meet most of the standardized tests by electronic programming. Even the reproducibility of those cabinets in terms of temperature, humidity, and salt fog is quite good the variation coefficient of one parameter set, that is standardized panel, electrocoating procedure, scribes accuracy, and test conditions is about 30 to 50%. This results in some statistical expenditure of many months to find out respective improvement of the corrosion performance of a new electrocoat material. In other words, one panel and even one single test set is not enough to find out a significant improvement.

Some of the car manufacturers require outdoor exposure tests. As a result of the different conditions on different locations at different times around the world, these tests can only find out differences in a single series. They are carried out with weekly salt spraying and evaluation procedure according to DIN 55665:2007-01. Both type of panels with and without top coats are tested. Only electrocoated panels are sometimes covered against UV radiation by an open box. Reasonable results are given after 2 years depending on the location. The outdoor tests best duplicate the real behavior of galvanized panels with the total automotive coating system, which normally does not corrode after more than 12 years under an untouched surface.

From a research standpoint, the electrochemical impedance spectroscopy (EIS) has gained acceptance for testing the potential of corrosion protection of an electrocoated panel [58, 59]. It helps to understand corrosion mechanism and influence of film properties like degree of cross-linking in a period of testing time of about 10 days. (see 12.7).

Table 12.7 Comparison of typical length of different corrosion tests

Test	Standard time (d)
Electrochemical impedance spectroscopy	10
Salt spray test	42
Climate change test (VDA 621-415)	70
Outdoor testing	365–730

12.3.3
Special Tests for Edge Protection, Contact Corrosion, and Inner Part Protection

Special specimen are designed sometimes in case of significant development steps of electrocoat materials to duplicate the flange corrosion [60], contact corrosion, corrosion in inner parts, and edge corrosion by the standard tests. These specimens are small parts welded or clinched together, then cleaned and pretreated, primed by cathodic electrode position process, and then tested.

As seen in Figure 12.57, the delamination occurs on the steel side (left) of those panels where also edge protection is less than on the galvanized side (right).

Throwing power panels (see Figure 12.58 and Chapter 4) according to the Ford test are also tested to evaluate the corrosion protection in inner parts by finding the lowest film build to pass the different tests. The line at which corrosion in the surface of the film starts gives two types of information: the film thickness and the throwing for power for corrosion resistance. For standard cathodic electrocoats, the thickness of this film is about 5 μm.

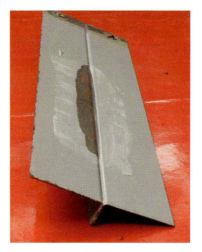

Fig. 12.57 Typical test panel for contact corrosion and sealer testing after 10–week climate change test (left panel: Steel).

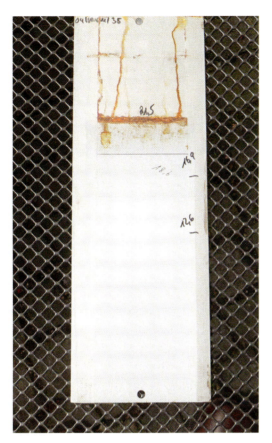

Fig. 12.58 Corrosion test result on a throwing power panel.

Special emphasis has been put to the fact that edge corrosion cannot be measured quite accurately. For development purposes, paint manufacturer sometimes uses specially developed test like the measurement of edge protection by electrical discharges in a specific device [61]. Main result is the fact that the viscosity level and profile of cathodic electrocoats during the baking process plays an important role as the lower the viscosity, the better is the flow and lesser the edge covering (see Chapter 4).

12.3.4
Total Body Testing in Proving Grounds

In case of complete new generations of electrocoat materials, the risk of unexpected weaknesses on the complex body of a car will be minimized by adding the so-called proving ground test to all the results on panels and specially designed parts in salt spray and climate change tests. More or less, all the car manufacturers have test areas where they can test new cars in terms of their driving properties and long-term behavior. This includes the overall corrosion resistance of the completely built vehicle.

The car undergoes mostly a daily cycle similar to the climate change test. This includes cool down periods with mechanical stress–like torsions. This is followed by a driving period on the streets of the proving ground with chipping impact and salt sludges.

After all this, the car is evaluated by complete knock down. The results are very important hints for construction methods of the body-in-white.

Owing to the fact that these tests are very costly and time-consuming, emphasis is made to calculate throwing power, film build of all coatings as well as constructive factors to reduce the number of tested cars [62].

12.4
Mechanical Properties

Gerhard Wagner

12.4.1
General Remarks

All mechanical examinations test the properties of organic coatings under simulated lab conditions are representative of conditions, which the car could encounter during its lifetime. Examples include stone-chip examinations, steam jet tests, abrasion, and scratch resistance tests. In addition, very simple examinations such as bending tests, impact tests, and pull-off tests fall into this category.

All specifications of automotive manufacturers, independent of their region, include mechanical test methods to prevent and/or avoid later failures of the coating during service life. Most tests are carried outduring development to first get a system approval, which is more fundamental, and then complicated tests are performed, for example, scratch resistance tests. Some of them like hardness and adhesion tests are later required for batch controls. The examinations are dominantly carried out at the full paint layer system.

Guidelines for the examinations are national and international standards, as well as supply and laboratory test specifications, which are agreed upon between paint supplier and carmaker. Looking at the history of paint inspection technology over the last decades, it has to be realized that even for the examination of the same properties, often a variety of different test methods is available. However, the effort for leaner processes and the cost pressure should force the partners to reduce the number of methods.

According to ISO 15528, ISO 1513, and ISO 1514, a representative sample of the coating material to be tested has to be chosen. The appropriate selection of the substrate and the specimen has to be made and then be coated. The time intervals and temperatures for drying, baking, and aging are to be carefully standardized. For the dimension of the samples, a size of approximately 10×20 cm is proved to be useful. The car body sheet thickness of approximately 0.7 mm is preferred on metals. In case of plastic substrates, the right materials have to be defined for the samples.

Before testing, the samples are usually preconditioned at 23 ± 2 °C and 50 ± 5% RH for a minimum of 16 hours. The mechanical properties of coatings are generally very strongly dependent on temperature and humidity. So it is necessary – at least in arbitration cases – to carry out the actual test under these standard conditions of 23 °C and 50% RH in air-conditioned rooms or chambers.

Usually the testing is carried out in duplicate or in triplicate form. The examinations are carried out as either of the following:
- 'pass/fail' test – that is by testing whether the sample withstands specified conditions to assess compliance with a particular requirement;
- a physical value is determined, where destruction is observed or which generally describes the material behavior.

In many cases, precision values like repeatability and reproducibility, which characterize the accuracy of the test method have already been determined in interlaboratory tests, so-called 'round robin tests'.

12.4.2
Hardness

The term *hardness* is not clearly defined with organic coatings and often leads to misunderstandings as nearly all coatings as a type of plastic films show a viscoelastic material behavior. The hardness is defined as follows in the DIN 55945: Hardness is the resistance of a coating against a mechanical impact like pressure, rubbing, or scribing. Dependent on the test method, the hardness allows drawing conclusions on different characteristics such as cross-linking, wearing characteristics, scratching resistance, or elasticity. Essentially, there are three principally different hardness-testing methods: the pendulum damping test, the indentation hardness tests, and the scratch hardness tests.

12.4.2.1 Pendulum Damping
The pendulum damping test method is based on the principle that the oscillation amplitude of a physical pendulum resting on a coating surface decreases the faster, the softer, and less elastic the surface is. The interaction of pendulum and coating is very complex and depends among other things on the viscoelastic properties of the coating film.

For carrying out respective tests, a pendulum of defined mass, geometry, and oscillation duration is applied upon the coated surface. After deflection to a start-amplitude of 6° angle, it is brought to oscillation. The duration time or number of oscillations respectively, up to the decrease of the oscillation amplitude to a defined final value of 3° angle is determined. The shorter the damping time or the number of oscillations, the lower is the hardness.

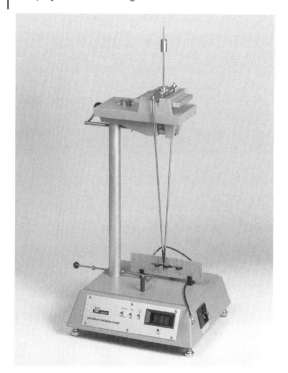

Fig. 12.59 König Pendulum damping device (source: Byk–Gardner).

This test method goes back into the beginning of the paint inspection technique and it is used till today for the characterization of coatings. Two different types of devices are available, the more frequently used König pendulum (see Figure 12.59) and the Persoz pendulum. The test method is described in ISO 1522. The calibration of the devices is carried out on a polished, flat glass plate. The deflection of the pendulum is made automatically with new devices. The measurement of the number of oscillations or the oscillation time takes place automatically in a similar manner by light barriers.

12.4.2.2 Indentation Hardness

A relative simple hardness test method is the Buchholz indentation test. It is a procedure, which can be carried out relatively quickly and inexpensively as field–test, however, only on horizontal surfaces. The precision of this test is not very high. In particular, at nonprofessional application for example, below the minimum thickness of the coatings chances of error are very high.

The indentation apparatus consists essentially of a rectangular block of metal with two pointed feet and a sharp-edged metal wheel of hardened tool steel as indenter (see Figures 12.60 and 12.61). The indenter and the two feet are positioned on the car surface in a way that the instrument is stable, its upper surface is horizontal, and the effective load upon the indenter is 500 ± 5 g. The indenter is allowed to affect

Fig. 12.60 Buchholz test equipment (source: Byk–Gardner).

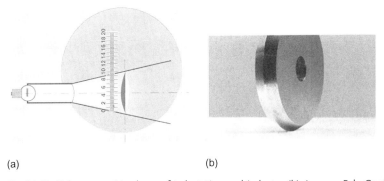

(a) (b)

Fig. 12.61 Light source (a), shape of indentation and indenter (b) (source: Byk–Gardner).

onto the coating for a defined duration time. In the coating surface, a remaining shape of the indentation is produced. Consequently, this test procedure is also not suitable for coatings with high flexibility as the indentation mark might heal away. The indentation depth depends on the layer thickness. The results are useful only if the layer thickness exceeds a minimum value indicated in the standard.

As test result, only the length of the produced indentation in the coating surface is taken. A conversion of the indentation length into the prior common Buchholz indentation resistance α (B) can be calculated additionally according to formula (1).

$$\alpha(B) = 100/l \quad \text{where } l = \text{indentation length in millimeter} \quad (1)$$

For the determination of the indentation length, the device is placed gently onto the test panel. The indenter is left in position for 30 ± 1 seconds.

The length of the indentation shadow is measured 35 ± 5 seconds exactly with a microscope in millimeters to the nearest 0.1 mm after removal of the device (see Figure 12.61). Five tests are carried out on different parts of the same test panel and the mean value is calculated.

The determination of the indentation length and/or the indentation resistance is affected by a large margin of error particularly at small indentation lengths induced on hard coatings. The test procedure is described in ISO 2815.

The microindentation hardness test is a matter of the load/indentation depth method according to ISO 14577-1. With this method, an indenter, typically a Vickers pyramid in the area of coatings, is continuously pressed into the tested material with an increasing test load and then unloaded. The indentation depth is measured at the same time under load. Taking into account the geometric relationship between the indentation depth and the shape of the indenter, this measurement produces the physically meaningful Martens hardness HM according to equation (2).

$$HM = F/A \qquad (2)$$

where F = test load in N and A = surface of the indentation in mm^2.

In consideration of the increasing/decreasing test load, one receives a hardness profile in dependence of the indentation depth. The indentation depths are situated thereby in the range of some μm, so that it is a matter of an absolutely nondestructive test method. Furthermore, the so-called indentation hardness H (IT) can be determined, which is calculated at maximum test load according to equation (3).

$$H(IT) = F_{max}/A \qquad (3)$$

Further, important technological characteristics can be obtained from the result of the load/unload cycle, for example the elasticity modulus of indentation. A plastic or elastic proportion of indentation work can be calculated. Additionally, the so-called indentation creep is an important value, which can be determined from the change of indentation depth at a constant maximum or minimum test load.

While for the examination of coatings, usually empirical test methods are used, with this test method a measuring procedure is available, which supplies a lot of physical parameters that are particularly of great importance for the development. Conclusions from important quality criteria can be drawn such as surface hardness, degree of cross-linking, and flexibility as well as scratch resistance with consideration of the Reflow behavior. The method is however, not suitable for field measurements. The execution of the measurements has to be carried out absolutely vibration-free. The measured values also depend strongly on the ambient temperature and the RH. With the help of a programmable measuring stage, the execution of the measurements can be automated. An accurate selection of special measurement positions can also be made with the help of a video microscope. By this way, for example measurements of defects like craters, inclusions of dirt,

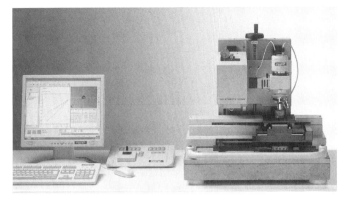

Fig. 12.62 Total hardness measurement system (source: Helmut Fischer GmbH & Co. KG).

and etchings are possible for failure analysis. In Figure 12.62, total hardness measurement system is shown.

12.4.2.3 Scratch Hardness

The scratch hardness tests represent a completely different type of impact. Here, test methods are specified and the resistance of a coating against the indentation of a scriber is determined under defined conditions. At this, either the coating is only scribed on the surface or it is scribed down to the substrate.

According to ISO 1518, the coating is scribed down to the substrate. A scriber with a hemispherical tip is used. The diameter of the scriber tip is 1 mm. The examination can be used as 'pass/fail'- test – that is after scratching with a defined constant test load the scribing down to the substrate is evaluated. In the other case, the minimum load is determined, which generates the scribing through the coating. In this case, the load is constantly increased. For the execution of the test, automatically motor-driven devices are also available. The scribing speed is 30–40 mm s^{-1}. This test method is well applicable for metallic substrates. Here, over an electrical contact of the scriber to the substrate, the penetration is indicated. Also, scribers with other geometry, for example according to Clemen, can be used for deviating from the standard.

The hardness meter (DUR-O-Test) (see Figure 12.63) is a pocket instrument allowing simple hardness tests on flat, and curved coating surfaces. It evaluates, principally, only the scratching of the coating surface. The instrument consists of a sleeve with a pressure spring that can be bent to various tensions by using a slide. The spring acts on a tungsten carbide needle with its tip extending out of the sleeve. A locking screw fixes the slide, thus maintaining constant spring tension. Three pressure springs with varying strengths of 0–3, 0–10, and 0–20 N are available to cover a large range of hardness. For different applications, needles with diameters of 0.5, 0.75, and 1.0 mm are used.

For the examination, the hardness meter is drawn evenly with the selected spring strength over the sample surface. The spring strength, which produced barely

Fig. 12.63 Hardness meter (DUR-O-Test) (source: Erichsen).

a visible scratch indicates the measurement reading. Often however, the visual evaluation is not simple and depends, in many cases, on the experience of the examiner, particularly with soft coatings. For field tests, however, this method is readily used to have a prompt evidence of the curing.

An old, well-proven, but not very exact hardness test method is the so-called pencil test. After many years of critical consideration, the method was published 1998 as an international standard ISO 15184. This documents an international interest in this nearly forgotten test method. The advantage is the simplicity of the method in any case. Quick evaluations of the surface hardness can also be accomplished in the field.

For the execution, a set of wooden drawing pencils with 20 different hardness levels between 9 B (soft) and 9 H (hard) are available. After removing approximately 5–6 mm of wood, the tip of the lead shall be sanded maintaining an angle of 90° until a flat, smooth, circular cross-section is obtained free of chips or nicks in the edges. For the determination, the pencil is pushed by hand at an angle of 45° over the coating surface in the direction away from the operator at a speed of approximately 1 mm s^{-1}. Also, a mechanical test instrument is helpful into which the pencils can be clamped. A defined constant contact load of 750 g is assured here.

If the pencils are softer than the coating, they are sliding easily over the surface. With the transgression of certain pencil hardness – if the pencil lead is harder than the coating – the pencil scratches the surface and a remaining mark occurs. The hardness of the hardest pencil, which does not mark the coating, is the so-called pencil hardness.

The disadvantage of this test method is the high inaccuracy. As a function of type, age and manufacturer of the pencil leads, and in addition, in dependence on the experience of the operator, large differences in the test results are obtained. The application is particularly difficult with soft or flexible coating systems.

12.4.3
Adhesion and Flexibility

12.4.3.1 Pull-Off Testing

The pull-off test describes a procedure for assessing the adhesion of a single coating or a multilayer system of coating materials. The minimum tensile stress is determined, which is necessary to detach or rupture the coating in a direction perpendicular to the substrate. The examination may be applied on a wide range of substrates such as metal, plastics, wood, or concrete. With small mobile devices, this test method is used also in the field, for example for an acceptance procedure in the severe corrosion protection or expertise in the automotive industry.

Onto the cured coating system, test dollies (test stamps) with a preferential diameter of 20 mm are glued directly using an adhesive. Special attention is required in selecting suitable adhesives for the test. In most cases, cyanoacrylates, two-component solventless epoxies, and peroxide-catalyzed polyester adhesives have been found suitable. In no case cohesive failures in the adhesive or adhesive failures between dolly and adhesive may be generated. The adhesion of the adhesive on the coating can be improved by slight sanding also.

The test is carried out by a tensile testing machine or a mobile hand-driven pull-off apparatus (see Figure 12.64). The tensile stress shall be increased at a substantially uniform rate, which may not exceed 1 MPa s^{-1}. If specified or agreed between the interested parties, the coating can be cut around the circumference of the dolly down to the substrate with a suitable cutting device. The test dollies can be bonded on the coating from one side only. More suitable, however, is a procedure where also a second dolly is bonded on the backside coaxially aligned. Then both test dollies are clamped into a tensile testing machine and pulled apart. In this way, the dollies are aligned optimal coaxially so that the tensile force is applied uniformly across the test area to avoid deflections and shear stress, which can easily falsify the test result. At least six determinations have to be carried out because the results show a high statistical variation.

During the test procedure, the tensile stress is increased evenly up to the pull off of the test dolly.

The maximum tensile stress required to break the test assembly (breaking strength) is recorded and reported in MPa. Further, the fracture surface is examined visually to determine the nature of fracture. The proportion of cohesive and adhesive failure as well as the failure area is estimated. The test method is standardized in ISO 4624.

12.4.3.2 Cross Cut

Although this empirical test method for the examination of adhesive strength is frequently disputed, it is used very often due to its simplicity. The advantage consists in the fact that it is also applicable as field check. According to ISO 2409, a right-angle lattice pattern is cut into the coating penetrating through to the substrate. With a lack of flexibility and adhesive strength, both at the substrate and in the intermediate layers, detachments and/or breaking-off of particular fragments in the coating can occur.

Fig. 12.64 Hand-driven pull-off apparatus (source: Erichsen).

As cutting tools, the so-called single-blade cutting tools as well as multiblade cutting tools are available. Because of their higher reproducibility and their better cut quality, the single-blade cutting tools are preferred. The test is to be carried out at least three different places on a test panel. For the preparation of the cuts in addition to simple handsets (see Figure 12.65), motor-driven instruments also are available (see Figure 12.66). In Figure 12.67, a fully automatic cross-cut equipment is shown. The number of cuts in each direction of the lattice pattern according to ISO 2409 should be six – that is a lattice with 5 × 5 squares is formed. In the automotive industry, however, different standardized variants are possible, for example 10 × 10 squares or additional diagonal cuts. The spacing of the cuts is chosen mainly depending on the total layer thickness.

$$\begin{aligned}
0\text{–}60\,\mu m &\quad : 1\,\text{mm} \\
60\text{–}120\,\mu m &\quad : 2\,\text{mm} \\
120\text{–}250\,\mu m &\quad : 3\,\text{mm}
\end{aligned}$$

The loose paint particles are removed before the evaluation takes place with a soft brush or better with an adhesive tape. Here the requirements of the users

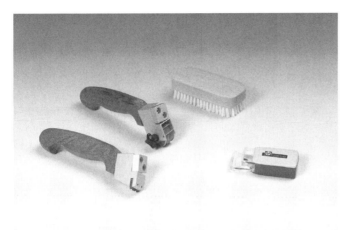

Fig. 12.65 Hand cutting tools (source: Byk–Gardner).

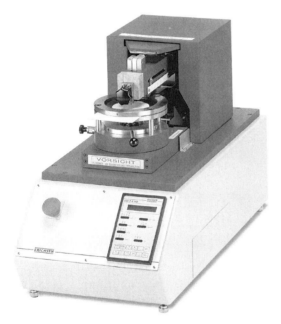

Fig. 12.66 Automatic cross-cut equipment (source: Erichsen).

deviate strongly from each other. In the ISO 2409, a transparent pressure-sensitive adhesive tape with adhesion strength of 10 ± 1 N per 25 mm width is required.

For the evaluation of the examination, the cut area of the coating must be assessed carefully. With reference standard pictures, a six-step classification is given. By comparison with the standard pictures these values can be defined (see Figure 12.68).

Fig. 12.67 Fully automatic cross-cut equipment (source: DuPont Performance Coatings).

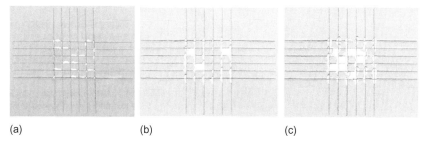

Fig. 12.68 Test panels with classification 2(a), 3(b), and 4(c) (source: Byk–Gardner).

12.4.3.3 Steam Jet

Meanwhile, the steam jet tests were established over the past years in the automotive industry. Unfortunately today, approximately 15 different test procedures are still known because no standardized method is present yet. However, first suggestions for a standardization have already been submitted at DIN (German institute for standardization).

These test methods are procedures for the determination of the paint adhesion both on metallic and on nonmetallic substrates, in particular, on plastics. In the process, the exposure of coated surfaces to a steam jet equipment during cleaning is simulated. In Figure 12.69a, steam jet test equipment is shown.

A specimen is usually treated with a cross cut– that is St. Andrews cross according to ISO 2409 and several specifications in the automotive industry, where the cut has to go down to the substrate. This scribe is stressed afterwards by a steam jet of high-pressure cleaner with specifically defined parameters.

Fig. 12.69 Steam jet test equipment : (a) = total view; (b) = panel positioning (source: DuPont Performance Coatings).

The usual test parameters are as follows:

Water temperature at the nozzle	: 50–90 °C
Pressure at the nozzle	: 30–120 bar
Distance nozzle to object	: 20–250 mm
Nozzle	: Fixed, oscillating
Angle of nozzle to object	: 25–90 °
Duration	: 15–300 s

After exposure to the steam jet, the sample is examined visually for possible delaminations of the coating. Delaminations within the region of the scribe are measured in millimeter perpendicular to the scribe. Normally, delaminations < 1 mm are tolerated. Multiple determinations are necessary because statistic fluctuations can appear, in particular, at high pressures and at small distances.

For the calibration of the testing device, a spray pattern on an expanded Styrofoam block (Styrodur) is produced under defined conditions. The generated spray pattern is measured by length, width and depth.

12.4.3.4 Bending

If a coated test panel with the painted side outside is bent over a cylindrical mandrel, the surface is stretched depending on the mandrel diameter. If a critical relative stretch rate is exceeded, cracks on the surface or detachments from the substrate are provoked. Small mandrel diameters with the highest relative stretch rates are most critical.

The examination can be performed as 'pass/fail'- test. Either it is examined if the coating can resist the bending over a mandrel with a specified diameter or, on the other hand, exactly the mandrel diameter can be determined at which the first defects arise with several examinations. The diameter at which the coating fails is the measure for flexibility. The smaller the diameter without failure, the better is the

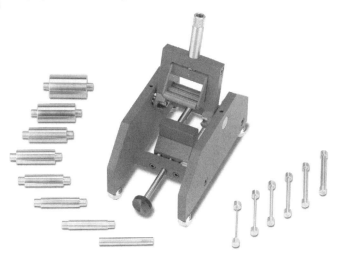

Fig. 12.70 Cylindrical mandrel (source: Erichsen).

quality of the coating regarding flexibility and detachment. For this type of testing, cylindrical mandrels with 2–32 mm diameter are used (see Figure 12.70). As per ISO 1519, the test panels shall be flat and free from distortions and the bending must be carried out within 1 to 2 seconds to a 180° angle around the mandrel.

In order to cover all mandrel diameters, which are applicable for the determination of a characteristic value with only one examination, a conical mandrel (see Figure 12.71) has been developed and standardized in ISO 6860. With this testing device, it is possible to cover the complete range of different bending diameters.

The examination with the conical mandrel allows indeed a fast result, but there is however, the danger that – particularly with brittle coatings – a crack can propagate from the small to the large diameter and consequently produces a false negative result.

In order to prevent this, nowadays, cuts in distance of 20 mm perpendicular to the propagation direction are applied into the coating down to the substrate. The bending rate is 2 to 3 seconds.

12.4.3.5 Cupping

The elastic properties of a coating are determined with the cupping test by slow bulging of a coated test panel. During the test, the sample is clamped firmly between a drawing die and the retaining ring. A half spherical indenter with a diameter of 20 mm is then pushed with a steady rate of 0.2 mm s^{-1} from behind into the test panel. A dome shape is formed with the coating on the exterior.

Using normal corrected vision or a microscope or a magnifying glass of tenfold magnification, the coating is examined for beginning cracks on the surface or

Fig. 12.71 Conical mandrel (source: Erichsen).

delamination from the substrate. The examination can be carried out in two different ways:
- on one hand, as a 'pass/fail' test for testing a specified depth of indentation to assess compliance with a particular requirement, one has to examine if cracks in the coating or detachments from the substrate appear or not (good/poor);
- on the other, by gradually increasing the depth of indentation to determine the minimum depth at which the first defects in the coating are observed while simultaneously observing the sample. The test result is the maximum depth of indentation at which the coating is passed. This is obtained from the calculation of the mean of three valid results and to the nearest 0.1 mm. The test method is described in ISO 1520.

In Figure 12.72a, cupping test equipment with stereomicroscope of the company Byk–Gardner as well as a tested and deformed specimen with cracks in the surface is shown.

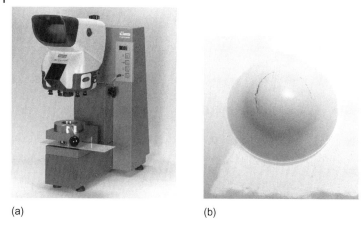

(a) (b)

Fig. 12.72 Cupping test equipment (a) and a sample surface with cracks (b) (source: Byk–Gardner).

12.4.3.6 Impact Testing by Falling Weight

In ISO 6272 and ASTM D 2794, the falling-weight test is described. Similar to the cupping test, the resistance of a coating against cracking (flexibility) and detachment from the substrate is evaluated. The substantial difference consists however, of the fact that the impact in this test takes place via an abrupt dropping of a spherical indenter to the specimen yielding a dynamic deformation.

Figure 12.73 represents a falling-weight apparatus of Erichsen. It consists essentially of a vertical guide tube. At its exterior, a scale for the drop height is fixed. Inside the pipe, a falling weight is led perpendicularly onto the test panel. The head of the falling weight has the shape of a spherical segment. In a solid base plate, a circular die is fixed at which the deformation of the test panel takes place.

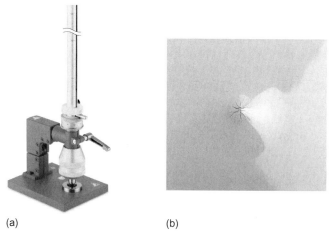

(a) (b)

Fig. 12.73 Falling-weight apparatus (a) with a typical result (b) on a test panel (source: Erichsen).

Table 12.8 Standard parameters for falling-weight test

Standard	Head	Die	Mass	Drop height expressed in	Result expressed in
ISO 6272	20 mm	27 mm	1 kg (2 kg)	m (mm)	kg × m
ASTM D 2794	0.625 in. (0.5 in.)	0.64 in.	2 lbs (4 lbs)	in.	in. × lb

Alternative values in brackets
1 lb = 453 g
1 in. = 2.54 cm.

The deformation can be caused in denting or bulging (coating inside or outside) depending on requirement (see Figure 12.73b for sample surface with a bulge). The deformation energy results from the falling height multiplied with the mass of the falling weight. Depending on the standard, an allocation results in head diameter, die diameter, and the mass of the falling weight (see Table 12.8).

The examination can likewise be applied again as a 'pass/fail' test. Then, the test is carried out with a specified impact energy, that is, from one drop height and with a defined mass, repeating the test a further four times at different positions, giving a total of five drops. The coating passes the test if at least four test positions show no cracking or peeling from the substrate. A classification test can also be applied to determine the minimum mass and/or drop height for which the coating cracks or peels from its substrate by gradually increasing the drop height and/or the mass (impact energy). The ASTM offers additional aids for the determination of explicit cracks. For example, cracks in the coating film can be more visible by immersion in an acidified copper sulfate solution. In the cracks of the coating, a deposition of copper appears on steel panels. Furthermore, cracks can be detected by the measurement of the electrical contact with a so-called pinhole detector on metallic underground. These simple porosity testers work with a 9-V electrical power source and a moistened probe sponge.

12.4.4
Stone-Chip Resistance

In addition to the corrosion damages, injuries by stone chips are an important factor in automotive finishing. On one hand the optical appearance is affected very strongly by defects in the top coat, and on the other, defects down to the substrate lead to corrosion and must be repaired as soon as possible. Particularly at low temperatures in wintertime, strong chipping of the coating film is caused by the impact of street gravel due to the high brittleness of the film below the glass transition temperature. This is the reason why stone-chip tests in the laboratory are often applied at low temperatures down to −30 °C. This can be done either via a freezer in which the test panels are cooled before they are tested in the gravelometer, or better in a special low temperature chamber in which the gravelometer and the test panels are maintained at the specified temperature of testing. However, this equipment is very difficult and is not present in all testing labs.

The task of the paint developer is to balance out the mechanical properties of an automotive coating, consisting of three to four layers, in such a way that defects in the coating film by stone chips are avoided as much as possible. In case of very high exposure to gravel, for example in off-road areas; this is, however, not yet perfect. In such cases it must be ensured that the impact energy is, at the latest, absorbed by the filler layer and damages to the substrate can be avoided.

Stone-chip resistance tests in the laboratory can be differentiated between the multi-impact tests, which applies many impacts on a surface of approximately 10 × 10 cm, and the single-impact test, where only an individual defect is produced.

The evaluation takes place either on ratings from zero to five or a percentage-affected area of surface, chipped by comparison with reference standards, is determined. Inmany cases, a differentiated evaluation is required, specifying the main separation level or the layers of the paint system between which loss of adhesion occurred. Modern techniques of digital image analysis can also be very useful here.

Today, there are still up to approximately 20 different internal test methods generally applied for the examination of stone-chip resistance by the automotive industry. As impact material, different kinds of stones like large, small, round, and sharp-edged are used as well as iron-grit in different sizes. In fact, in standardization-groups it is tried to limit the multiplicity of the methods to approximately 4.

12.4.4.1 Standardized Multi-Impact Test Methods

A simple and most meaningful test method is applied according to ISO 20567-1. The stone-chip resistance of the coating is examined by exposure to many small sharp-edged impact bodies, which impact in rapid sequence and are predominantly independent. As impact material, a defined iron-grit with a particle size of 4–5 mm is used, which is projected via air pressure at an angle of 54° onto the coating. Three different procedures are provided regarding working pressure, mass of the steel grit, and time used to project the grit. In Figure 12.74a, multi-impact tester as well as typical test results on test panels are shown.

A problem with all these methods is the choice of the adhesive tape with which delaminated pieces of paint that have not been completely separated from the panel must be removed. The adhesive strength should be 6–10 N/25 mm width.

Usually, in dependence of the affected area in percent, a characteristic rating from zero to five is estimated by comparison with reference standards (see Figure 12.74a,b for rating one and two).

A draft version of the ISO 20567-4 has the intention to standardize and harmonize the requirements of SAE J 400, ASTM D 3170, as well as internal specifications of well-known car manufacturers such as Volvo, GM, and Toyota. A special water-worn road gravel with a size of 9.5–16.0 mm is to be used for projecting. To get a better constancy in the size distribution, the gravel from three defined size fractions

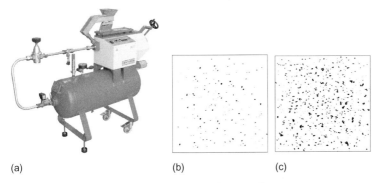

Fig. 12.74 Multi-impact tester (a) (source: Erichsen) and typical test results (b = 1, c = 2) as reference standards according to ISO 20567-1.

is mixed together. For the evaluation of the affected area, the comparison with reference picture standards is favored.

12.4.4.2 Single-Impact Test Methods

Because of the fact that multi-impact tests are very empirical and subject to large errors both in execution and in evaluation, single-impact tests are suitable for more detailed investigations. On one hand, the size of an individual impact can be measured accurately, and on the other, the main separation level or the layers of the paint system between which loss of adhesion occurred can also be well identified.

The examination according to ISO 20567-2 refers back to a procedure, which was originally developed particularly, for the determination of the main separation level of multilayer coatings by BMW [63] as well as Byk–Gardner (see Figure 12.75). The stone-chip resistance of a coating is examined in this method by a single impact of a defined guided impact body (see Figure 12.76). The impact body has a wedge-shaped cutting edge to produce an impact that can be evaluated. The impact of the wedge-shaped body into the coating is affected by the transmission of a pulse of energy from an accelerated steel ball using compressed air of 300 kPa.

After removing the loose fragments of coating material using an adhesive tape, the degree of the damage is determined by measuring the total width of the damaged coating in millimeter. Each test shall comprise three test runs. The average value of the total width is indicated. In Figure 12.77, some typical examples of defects are shown. Additionally, the main separation level or the layers of the paint system between which loss of adhesion occurred can be indicated on request.

The impact body consists of hardened steel. There is a high risk that parts of the cutting edge break out. The impact body must be checked after a maximum of 500 test runs and replaced if necessary.

For the calibration of the test equipment, a special cast aluminum panel is available as calibration standard.

Contrary to the test method with a guided impact body, described before, in the test method according to DIN 55996-3 free flying impact bodies are used. For a better understanding of the test method the comparison with an air gun may be

Fig. 12.75 Single-impact tester (source: Byk–Gardner).

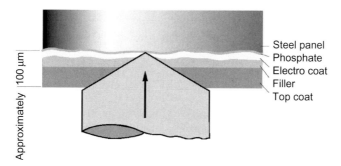

Fig. 12.76 Illustration of a test run of a single impact (source: Byk–Gardner).

allowed here, shooting a projectile with a defined velocity onto the surface. The field conditions, where relative low masses with high velocity can produce bad damages, can be well simulated here. The stone-chip resistance of a coating is examined by the impact of a single, free flying impact body, where the important parameters for the damages, angle of impact, velocity, mass, and geometry of the impact bodies as well as the well-defined working temperature have to be adjusted. In this way, all practically relevant kind of exposure can be simulated realistically.

Several years ago, this test procedure was developed at the Research Institute for Pigments and Paints (Forschungsinstitut für Pigmente und Lacke e.V., FPL)

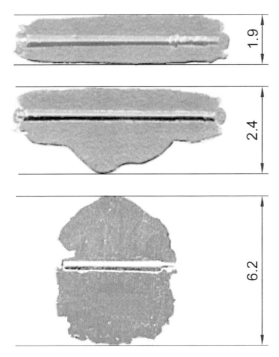

Fig. 12.77 Typical forms of defects on test panels caused by single-impact tester.

in Stuttgart, Germany [64]. A high-speed camera has been installed which records the effect of the impact body to the coating. From the difference of the measured impact velocity and the velocity of the reflected impact body, the energy absorbed by the coating could be calculated and the exposure could be evaluated scientifically. Further developments of this test method have been carried out by Vianova/Hoechst AG [65]. Daimler Chrysler has readily adopted this test method for the testing of the stone-chip resistance of high-quality automotive coatings.

In the current version the impact is a fast flying steel ball with a diameter of 2 mm and a mass of 0.033 g (see Figure 12.78a). The little steel ball impacts the surface of the test panel with a velocity of 250 km h^{-1} in an angle of 2°. Various impact bodies with different masses up to approximately 0.5 g are available for the examination (see Figure 12.78b). The impact velocity is 100 km h^{-1} for this application. The angles can be selected between 2° to 45° respectively 60° to the normal.

The impact bodies are accelerated in a defined manner by air pressure and the velocity is measured using two light barriers. Devices with different tube-inside-diameters are available. The test panels are fixed on a movable, angle adjustable sample rack, whose temperature can be adjusted within the range of – 20 °C to + 30 °C.

After removing loose particles of coating material using an adhesive tape, the evaluation has to be made visually by the help of a microscope or magnifying glass

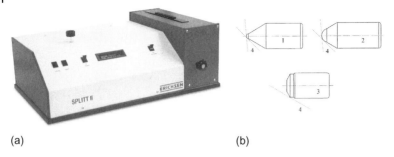

(a) (b)

Fig. 12.78 Single-impact tester (a) (source: Erichsen) and different types of impact bodies (b).

by comparison with special templates or by means of digital image processing. Measuring the total affected area in square millimeter indicates the degree of damage. Additionally, the main separation level can be determined for multilayer coatings.

For the calibration of the equipment a calibration standard in the form of a cast aluminum panel is available. After applying an impact with given velocity, the circular diameter is to be measured using a microscope and compared with the certified value of the standard panel. If the measured value differs from the certified value, the apparatus must be adjusted.

12.4.5
Abrasion

For the examination of abrasion or abrasion resistance, a series of empirical test methods is available. The most important methods, which are used in the automotive industry, will be presented in detail.

12.4.5.1 Taber Abraser
Probably the best-known and very sensitive method for the examination of abrasion resistance is the Taber Abraser procedure. While primarily developed for coated wood surfaces, it can also be used for coatings on metals and plastics. Many details are to be found in technical data sheets and customer specifications as well as in the largely used standard ASTM D 4060.

With this test method, rotating a plane rigid test panel under two turning abrasive wheels abrades the coating surface (see Figure 12.79). A circular mark is generated by the two rolling wheels. The rotating speed is 60 revolutions per minute. The load can be 250, 500, or 1000 g. As abrasive, defined sandpaper is normally used which is bonded onto the contact surface of the wheels. Further, large number of rubber wheels with different granulation and hardness is available.

For the evaluation, the mass loss can be measured as a function of the number of cycles or alternatively the number of cycles up to the degradation of the coating layer down to the substrate.

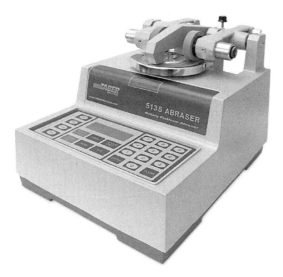

Fig. 12.79 Taber Abraser (source: Erichsen).

Furthermore, in the ASTM D 4060 two characteristic terms are defined:

$$\text{Wear Index}: I = (A - B) \times 1000/C \tag{4}$$

where: A = weight of test specimen before abrasion in milligrams
B = weight of test specimen after abrasion in milligrams
C = number of cycles of abrasion recorded

$$\text{Wear cycles per mil}: W = D/T \tag{5}$$

where: D = number of cycles of abrasion required to wear coating to substrate
T = thickness of coating in mils (1 mil = 25.4 µm)

12.4.5.2 Abrasion Test by Falling Abrasive Sand

The sand abrasion test, standardized in ASTM D 968, is merely an empirical test method. However, with careful application it often yields good interpretable results. Normally applied to flat test panels, the application is suitable especially for shaped samples, for example tube segments, with which 2D abrasion tests cannot be used. The abrasion is achieved by falling sand from a specified height onto the test panel exposed at an angle of 45° to the vertical. The sand is introduced into a filling equipment in vertical position consisting of a filling funnel and a guide tube of approximately 90 cm defined quantity.

As standard abrasive a natural silica with a particular particle size, consistency, and origin could be used for this examination. The particle size is adjusted by sieving through sieves of 600–850 μm. The sand comes from the central United States and is well-known as *Ottawa sand*.

For the calculation of abrasion resistance, the amount of sand in liter per unit film thickness is reported that is needed until a specified area of total abrasion of the coating is reached and the substrate becomes visible. For the execution of the test, 2000 ml is a suitable amount of sand to start with. The operation has to be repeated until an area of the coating of 4-mm diameter has been abraded through to the substrate. During the test, the painted panel must be constantly observed. As the end-point approaches, quantities of 200 ml sand may be introduced into the funnel for a better differentiability.

The abrasion resistance A has to be calculated: $A = V/T$

where: V = volume of abrasive (sand) used in liter
T = thickness of coating in mils (1 mil = 25.4 μm)

12.4.6
Scratch Resistance

Particularly the scratch resistance is very important for top coats in the automotive industry. The resistance against the strong impact of dirt and rotating brushes in a car wash line is of much interest for all end users. For testing the scratch resistance, a multiplicity of often internal test methods (brush methods, abrasive tests, sanding methods) exists, whose test results can often deviate explicitly from each other because of the different principles [66–68]. Here, the three most important test methods are described, which are used by different manufacturers in the domain of automotive . In all cases, a gloss measurement before and after exposure is carried out for the evaluation. As test result, either the loss of gloss or the remaining gloss of the coating is indicated. Moreover, it has to be considered for the evaluation that there are coatings, which show an extraordinary reflow effect. In this case, the scratching marks can recover more or less during aging due to the healing process.

Furthermore, a trend-setting, scientifically oriented test method is described which permits a significant characterization of the scratch behavior of organic coatings [69–71]. With the so-called nanoscratch test, single scratches are produced, which can be evaluated explicitly.

12.4.6.1 Crockmeter Test
The Crockmeter test originally has been developed by the textile industry for testing the color fastness or transference of color by either wet or dry rubbing. The test method is described in specifications of American association of textile chemists and colorists (AATCC). The applications and test parameters are very multifunctional. Over the past years, the method has also been used in the automotive industry. There it finds entrance into specifications for testing scratch resistance.

A round and flat rubbing finger with a diameter of 16 mm rests solidly upon the surface of the test panel for scratching with a defined load of 9 N. The test

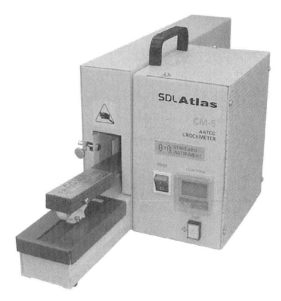

Fig. 12.80 Motor-driven Crockmeter (source: Atlas Textile Testing Solutions).

panel is clamped on the flat base of the equipment with a specimen holder. A test cloth is fastened to the rubbing finger. The sample is stressed by the moving of the rubbing finger back and forth. The stroke length as well as the number of double strokes is adjustable. Figure 12.80 shows a motor-driven crockmeter. This is to be preferred over the manual execution due to the higher test precision. This machine, fitted with an electric motor, carries out 60 revolutions per minute and has a digital automatic switch off cycle counter.

In the automotive industry, as scratching medium, a dry standard polishing paper of 2 and/or 9 µm grain size is used in most cases, which is fastened together with a soft piece of felt beneath the rubbing finger. Furthermore, other abrasives can be used such as feldspar/calcite powder for dry scratch tests, which is put directly in a thin layer onto the sample surface. Contrary to dry test methods, so-called wet scratching tests are known, where defined grinding pastes are applied.

For the evaluation of the degree of scratching, a gloss measurement before and after the exposure is performed. The loss of gloss or the remaining gloss in relation to the initial gloss in % is indicated as the measure for the scratch resistance.

12.4.6.2 Wet-Scrub Abrasion Test

Coated surfaces have already been tested for a long time with a special wet abrasion scrub tester (see Figure 12.81) for their stability against abrasion and scratching by brushes, sponges, or other materials. Originally used mainly for dispersions, this test method is, in the meantime, also applied for the testing of automotive solid color top coats and clear coats. A method according to ASTM D 2486 or ISO 11998 has gained general acceptance.

Fig. 12.81 Abrasion scrub tester (source: Byk–Gardner).

Moving of special brushes back and forth on the sample surface produces the scratching here. The bristles of the brush are slathered evenly before the exposure using a spatula with a prepared aqueous grinding paste as abrasive calcium carbonate is used. On the flat test panels, 20 and/or 200 double strokes are carried out. At the same time, two to four samples can be exposed depending on the testing machine.

After exposure, the test panels are detached from the machine, cleaned immediately with tap water and dried with soft, nonscratching paper tissues. For the evaluation, a determination of gloss must be carried out directly after scratching. As measure for the scratch resistance, the percentage of gloss loss relative to the initial readings is calculated as follows:

$$\text{Gloss loss (\%)} = 100 - (\text{residual gloss} \times 100/\text{initial gloss}) \tag{6}$$

12.4.6.3 Simulation of Car Wash

In order to replace the multiplicity of the applied test methods by a generally accepted method with good practice simulation and also reduce test and development costs, a new laboratory test method for the determination of scratch resistance using a car wash facility was developed some years ago. The main instrument is supplied from Amtec Kistler. Here, the automated car cleaning in a car wash with rotating brushes is simulated using quartz powder defined as synthetic dirt. The received scratching results show a good correlation with real life.

In ISO 20566, the test method is described for evaluation of scratch resistance of organic coatings, in particular, for paint surfaces used in the automotive industry. The specified test parameters are aligned to conditions in a car wash line as far as possible. As brushes, the original equipment of a car wash line are used (see Figure 12.82).

The principle of the method is as follows: The painted test panels are fixed on a sample table and moved back and forth under a rotating brush with defined distance and defined speed during the test. The washing solution, consisting of

Fig. 12.82 Car wash device from Amtec Kistler (source: DuPont Performance Coatings).

water and quartz powder with a defined particle size of approximately 24 µm, is sprayed alternately through two nozzles into the brush. After finishing the specified number of wash cycles, the test panels are cleaned with a suitable solvent, for example white spirit using nonscratching paper tissues to remove brush abrasion and quartz powder residues. After cleaning and drying, the measurement of gloss in 20° geometry is carried out. The measured remaining gloss in relation to the initial gloss provides the percentage measure for scratch resistance.

12.4.6.3.1 Test Conditions

Brush:
- Material : Polyethylene
- Diameter : 1000 mm
- Width : 400 mm
- Profile : x-shaped, spliced ends

Penetration depth : 100 mm
Speed of brush rotation : 120 rpm
Wash cycles : 10 washings (10 double strokes)
Water flow rate : (2.2 ± 0.1) l min^{-1}. at (300 ± 50) kPa
Feed speed : (5 ± 0.2) m min^{-1}
Concentration quartz powder : 1.5 g l^{-1}

It is important to note that the test results will not remain constant over time as a result of changes in the brush material. As the brush ages, the test will become more rigorous. As a consequence, the test method can only be regarded as a comparative examination. The advantage however, is the good correlation to field damages in a car wash. Further, also uneven samples can be examined.

12.4.6.4 Nanoscratch Test

In a car wash, scratches can be caused on the surface of the coating with a width of a few μm and a depth of a few hundred nanometer due to the dirt on the car body or the brushes. For a clear characterization in the laboratory, so-called nanoscratch experiments are provided with which single scratches can be generated with field-oriented appearance [72, 73]. The indentation depth is usually in the range of 1 μm and smaller. A diamond indenter with a radius at the peak of 1 to 2 μm is pushed onto the sample surface applying a defined force to generate the scratch while the sample is moved in a linear direction at constant velocity underneath the indenter (see Figure 12.83). The applied normal force can be constant (1) or progressive (2) during the scratching procedure.

During the scratching procedure, the tangential forces and the indentation depths are measured simultaneously. Deformations or damages can be observed with a microscope or additionally with a video camera. Using the atomic force microscopy (AFM) technology, the profile of the generated scratches can be measured. Cracks can be observed with a microscope at a magnification of approximately 1000. In Figure 12.84, images of different scratch tracks are shown in which some cracks can be clearly identified.

This test method was developed in the DuPont Marshall Lab. (Philadelphia). The method is particularly used for investigations along with development work for clear coats, and is commercialized by the Swiss company, Centre Suisse d'Electronique et de Microtechnique (CSEM, now CSM).

For the determination, the so-called critical load is calculated in this test method, where first irreversible cracks or fractures in the coating are generated. These indicate the transition from plastic deformation to significant/lasting damages. For this, the normal load is constantly increased and indentation depth and tangential

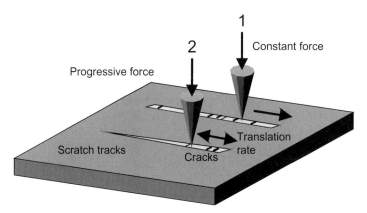

Fig. 12.83 Principle of Nanoscratch Tests.

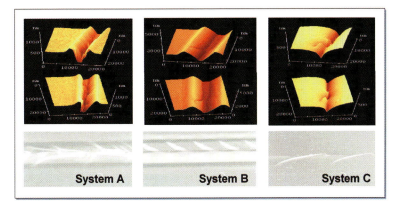

Fig. 12.84 Images of the results of different scratch tracks.

load are simultaneously recorded. The transition from plastic deformation to the fracture range is indicated, for example, by unsteadiness or fluctuations in the detected load flow and the indentation depth. This transition range can also be evaluated by additional optical or AFM analysis.

Besides the determination of the critical load, a typical measured value is the residual indentation depth after scratching – typically at a normal load of 5 mN – in the range of plastic deformation. In Figure 12.85, the schematic procedure of the test method is shown.

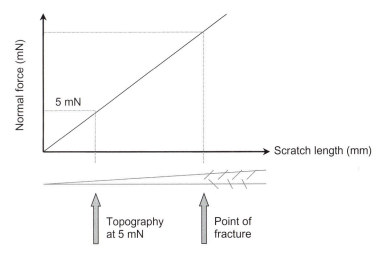

Fig. 12.85 Schematic process of the nanoscratch test.

12.4.7
Bibliography, Standards

Bibliography

Brock, T., Groteklaes, M., Mischke, P. (**2000**) *European Coatings Handbook*, Vincentz, Hannover.

Goldschmidt, A., Streitberger, H.-J. (**2007**) *BASF Handbook on Basics of Coating Technology*, (2. ed.), Vincentz, Hannover.

Kittel, H. (**2006**) *Lehrbuch der Lacke und Beschichtungen*, Band 10, Hirzel, Stuttgart.

Lückert, O. (**1992**) *Prüftechnik bei Lackherstellung und Lackverarbeitung*, Vincentz, Hannover.

Sadowski, F. (**2000**) *Basiswissen Autoreparaturlackierung*, Vogel, Würzburg.

Standards

ISO 1518: 1992	Scratch test
ISO 2409: 1992	Cross-cut test
ISO 1513: 1994	Examination and preparation of samples for testing
ISO 15184: 1998	Determination of film hardness by pencil test
DIN 55945: 1999	Fachausdrücke und Definitionen für Beschichtungsstoffe und Beschichtungen
ASTM D 2794: 1999	Test method for resistance of organic coatings to the effects of rapid deformation (impact)
ISO 15528: 2000	Sampling
ASTM D 3170: 2001	Test method for chipping resistance of coatings
SAE J 400: 2002	Test for chip resistance of surface coatings
ASTM D 4060: 2002	Test method for abrasion resistance of organic coatings by taber abraser
ISO 14577: 2002	Instrumented indentation test for hardness and materials
ISO 4624: 2002	Pull off test for adhesion
ISO 1519: 2002	Bend test (cylindrical mandrel)
ISO 6272: 2002	Rapid-deformation tests, falling-weight test
ISO 2815: 2003	Buchholz indentation test
ISO 1514: 2004	Standard panels for testing
DIN 55996-3: 2004	Determination of stone-chip resistance of coatings, single-impact test with free flying impact body (draft paint)!
ISO 20567-1: 2005	Determination of stone-chip resistance of coatings, multi-impact testing
ISO 20567-2: 2005	Determination of stone-chip resistance of coatings, single-impact test with a guided impact body
ISO 20566: 2005	Determination of the scratch resistance of a coating system using a laboratory car wash
ASTM D 968: 2005	Test methods for abrasion resistance of organic coatings by falling abrasive

ASTM D 2486: 2006	Test methods for scrub resistance of wall paints
ISO 1522: 2006	Pendulum damping test parameters
ISO 6860: 2006	Bend test (conical mandrel)
ISO 1520: 2006	Cupping test
ISO/WD 20567-4: 2007	Determination of stone-chip resistance of coatings, multi-impact test (Gravel Method) (draft)

References

1 Wicks, T.W., Jones, F.N., Pappas, S.P. (**2001**) *Journal of Coatings Technology*, **73** (917), 49.
2 Bloser, F. (**2005**) *Coatings World*, **10** (10), 52.
3 ASTM Standard E 284. (**1994**) *Terminology of Appearance, Annual Book of ASTM Standards*, Vol. 06.01, American Society for Testing and Materials, Philadelphia.
4 ASTM Standard D 523. (**1994**) *Test Method for Specular Gloss, Annual Book of ASTM Standards*, Vol. 06.01, American Society for Testing and Materials, Philadelphia.
5 DIN EN ISO 2813: 1999–*Paints and Varnishes – Measurement of Specular Gloss of Non-Metallic Paint Films at 20, 60 and 85*, International Organization for Standardization.
6 Landolt, E. (**1899**) *Archives d'Ophthalmologie*, **19**, 465.
7 Hunter, R.S. (**1937**) *Journal of Research*, **18** (77), 281.
8 Lex, K. (**1992**) In *Prüftechnik bei Lackherstellung und Lackverarbeitung* (ed. O., Lückert), Vincentz, Hannover, p. 70.
9 Czepluch, W. (**1990**) *I-Lack*, **58** (4), 149.
10 Loof, H. (**1966**) *Journal of Paint Technology*, **38** (501), 632.
11 Tannenbaum, P.H. (**1998**) New routes to surface appearance assessment – the ASTM E12.14 approach, *4th Wave-Scan User Meeting by BYK-Gardner GmbH*, Geretsried, September 1998.
12 ASTM Standard E 430. (**1994**) *Test Methods for Measurement of Gloss of High-Gloss Surfaces by Goniophotometry, Annual Book of ASTM Standards*, Vol. 06.01, American Society for Testing and Materials, Philadelphia.
13 Schneider, M., Schuhmacher, M. (**1999**) Untersuchung zur Entstehung des visuellen Glanzeindruckes aus den Eigenschaften der Lackoberfläche, DFO research report, March 1999.
14 Lex, K., Hentschel, G. (**1999**) Neues Verfahren zur Glanz- und Verlaufsstrukturbewertung, Berichtsband-Nr. 41: Jubiläumstagung 50 Jahre DFO, September 1999, Düsseldorf, p. 73.
15 Biskup, U., Petzoldt, J. (**2002**) *Farbe + Lack*, **108** (5), 110.
16 Schene, H. (**1990**) *Untersuchungen über den Optisch-Phsiologischen Eindruck der Oberflächenstruktur von Lackfilmen*, Springer, Berlin.
17 Lex, K. (**03/2006**) *Welt der Farben*, 14.
18 (a) Guild, J. (**1928**) *Guild and Wright Experiments*, National Physical Laboratory, Teddington; (b) Wright, W.D. (**1931**) Imperial College, Kensington; (c) Wright, W.D. (**1969**) *The Measurement of Color*, 4th edn, Hilger, Bristol.
19 (a) Publication CIE 15: 2004, 3rd Edition, Colorimetry (b) Kittel, H. (**2003**) *Lehrbuch der Lacktechnologie*, 2nd edn (ed. J. Spille), S. Hirzel, Stuttgart, Vol. **5**, p. 292.
20 Billmeyer, F.W. Jr., Saltzmann, M. *Principles of Color Technology*, 2nd edn, John Wiley & Sons, Ltd, New York.
21 ISO 7724: 1984, 01-03, *Paints and Varnishes – Colorimetry*, International Organization for Standardization.
22 ASTM Standard D 2244. (**1994**) *Standard Test Method for Calculation of*

Color Differences from Instrumentally Measured Color Coordinates, Annual Book of ASTM Standards, Vol. 06.01, American Society for Testing and Materials, Philadelphia.
23 Sung, L.-P., Nadal, M.E., McKnight, M.E., Marx, E. (**2002**) *Journal of Coatings Technology*, **74** (932), 55.
24 Cramer, W., Gabel, P. (**7-8/2001**) *European Coatings Journal*, 34.
25 Baba, G., Kondo, A., Mori, E. (**1989**) Goniometric colorimetry, *Proceedings of the 6th Congress of the AIC*, Vol. II, Buenos Aires, p. 213.
26 Alman, D.H. (**1987**) Directional color measurement of metallic flake finishes, *Proceedings of the ISCC Williamsburg Conference on Appearance*, p. 53.
27 Schmelzer, H. (**1986**) Farbmessung und Rezeptberechnung bei Metallic- Automobillacken, *Proceedings of the 18th FATIPEC Congress*, Venice, Vol.I(B), p. 607.
28 Saris, H.J.A., Gottenbos, R.J.B., van Houwelingen, H. (**1990**) *Color Research and Application*, **15** (4), 55.
29 ASTM Standard E 2194-2003. Multiangle *Color Measurement of metal Flake Pigmented Materials*.
30 Rodrigues, A.B.J. (**1990**) Measurement of metallic & pearlescent colors, *Proceedings of the AIC Interim Symposium On Instrumentation For Colour Measurement*, Berlin, September, 1990.
31 Schulze, H.J. (**2001**) *Journal für Oberflächentechnik*, **41** (9), 86.
32 British Standard BS 6923 (**1988**) *Calculation of small color differences*.
33 ASTM Task force E12.12.06 –WK 1164. Standard practice for Multiangle Color Measurement, Identification, and Characterization of Interference Pigments (draft).
34 Pfaff, G., Huber, A. (**2005**) *Welt der Farben*, **3** (9), 14.
35 Kirchner, E.J.J., van den Kieboom, G.J., Njo, S.L., Super, R., Gottenbos, R. (**2006**) *The Appearance of Metallic and Pearlescent Materials*, Color Research and Applications, Wiley Periodicals, Inc., New York.
36 Cramer, W. (**2003**) *Farbe + Lack*, **109** (4), 132.
37 Hirayama, T., Gamou, S. (Kansai), (**2004**) JP2004020263A.
38 Voye, C. (**2000**) *Farbe + Lack*, **106** (10), 34.
39 Jacques, L.F.E. (**2000**) *Progress in Polymer Science*, **25**, 1337.
40 Gerlock, J.L. (**2001**) *Journal of Coatings Technology*, **73** (918), 45.
41 Gerlock, J.L., Kucherov, A.V., Smith, C.A. (**2001**) *Metal Finishing*, **99** (2), 8.
42 Wernstahl, K.M. (**1997**) *Surface Coatings International*, **12**, 560.
43 Schulz, U., Geburtig, A., Crewdson, M., Stephenson, J. (**2005**) *2nd European Weathering Symposium*, Gothenburg, p. 49.
44 Schulz, U., Trubiroha, P., Schernau, U., Baumgart, H. (**2000**) *Progress in Organic Coatings*, **40**, 151.
45 Adamsons, K. (**2002**) *Progress in Organic Coatings*, **45**, 69.
46 Wernstahl, M. (**1996**) *Polymer Degradation and Stability*, **54**, 57.
47 Henderson, K., Hunt, R., Spitler, K., Boiseau, J. (**2005**) *Journal of Coatings Technology*, **2** (18), 38.
48 Riedl, A. (**2/2006**) *Welt der Farben*, 16.
49 Boiseau, J., Campbell, D., Wurst, W. (**2003**) *1st European Weathering Symposium*, Prag.
50 Rauth, W., Nowak, S. (**2006**) *25. Jahrestagung der GUS (25th Annual GUS Conference), Proceedings*, p. 177.
51 Osterhold, M., Glöckner, P. (**2001**) *Progress in Organic Coatings*, **41**, 177.
52 Gerlock, J.L., Peters, C.A., Kucherov, A.V., Misovski, T., Carter C.M. III, Nichols, M.E. (**2003**) *Journal of Coatings Technology*, **75** (936), 35.
53 Bauer, D.R. (**1997**) *Journal of Coatings Technology*, **69** (864), 85.
54 Bauer, D.R. (**2000**) *Polymer Degradation and Stability*, **69**, 297.
55 Bauer, D.R. (**2000**) *Polymer Degradation and Stability*, **69**, 307.
56 Amirudin, A., Thierry, D. (**1996**) *Progress in Organic Coatings*, **28**, 59.
57 Bracht, U., Kurz, O. (**2005**) *wt Werkstatttechnik Online*, **95** (1/2), 38.

58 Ranjbar, Z., Moradian, S., Attar, M.R.M.Z. (2004) *Progress in Organic Coatings*, **51**, 87.
59 Loveday, D., Peterson, P., Rodgers, B. (2004) *Journal of Coating Technology*, **1** (8), 46.
60 Stellnberger, K.-H. (2006) *Journal für Oberflächentechnik*, **46** (5), 60.
61 Dirking, T., Große-Brinkhaus, K.-H. (1996) *Proceedings XXIII Fatipec*, B-260, Bruxelles.
62 Rother, K. (2006) *Journal für Oberflächentechnik*, **46** (3), 26.
63 Harlfinger, R. (1988) *Farbe + Lack*, **94**, 179.
64 Zorll, U. (1975) *Farbe + Lack*, **81**, 505.
65 Ladstädter, E. (1984) *Farbe + Lack*, **90**, 646.
66 Wagner, G., Osterhold, M. (1999) *Materialwissenschaft und Werkstofftechnik*, **30**, 617.
67 Osterhold, M. (2006) *Progress in Colloid and Polymer Science*, Springer-Verlag, Berlin, Vol. **132**, p. 41.
68 Osterhold, M., Wagner, G. (2002) *Progress in Organic Coatings*, **45**, 365.
69 Blackmann, G.S., Lin, L., Matheson, R.R. (1999) *ACS Symposium Series* .
70 Lin, L., Blackmann, G.S., Matheson, R.R. (2000) *Progress in Organic Coatings*, **40**, 85.
71 Klinke, E., Eisenbach, C.D. (2001) *Proceedings of the 6th Nuremberg Congress*, p. 249.
72 Shen, W. (2006) *Journal of Coating Technology*, **3** (3), 54.
73 Mi, L., Ling, H., Shen, W., Ryntz, R., Wichterman, B., Scholten, A. (2006) *JCT Research*, **3** (4), 249.

13
Supply Concepts

Hans-Joachim Streitberger and Karl-Friedrich Dössel

13.1
Quality Assurance (QA)

Following the success story of the Japanese car manufacturers, boosted by the impressive image of their product quality, the rest of the car manufacturers mainly in North America and Europe have improved on their quality assurance (QA) system by demanding and controlling quality management systems at the supply industry over the last 20 years. These systems, like the QS 9000 for GM, Ford, and Chrysler in North America, the ISO 9000 series, EAQF (Evaluation d'Aptitude à (la Qualité pour les Fournisseurs), and VDA 6.1 (Verband der Deutschen Automobilindustrie) are all based on general management systems for quality enhancements with continuous improvement efforts. They also apply to the paint suppliers who enter into basic agreements with the manufacturers on batch to batch variations and product consistency. Certifications based on audits conducted by the car manufacturers or by independent associations have become mandatory for a supplier to be approved.

One of the standards accepted almost worldwide, which covers all sectors of the economy, is described in the DIN ISO 9000 series of standards developed by the TC 176 committee of the International Organization of Standardization (ISO). It contains an instruction manual for various standards that must be observed, and guidelines for the installation of QA-systems describing detailed criteria that must be followed. It deals with all business segments of the supplier, including strategy, development and design, manufacturing, the supply chain, and customer service. Furthermore, it addresses the management structure, tasks and responsibility, and personnel training and education. Other notable sections include testing, monitoring of test and inspection equipment, contract agreements, and customer satisfaction. The supplement ISO DIN 10001:2006 focuses on management systems for customer satisfaction and ethics compliance.

Additionally, environmental management system requirements, according to DIN ISO 14000 series, were included about 10 years ago. The basic philosophy of these standards is to provide the suppliers with assistance and incentives to

Automotive Paints and Coatings. Edited by H.-J. Streitberger and K.-F. Dössel
Copyright © 2008 WILEY-VCH Verlag GmbH & Co. KGaA, Weinheim
ISBN: 978-3-527-30971-9

achieve sustained improvements in the environmental quality of manufacturing and management, making environmental compliance their responsibility.

The most accepted variant of the series of standards seems to be the norm ISO/TS 16949:2002, which concentrates on special requirements for the automotive supply industry and tries to bring into consonance all quality standards on a worldwide basis. The three big automotive companies in North America accept only those suppliers who have complied with this norm.

To fulfill the requirements of these standards and programs, certain tools like the balanced score card (BSC) have gained importance in the paint and coatings industry. The BSC takes into account financing and corporate goals as well as human resources, processes, and customers, thus giving all systems used, including the quality management system, the same strategic direction. It helps to transform the company's vision and strategic goals into actuality. This is achieved by determining key factors that are specific to the company and which are referred to as *key performance indicators* (KPIs).

Furthermore, 'Kaizen', '6 Sigma' and other concepts are widely used in the automotive supplier industry to fulfill the continuously increasing quality targets of the car manufacturers [1].

Submissions of new products are based on the specification sets of the respective automobile manufacturer. However, there is a differentiation between new products with new or increased performance profiles and new colors. In general, the specifications are the backbone of a submission, but quite often a mutual risk assessment is made by the manufacturer together with the supplier before developing the submission profile. This is especially the case in a supply situation with the paint supplier as a single source or system supplier (see Chapter 13.2). Most of the testing has to be carried out by the supplier.

New colors are based on the approval of the relevant top-coat technology and additional workability including repair issues and short-term weathering test results. The replacement of a color is connected with many changes in the car and marketing. Technically a new color needs 1–2 years of development and testing before it is introduced in the coating line.

13.2
Supply Chain

13.2.1
Basic Concepts and Realizations

The traditional relationship between car manufacturers and paint suppliers is based on carefully selected and detailed specified products as delivered until the completion of final coating process. Suppliers are chosen on the basis of technical performance, service, and product prices. Purchase departments follow the strategy of certifying as many suppliers as possible to ensure a reliable supply situation and

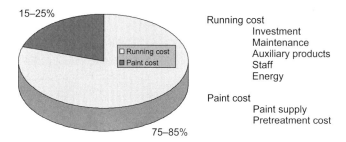

Fig. 13.1 Typical share of paint costs and running costs of a car body paint shop.

create technical and price competition. This type of business is simply based on price per volume or kilogram of paint.

There is room for improvement especially when cost per paint has reached the bottom line. In general, the cost structure of a car body paint shop is dominated by the running cost according to Figure 13.1. This includes capital cost, that is, investment and depreciation, maintenance, auxiliary products, energy, QA, and respective staffing. This simply means that the cost saving potential of the running expenditures is higher especially considering the following factors.

First in the case of a complex painting problem, owing to the many layers including pretreatment of an automotive coating, the suppliers involved have to cooperate in spite of their situation as competitors, for example, to solve the problems in top-coat colors. Under these circumstances, it is hard for the car manufacturer to combine the right expertise and define accountability to make troubleshooting activities very effective. The 'political' as well as marketing aspects of the supplier are often counterproductive. Furthermore, the evaluation and testing procedures have to cover all potential products similar to the delivered products to keep the competitive situation of the suppliers alive.

Second under any circumstances, two expert groups exist at the customer and the supplier levels dealing with specifications, testing, process, and application technology, which may create unnecessary communication and decision problems. Finally, the consumption of paint is viewed differently by the supplier and the car manufacturer – the supplier wants to sell as much as possible, but the car manufacturer wants to use as little as possible.

Table 13.1 Steps in the supplier–customer relationship in automotive OEM coating business

Business type	Year	Region
Traditional (cost/paint)	until today	Worldwide
Single sourcing	1985–today	North America
Single sourcing (cost/unit)	1990–today	North America
System supplier (cost/unit)	1995–today	Germany/Europe
Paint facility ownership (cost/unit)	2000–today	Europe/South America

The traditional relationship changed slowly, but steadily, during the mid-1980s. The first step was the concept of 'single sourcing'. This meant that one paint manufacturer would deliver all pretreatment chemicals and paints for the specific processes, whereas, earlier on the pretreatment used to be excluded, quite often. The basic advantage of single sourcing was the simple responsibility structure of such a concept. There was only one supplier who had to deal with troubleshooting in case of coating problems. To keep competition in place, the contracts for delivering total paint shops was often restricted to 2–3 years. Then the customer would call for bids from the paint suppliers. This would result in strong price competition, which was increasingly combined with cost targets during the contracted supply time in the form of annual cost reductions. Running the paint shops, investing, maintenance, and staffing was still the complete responsibility of the car manufacturers. Support was given by the supplier for the daily painting process. However, with time, the paint material supply chain was transferred gradually to the paint supplier to include the management of the paint kitchen. (See Table 13.1).

To overcome delays in the delivery all paint products, to keep leading suppliers of technologies in the business, and to be flexible and fast in introducing new technologies and paints with improved properties, the paint suppliers started taking over the responsibility of supply chain management (SCM), confining it to a certain degree to one key supplier. The so-called tier1- and tier2-suppliers were defined, where the tier1-supplier is responsible for the stock management of all the products and manages the business with a service team, whereas some tier2-suppliers delivers products through the tier1-supplier to the customer. This could be the pretreatment chemicals, auxiliary products, and special paints. These concepts are defined by the terms *system supply* or *lead supply system* [2]. The partnership between the customer and a single supplier to improve efficiency in cost and quality of the paint shop is the basis of these supply concepts [3]. To achieve this, an open book philosophy where both partners inform each other about cost structures and cost factors is mandatory. Detailed rules have to be agreed upon between the car manufacturer, who will still be the owner of the paint shop, and the paint supplier regarding the interfaces of the areas of responsibility. Payment to the paint supplier is increasingly made taking into account the 'cost per unit (CPU)', that is, the fully painted car, once accepted, will be charged to the car manufacturer (see Table 13.2).

There are many ways of sharing responsibility in the complex task of running a paint shop for car bodies. Figure 13.2 shows the basic functions or modules of a paint shop in the automotive industry. The tasks under Figure 13.2(a) are normally in the realm of responsibility of the paint shop owner, whereas, responsibility for the tasks shown in Figure 13.2(b) are transferred to the paint product supplier. To manage perfect product supply and quality control (QC) paint warehouses together with supply rooms, the so-called 'paint kitchens', and laboratories for measuring basic physical data for appropriate product performances are put in place and taken over by the paint supplier.

Table 13.2 Actual supply concepts and their characteristics

Concept	Responsibility of paint supplier	Risk	Comment
Paint facility ownership (PFO)	Investment Auxiliary product supply Paint product supply Process control QC Repair	Capital costs	Cost per unit (CPU) tier1 and tier2 suppliers
System supply management (Lead supply system)	Auxiliary product supply Paint product supply Process control QC Repair	Deficient contract	Cost per unit tier1 and tier2 suppliers
Total fluid management	Auxiliary product supply Paint product supply QC	CPU: process control	Cost per paint tier1 and tier2 suppliers
Single sourcing	Paint product supply QC	Only one supplier Deficient responsibility for CPU	

The main areas for mutual cost reduction are as follows:
- paint consumption per car
- energy consumption of the process
- first time capability.

First time capabilities especially gain in importance. The definition may be different from one car manufacturer to another, but commonly included in this number are the bodies with 'no touch', that is, no corrections, and those cars that needed minimal polish or repair work, manageable at the inspection desk. The numbers for an excellent paint shop reach a level of >90%. (See Chapter 6).

Table 13.2 shows the basic models of supply concepts, their risks, and their respective potential for cost sharing.

In reality, there are a large number of concepts in addition to those described in Table 13.2 [4]. Even total toll manufacturing concepts are in place like the coating process of the first smart cars. In those cases, the equipment manufacturers invest in, build, and run the paint shop. This concept is known as the *paint facility ownership* (PFO) model (see Table 13.1 and 13.2). It significantly reduces the fixed cost structure of the car manufacturer by shifting part of the costs to the variable costs.

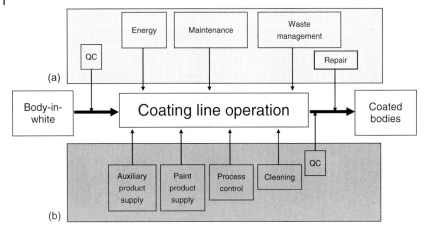

Fig. 13.2 Modules and potential responsibilities of a car body paint shop operation. (a) car manufacturer and (b) paint supplier.

The basic requirements for all the system supply concepts targeting CPU pricing areas follows:
- clear technical targets and QC for the car bodies entering the paint line ('body-in-white');
- clear accountabilities for all parameters and each process step of the paint process;
- clear technical targets and QC for the coated car body before entering the assembly line;
- CPU prices depending on manufacturing volume.

To manage the concepts, many electronic programs under the term *enterprise resource planning* (ERP) or *SCM* are available on the market [5]. Often, up to 30 employees of the paint supplier work at the paint shop of the respective car manufacturer. Today, it is estimated that about 20% of about 330 paint shops worldwide are running with one of these modern concepts.

A very demanding version of a system supply concept was the new paint shop line that started operating in 1997 at the Daimler plant in Rastatt, Germany. This paint shop had to be designed by a team consisting of the equipment manufacturer, paint supplier, and Daimler according to the basic design framework consisting of capacity numbers, Daimler product specifications, environmental compliance, and cost, also taking into consideration variations in the product portfolio and production numbers, which normally changes the CPU. A merit/demerit contract was signed so that all technical efforts to improve efficiency, cost, and risks were shared by the partners. At that time, not only the supply concept, but also the technology, was developed by the selected supply team. It became one of the best paint shops in terms of cost, quality, and operations and still represents an excellent example of the validity of the system supply concepts [6].

The technical approval system for products and suppliers will normally not be changed in these new supply concepts. Approval for tier2 suppliers is the joint responsibility of the car manufacturer and the tier1 supplier.

13.2.2
Requirements and Limitations of a System Supply Concept

For the car manufacturer, it is mandatory to provide a sufficiently stable business, including type of bodies, colors, and area to be coated. It is not easy to generate rules in case of changing production numbers of more than 20%, which would lead to different CPU. It is essential that the car manufacturer is able to monitor the number and the type of units. This is almost state of the art in the automotive industry. Furthermore, competition is somewhat restricted by the fact that total paint shops around the world are now the products on the market.

The supplier needs to establish an internal accounting system from CPU to cost per kilogram based on each color for reasons of control. In some cases, separate business units have been created by the paint supplier to get a better handle on the accounting. Trust in the technical and management capabilities of the supplier is supported by the certification processes according to the standards (see Section 13.1).

From a contract standpoint it is absolutely mandatory to carefully define the quality level of the 'body-in-white' because dirty and rusty bodies reduce the first time capability and increase usage and cost of auxiliary products (see Chapter 13.2.1). At the same time, the quality level of the coated car needs to be defined. Cooperation is crucial in the decision processes for parameter settings as well as other correction steps.

References

1 Koppitz, F.-D. (**2005**) *Metalloberflache*, **59**(4), 38.
2 Cramer, W.R. (**2005**) *Fahrzeug+Karosserie*, **58**(5), 34.
3 Bloser, F. (**2004**) *The Coatings Yearbook 2004*, Vincentz, Hannover, p. 103.
4 Agosta, M. (**2002**) *Coatings World*, **7**(3), 28.
5 Challener, C. (**2005**) *Journal of Coatings Technology*, **2**(19), 30.
6 Klasing, J. (**2000**) *Fahrzeug+Karosserie*, **53**(2), 10.

14
Outlook

Hans-Joachim Streitberger and Karl-Friedrich Dössel

14.1
Status and Public Awareness of the Automotive Coating Process

The performance of an automotive coating system has reached a level that satisfies most customers around the world. Compared to 30 years ago, the problem of corrosion is almost nonexistent, and durability and appearance of the top coats are acceptable for the lifetime of a car. With worldwide introduction of the two-layer top coats, color, gloss, and chip resistance stay excellent during the first 7–10 years. Almost any color and color effect is available, and technology is able to keep pace with changing fashion trends. This performance, generated over the last 20 years, presents a positive image of automotive coatings. Furthermore, the increasing trend of outsourcing of parts like bumpers, mirrors, fenders, grills, hoods, and roofs stresses the fact that color match problems of all assembled parts are also being overcome.

The technological leaps in the application processes involving abatement technology, the use of robots, the emergence of waterborne and powder paints, and a wide range of colors support the high-tech image of this industry. In other words, nobody will accept Henry Ford's words: 'Any customer can have a car painted any color that he wants, as long as it is black!' The color of the car today is strongly connected with the expression of individuality of the customer and has a relatively low cost potential for mass customization. The environmental impact of an automotive paint shop has been reduced significantly over recent years. The market penetration of waterborne coatings as base coat or primer surfacer, and powder as primer surfacer and clear coat will sustain this trend on a worldwide basis. Recently, Japan has started to switch over to waterborne technology.

According to a study of the Federation of German Paint Manufacturers (VdL), the image of automotive coatings ranks first among all coating segments, on the basis of customer awareness and acceptance in Germany [1]. In this respect, it

Automotive Paints and Coatings. Edited by H.-J. Streitberger and K.-F. Dössel
Copyright © 2008 WILEY-VCH Verlag GmbH & Co. KGaA, Weinheim
ISBN: 978-3-527-30971-9

ranks superior to the general paints industry, which does not have an image of creativity, innovation, high-tech finishes, and environmental compliance.

Costs of a paint shop in terms of investment and running costs have been the main focus for many years, but it is very difficult to monitor improvements owing to the overlapping and increasing regulatory demands of environmental and safety issues over the last 30 years. Nevertheless, with paint shops becoming more and more standardized, the workforce has been drastically reduced. Pretreatment and electrodeposition processes run automatically, robots have taken over almost all the application and other handling requirements in the spray booth, and repair work has become less frequent owing to better first time capabilities of more than 90% in a well-maintained paint shop.

The total cost of car painting in the 1970s was € 500–800 per unit, including all running costs as well as depreciation. Today, with improved quality and less environmental impact, the total cost is estimated at a level of € 300–400 per unit. Further improvements in the traditional paint shops are expected with reduction in energy and investment costs. Both can be achieved by wet-on-wet-on-wet processes, where the primer surfacer, the base coat, and the clear coat are baked together [22]. Again, low bake coatings are considered and tested for inline coating processes, including all kinds of plastic parts [3]. The success is heavily dependent on the achievable first time capabilities and the performance level of the car appearance.

The introduction of waterborne and powder technology to maximize the total potential of environmental compliant paints (see Figure 1.8) is expected to continue. In Europe, waterborne primer surfacer and waterborne base coats have reached a market penetration of 50 and 80% respectively. In the clear coat segment, high solids, powder, and waterborne technology will continue to coexist. Ecoefficiency evaluations (life cycle analysis) have gained more and more importance in the decision processes, and technological and performance topics are no longer the key factors. [4].

In North America, powder primer surfacer is used at Daimler and GM, whereas waterborne primer surfacer is rarely used. Chrysler and the European, Japanese, and Korean transplants use more of waterborne base coats, whereas GM and Ford mainly use the regionally specific high solid paints. Among the Asian manufacturers, the Japanese, Koreans, and Chinese have started to change over to waterborne base coat technology after 30 years of 'experience' in Europe and North America.

14.2
Regulatory Trends

Three main legislatory trends will have an impact on the automotive and automotive coatings industries in Europe. The first is the European Unions' Directive 2000/53/EG, which requires an 85% level of reuse of cars with 80% recycling in 2006, and 95% and 85% levels respectively, by 2015. This has to be seen together with increasing considerations and model development of life cycle analysis of products like paints (see chapter 14.1). The second is the introduction of mandatory risk assessments of each chemical in the European industry, the so-called Registration, Evaluation, Authorisation and Restriction of Chemicals (REACH) process [5]. Finally, both the Biocidal Products Directive 98/8/EC and the US Environmental Protection Agency (EPA) deal with antioxidants, preservatives, and similar compounds, which became important for the paint industry owing to the significant increase of waterborne coatings and reduced solvents in them [6]. Together with the ban on heavy metals, this makes the waterborne coatings prone to microbiological attack.

The EU directive for recycling cars at the end of their lifetime has an impact on the painting process in terms of toxic substances. It is very understandable and almost mandatory for the required level of car recyclability that toxic materials have to be banned. The focus is on lead, cadmium, mercury, and chromium(VI) compounds. These compounds have to be avoided in coatings, where ever technically possible. These chemicals could be found in paints used as pigments for top coats and electrodeposition coatings. Since the beginning of 2005, they are being replaced in the respective formulations (see Chapters 4 and 7). This also holds for the pretreatment processes, in which chromium(VI) played a role as a passivation agent during the 1980s and 1990s. This has either been abandoned or replaced by less toxic compounds, almost completely [7]. The real world of recycling starting around 2007 will probably encounter other problems and may cause other replacement necessities, which are not foreseeable today. As already mentioned, the European government replaced the European Inventory of Existing commercial Chemical Substances (EINECS) list by a legislation that makes the registration of chemicals mandatory, in Europe. All chemicals need to undergo a risk assessment in the direction of their intended use. This seems to be a costly measure for both the manufacturer and the user, and may lead to the withdrawal of some low volume chemicals [5]. Paint companies will then have to reformulate their products while maintaining performance.

The Biocidal Products Directive may have a similar effect on the use of preservatives. All new biocidal products, both active substances and preparations, will need formal approval for marketing. Existing biocides will be reviewed retrospectively during a 10-year transitional period [8]. It will become difficult to introduce new products in the market and costs will be higher for those that enter the market.

The actual legal limits of volatile organic compounds (VOC) in Europe and the United States [9] will follow the state-of-the-art application development. On the basis of emission numbers in g VOC m^{-2} coated car, the best available technology today yields numbers in the range 10–35 g m^{-2}. All paint shops have to chart out strategies to achieve this range by 2012. The constantly increasing knowledge of health and environmental impact of substances will further influence the exposure limits of toxic products mainly in organic solvents, resins, and additives.

Emission into the air is strongly in focus, but waste and water are also important outflows of car painting processes. Zero waste processes and closed loops for water are also targets to be achieved. Under these considerations, the spray processes are not deemed satisfactory.

Future legislation is likely to stress on fuel consumption of cars to protect the world's climate. This will also have an impact on the coatings process and will boost the integrated processes together with lower film builds, without detrimental effects to the coating performance.

14.3
Customer Expectations

The success of a car in the market depends on pricing, technical features, image of the manufacturer, and design, that is, the looks of the car. An appropriate or trendy color is one of the important success factors that supports the design of a car model. In mature markets, the function and technical features of a car coating are also part of customer's expectations, as demonstrated by a study of Toyota in 2002 and 2003. (see Figure 14.1). The main focus of all car manufacturers is still set on colors. Therefore, the paint shop and the paint supplier should be able to follow color and fashion trends quickly. This is standard for a color year basis, but some car manufacturers are marketing their ability to produce cars with colors according to the individual demands of customers. However, only a few car manufacturers are able to fulfill such demands by having made the necessary investments, but at high costs. To provide this feature on a low cost basis is one of the customer expectations.

Interference colors are gaining some popularity. Their penetration into the market depends also on the costs of the necessary pigments.

Color harmony on future cars, made with an increasing number of different substrates and fully painted components like bumpers, roofs, fenders, and trunks, is set to become a customer expectation. The technical approaches of the car manufacturer, at the moment, are varied – from relying on suppliers on the basis

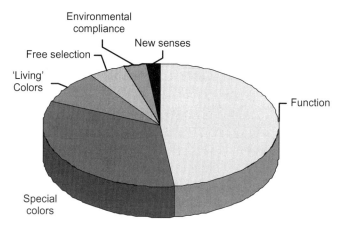

Fig. 14.1 Customer survey of Toyota in 2002/2003 [10]. Function = 80% scratch resistant, 20% maintenance-free

of their specifications, up to taking over the painting work themselves. This may be one of the driving forces in the introduction of new coating strategies making use of innovative processes.

Long term durability of color and gloss of the coating has been mostly achieved. Nevertheless, a 10-year warranty remains a target to be worked on by improving resin and additive performances in optimized basecoat and clear coat formulae.

Maintenance-free coatings have been prevalent in the automotive market for some time. They were based on fluorinated resins providing very low surface tensions and high gloss retention. The cars so protected stayed clean, but durability was not sufficient and after some time they became dirty and blackened. The cause was the hydrophobic absorption of carbon particles from gas–oil fueled cars. A new self-cleaning property approach is based on an example from nature and named after it – the lotus effect.

Customers are aware of the scratch problems caused by car wash. Furthermore, scratch-free cars have a higher attraction potential in dealer show rooms. Solutions are difficult to find owing to the fact that a coating has to pass a number of tests and has a broad spectrum of performances. One approach for improvement in this respect without detrimental effects on other properties, is the evolution of new ideas based on nanotechnology [11, 12]. Other solutions may be based on sol–gel technology and an additional coating process for a second, thin, clear coat layer. They result in easy-to-clean surfaces, especially if the resins are based on fluorinated compounds [13].

Very high potential is expected from UV-coating technology [14] for UV-curable clear coats [15, 16]. Their hardness can reach very high scratch resistance, but the complicated body shape does not allow all surfaces to be cured by radiation. New technical approaches as well as formulation approaches in the direction of dual-cure systems are under development [17].

Self-healing coatings may also gain interest. These coatings are made of microcapsules of 60–150 µm that contain film-formers and corrosion inhibitors. They are incorporated into the paint at the time of coating application. When the coating is damaged or scratched, the microcapsules break and spill their ingredients into the defect and repair the film [18]. This idea is actually directed toward corrosion protection, but has potential for other coating films, especially if the particle size of the microcapsules can be lowered.

Other properties of coatings like IR reflection to keep the car interior as cool as possible, personal identity inscriptions in the form of holograms, and color change features are based on material developments outside the automotive industry.

14.4
Innovative Equipments and Processes

The global standardization of the paint shop technology allows step-by-step improvements toward a smaller environmental footprint on the basis of better cooperation between the involved parties. These steps include the complete introduction of electrostatic application of waterborne coatings [19] and of robots equipped with very small bells for improved transfer efficiency, including the virtual development of spray booths based on spray patterns [20], the RoDiP® process that uses less space in pretreatment and electrodeposition [21], better cleanliness by keeping all nonpainting operations out of the paint shop, and improved pipe and gun rinsing [19, 22].

Real 'quantum leap' steps only can be realized by a complete consideration not only of the painting process and paint, but also of the car design and manufacturing process, and needs the commitment of all involved parties: the car manufacturer, the component supplier, the equipment manufacturer, and the coating industry. The driving forces are mainly the decrease of fixed cost in car manufacturing and better control of color harmony of the increasing number of components supplied by specialized manufacturers.

There are many ideas for new and somewhat revolutionary coating processes focusing on two main aspects. The first focus is to reduce energy, space, and time and eliminate spray application as much as possible in the traditional painting process as described before. The second is to transfer coating operations to suppliers.

There has been considerable interest in the partial elimination of spray application processes, space, and baking ovens to save material and energy. One approach is already established on the basis of the wet-on-wet-on-wet-system [23], which mostly eliminates the primer oven in the coating process. A step further is the elimination of the primer surfacer, its function being taken over by the first base coat layer [24]. This process has been introduced recently [25]. Both are known as the *integrated paint process* (*IPP*) or *eco concept*.

New techniques of drying or curing of waterborne coatings by microwaves [26] and UV clears (see Section 14.3) are likely to enter the scene.

A two-coat-electrodeposition process [27] is another alternative to reduce the amount of spray coating. In this process, the primer surfacer is applied by the electrodeposition process on a conductive electrodeposition primer. The conductivity could be achieved by using carbon black, which provides the necessary conductivity in the film after the shrinking caused by the baking process.

For many years, coil coating technology has been considered as the basis for a completely new coating process, offering the potential for reduction in fixed costs in the automotive manufacturing process [28] and for achieving color harmony. Coil coating is the most efficient coating technology in terms of coating speed and cost. Strategies for introducing this technology for car painting follow a three-step road map. First, weldable coil coatings are used to reduce the amount of additional corrosion protection steps like sealing and waxing. This has already been in use for some time, and to a certain degree, in car manufacturing [29]. The second step provides coil coatings that may or may not be weldable, and are designed to replace the pretreatment and electrodeposition process in a car manufacturing paint shop. The critical issue here is weldability. In case the coil coating is not weldable owing to the high mass of organic compounds in it, the car manufacturer would need to assemble the car differently. Technology is available for this, but many obstacles have to be overcome like creating a completely new body assembly line, and providing for crash resistance and protection of the edges after cutting the coils [30]. Trials with weldable coil coating primers were conducted on a few hundred cars at Chrysler in the 1990s. They were finished with powder top coats. The most attractive step would be to eliminate the complete paintshop by using top-coated coils for car assembly. To this point of time, cars have been assembled by clinching, riveting, gluing, and other similar techniques, but without welding. Some cars have been manufactured in this manner, based on coil coatings and have undergone tests for feasibility studies, but the process is quite far from its technical realization [31].

A similar idea is based on foil coatings [32]. Besides the increasing use of vehicle wraps for fleet cars, vans, and light trucks [33], foil coatings have the same potential as coil coating for eliminating paintshops in car manufacturing. The first application has been seen in the bumper-coating process, where foils are used for the thermoplastics in the molding process [34].

It is not easy to foresee the success of these technologies owing to the fact that revolutionary progress can be achieved only by strong engagement of all parties. The traditional coating process is integrated into several industries and the shift of the value added has a big barrier and involve some risk. Support can be expected by increasingly tighter cooperation between the automotive industry and the suppliers (see next section).

14.5
New Business Ideas

The change in business trends for automotive coatings has started with the strategy of the car makers to concentrate on core competencies, followed by a strategy to outsource nonstrategic activities. Up till now, many component manufacturers have grown into big companies not only manufacturing complex parts, but also developing and designing them [35].

In the coating process, the single sourcing concept was the starting point for leaving behind the traditional procurement concepts, in the early 1990s. The first target was to transfer the responsibility for the coating process and the coating film to one supplier, avoiding discussions with many suppliers, for the pretreatment stage, and for the different paints required for various applications in cases of disturbances and defects. According to the general strategy outlined, the next step was to establish year-by-year cost targets with the paint supplier participating in the success using bonus regulations. Next, the complete handling of the paintshop was taken over by the paint supplier (see Chapter 13). There are several steps between the traditional supply of paint being paid by volume or kilogram and the total responsibility for the coating process by the paint supplier, like total fluid management or system supplier [36]. The system supplier concept also allows the use of so-called tier-2 suppliers for supplying the necessary materials via the tier-1 supplier.

Many automotive paint shops are already run on the basis of these concepts. In future, these concepts may go further by outsourcing the paint shop totally to the paint or equipment supplier when they will additionally take over the investment risks as well.

The only drawback is that all involved parties are focused on efficiency, cost, and environmental targets. Consequently, the risk of introducing new technologies is transferred to companies with lower market capitalization. This situation is not very supportive for implementing innovations.

References

1 Eichstädt, D., Bross, M. (2002) European Coatings Journal. **10**(3), 32.
2 Valero, G. (2002) Modern Paint and Coatings, **92**(9), 26.
3 Stranghöner, D., Dössel, K.-F. (2000) European Coatings Journal, **6**, 39.
4 (a) Pappasavva, S., Kia, S., Claya, J., Gunther, R. (2002) Journal of Coatings Technology, **74**(925), 65. (b) Engel, D., May, T., Matheson, R.R., Nguyen, P.Q. (2004) European Coatings Journal, **1–2**, 40.
5 Scott, A. (2007) Chemical Week, **169**(11), 17.
6 Shaw, A. (2003) Modern Paint and Coatings, **93**(3/4), 28.
7 Gehmecker, H. (2001) Journal für Oberflächentechnik, **41**(9), 70.
8 Freemantel, M., Backhouse, B. (2001) Coatings World, **6**(9), 43.
9 Drexler, H.J., Snell, J. (2002) European Coatings Journal, **4**, 24.
10 Umeruma, S., Suzuki, T. (2005) Proceedings of the 5th International Strategy Conference on Car Body Painting, Bad Nauheim.
11 Fernando, R. (2004) Journal of Coatings Technology, **1**(5), 32.
12 Valero, G. (2002) Modern Paint and Coatings, **92**(9), 36.
13 Weigt, W., Auer-Kanellopoulos, F. (2004) Farbe+Lack, **110**(10), 20.
14 Bruen, K., Davidson, K., Sydes, D.F.E., Siemens, P.M. (2004) European Coatings Journal, **4**, 42.
15 Schwalm, R. (2000) Farbe+Lack, **106**(4), 58.
16 Seubert, C.M., Nichols, M.E. (2004) Progress in Organic Coatings, **49**, 218.
17 Skinner, D. (2002) Polymers Paint Colour Journal, **192**(4456), 14.
18 Kumar, A., Stephenson, L.D. (2004) Coatings World, **9**(6), 24.
19 Gamero, J. (2003) Journal für Oberflächentechnik, **43**(5), 50.
20 Eickmeyer, D. (2004) Journal für Oberflächentechnik, **44**(6), 34.
21 Klocke, C. (2002) Journal für Oberflächentechnik, **42**(9), 32.
22 Wolf, W., Scholz, A. (2003) Journal für Oberflächentechnik, **43**(1), 16.
23 Wegner, E. (2004) Coatings World, **9**(10), 44.
24 (a) Dössel, K.F. (2006) Journal für Oberflächentechnik, **46**(9), 40. (b) Wright, T. (2006) Coatings World, **11**(3), 26.
25 Svejda, P. (2006) Metalloberflache, **60**(11), 12.
26 Harsch, M., Schüller, P. (2002) Metalloberflache, **56**(3), 42.
27 Schnell, A. (2000) Proceeding of the 7th DFO Automobiltagung, Strasburg.
28 Stewing, T. (2006) Journal für Oberflächentechnik, **46**(8), 42.
29 Alsmann, M., Reier, T., Weiß, V. (2000) Journal für Oberflächentechnik, **40**(9), 64.
30 Kutlu, I., Reier, T. (2005) Journal für Oberflächentechnik, **45**(5), 48.
31 Pilcher, G.R. (2001) European Coatings Journal, **4**, 134.
32 Bannwitz, P., Gleich, H. (2004) Journal für Oberflächentechnik, **44**(9), 40.
33 Vaughan-Lee, D. (2004) Polymers Paint Colour Journal, **194**(4478), 24.
34 Barth, C. (2001) Kunststoffe, **91**(8), 92.
35 Schmidt, K.J. (2002) Automobil Production, **4**, 120.
36 Bloser, F. (2005) Coatings World, **10**(10), 52.

Index

a

abrasion evaluation
– falling abrasive sand test 455–456
– Taber abraser procedure 454–455
absorption heat exchanger 268
abstraction bionics 23
accelerated outdoor corrosion test using salt spray 413
accelerated weathering tests 134
acetaldoxime 71
acetone process 235–236
acid etch phenomenon 198–199
acrylic resin 3, *see also* polyacrylic resins, as coating materials
activation process 69–70, 84, 316–317
additives
– in plastic coatings 319–320
– in primer surfacer
– defoaming and deaerating agents 148
– for substrate wetting 149
– for the resins or resins' mixture like cross-linkers 147–148
– in waterborne primer surfacers 147
– pigment wetting and dispersion 148
– rheology 149–150
adhesion and flexibility examinations
– bending 445–446
– cross-cut test 441–443
– cupping test 446–447
– electrocoat primer 133
– falling-weight test 448–449
– pull-off test 441
– steam jet tests 444–445
adhesive bonding
– adhesives as process materials 355–356
– advantages 356–359
– application 357
– body shop 367–370
– in car production 366–374
– paint shop 370–371
– principles of bonding and materials 351–359
– surface preparation 359–366
– trim shop 371–374
alkaline cleaners 65
alkaline degreasers 65
alkyd-based monocoats 180–181
alkylphenolethoxylates (APEOs) 69
aluminum
– alloys 38–39
– application potential of 36
– extraction 34–35
– as light-construction material 38–40
– pretreatment process 75–76
– properties 35
– texturing of surface methods 36–37
– Ti-/Zr-oxide/hydroxide layers on 64
– treatments 35–37
aluminum or sheet molding compounds (SMCs) 5
analogy bionics 23
anodic deposition coatings 3
anodic electrodepostion coating 99
antiflutter adhesives 367–368
appearance measurement techniques 381–393
applied coatings 1
araliphatic isocyanates 233
Arizona, outdoor weathering in 406, 408
atomizer 272–274
atom transfer radical polymerization (ATRP) 219
attributes, of color 176

Automotive Paints and Coatings. Edited by H.-J. Streitberger and K.-F. Dössel
Copyright © 2008 WILEY-VCH Verlag GmbH & Co. KGaA, Weinheim
ISBN: 978-3-527-30971-9

Audi space frame (ASF)   18
Audi–Volkswagen   207
Austin Martin car   18
autoclave test, for stresses   134
automated quality assurance   293–297
automatic cleaning, of the body surface   266
automation, in paint application   269–271, 298
azeotrope-forming solvents, for water removal   230
azeotropic process   230

b

base coats   3, 4, 181–184, 263, 323
  – drying of   189
  – high solid   186
  – low and medium solid   185–186
  – rheology   184–185
  – waterborne   186–188
batch and semibatch processes   224
bead patterning   21
'Bell–Bell' application, of base coat   300–301
benzoguanamines   246
bionics   23–26
bis(4-hydroxycylohexyl)propane   226
black box   412
black standard thermometers   407
blocked isocyanates   100, 250
blow molding compounds (BMC)   5
body construction, of automobile, *see also* manufacturing methods, of automobile bodies
    – functions of body structure   13
    – joining methods
      – bonding   53–54
      – clinching   55–56
      – evaluation   53
      – laser welding   54–55
      – riveting   56
      – roller hemming   56–57
    – manufacturing methods
      – hydroforming   49–50
      – metal foam   50–51
      – press hardening   50
      – roll forming to shape   52
      – sandwich structures   51
      – tailored products   47–49
    – materials
      – aluminum   34–40
      – magnesium   40–42
      – nonmetallic parts   43–47
      – sandwich structures   51
      – sheet   61–63
      – steel   26–34
      – titanium   42–43
    – materials and manufacturing methods for advanced   14
    – methods
      – hybrid type of construction   19
      – modular way of construction   19–20
      – monocoque design   15–17
      – space-frame concept   18–19
    – milestones   14
    – principles of design
      – bionics   23–26
      – conventional   20–21
      – light-weight vehicles   22–23
    – requirements in   13–14
    – studies and concepts on light construction   57–59
    – substrates used for   62
    – surface protection
      – corrosion prevention in design phase   59–60
      – precoating of sheets   59
body-in-white   13, 20, 63
body on frame (BOF), mode of construction   13
body shop bonding   367–370
bonding technique   53–54
Buchholz penetration hardness   133
2-butyl-2 ethyl propane 1.3-diol   226

c

car body pretreatment lines
  – continuous horizontal spray/dip   80
  – cycle box spray   82
  – cycle dip   82
  – cycle immersion   82
  – RoDip3   80–81
  – schematic outline   79
  – sequence of treatment
    – for continuous conveyorized line for plastic cleaning   78
    – in a spray/dip   78
  – spray   80
  – vario shuttle   81
car painting industry
  – evolution   1–7
  – legislative acts   7–9
  – market   9–11
Carrozzeria Touring Company   18
car wash simulation test   458–459
Cataplasma testing   362
cathodic electrodeposition paints   6, 99–101
  – design of
    – baking oven   118–119

– general functions and equipment 107–108
– integration process 106–107
– pretreatment 107
– replenishment and anode cells 111–114
– tanks, filters, heat exchanger, and power supply for 108–111
– ultrafiltration and rinsing zones 114–117
– film performance of 101–106
cationogenic polyurethanes 233
cavity preservation 265
Chromic-6 5
circulation line system 282
clarity preservation 263–264
clear coats 3, 4, 323
– durability 198–203
– epoxy acid chemistry 192
– liquid
– acrylic melamine silane 192
– carbamate-melamine-based 1K coat 192
– one- (and two-) component epoxy acid coat 192
– one-component (1K) acrylic melamine 190–192
– one-component polyurethane (PUR) coat 192
– scratch resistance 201
– two-component (2K) polyurethane coat 193–194
– UV curving 206
– waterborne coat 194–195
– market 189–190
– powder 195–198, 264
clinching technique 55–56
coating processes, *see also* specification and testing, for paint finishes; surface protection
– automotive specifications for exterior appearance 391
– business trends 481–482
– customer expectations 478–479
– film performances 7
– innovative equipments and processes 479–481
– milestones and driving forces in 2, 6
– for plastic parts
– exterior, *see* exterior plastic part coating
– interior plastics, *see* interior plastic coating
– quality standards 6

– regulatory trends 476–478
– status and public awareness of the automotive 473–474
– of the twentieth century 1
coat performance, requirements for
– application conditions 204–206
– durability 199–201
– environmental etch resistance 198–199
– integrated paint processes
– primerless coating process 209–210
– wet-on-wet application 208–209
– scratch resistance 201–204
– UV-curing technology 206–208
cold check test 134
colorchanger 275
color and appearance examination
– appearance measurement techniques 392–393
– applications of color control in the automotive industry 400–401
– distinctness-of-image 385–386
– dullness measurement 385–386
– evaluation of 'orange peel' 386–390
– measurement of colors 398–405
– metallic and interference colors, visual evaluation of 397–398
– optical design of sphere geometry for 398
– outlook 403–405
– solid colors, visual evaluation of 393–397
– specular gloss measurement 382–384
– visual evaluation of appearance 379–380
– visual perception of a surface 390–392
color changes 275
color control, in the automotive industry 402–403
color flop 177
color of cars 6
computer aided design (CAD) systems 267
computer aided optimization (CAO) method, of body construction 24
Concentrated Radiation Using Fresnel Mirrors (Q-Track or EMMAQUA) 412
construction methods, of automobiles, *see* body construction, of automobile
contrast sensitivity, of the human eye 387–388
convection ovens 288

conventional design, in body construction 20–21
conveyor systems 287–288
copper accelerated salt spray test (CASS) 74
corona discharge method 317
corrosion prevention 59
corrosion protection 2–3, 5, 358
– oils 63
– primer 62
– tests for edge, contact and inner part protection 432–433
– tests for surface protection 429–432
– total body testing 433–434
– types 428–429
– use of zinc-coated steel 61
crash-management systems 40
crash-resistant adhesives/bonding 369
craters, in amorphous films 120–121
Cr-containing conversion treatment 77–78
crockmeter test 456–457
cross-cut test 156, 441–444
cross-linking agents 211–212, 214, 222, 227
– for liquid coating materials
– melamine and benzoguanamine resins 245–248
– other 252
– polyisocyanates and blocked polyisocyanates 249–252
– tris(alkoxycarbonylamino)-1,3,5-triazine (TACT) 248
– for powder coatings 252–254
crosslinking of paints 1
cupping test 446–448
cyclic corrosion tests (CCT) 421–422
cycloaliphatic diisocyanates 233

d

defects, in electrodeposition coatings
– craters 120–121
– dirt 119–120
– film thickness/throwing power 122–123
– hash marks 123
– pinholes 123–124
– surface roughness 121–122
– water spotting 123
degreasing process 65–69
deposition process 92
– coulomb-efficiency 94
– voltage effects 96
dibutyl tin laurate (DBTL) 153
diffusion and migration of dispersion particle, in deposition process 92

dimeric fatty acids 226
dimethylolpropionic acid (DMPA) 322
dip coating 1
dip degreasing 83–84
diphenylethene (DPE) 219
direct glazing adhesives 371–372
dirt cleaning 267
distinctness-of-image (DOI) measurement 385–386
dosing technology 277–278
dry film lubricants 64
Dullness measurement 383–384

e

EcoConcept 210, 263, 480
elastic bonding 353
elasticity, of primer surfacer coat 133
electrocoating process 261–262
electrodeposition coatings 2
– in automotive supply industry 124
– defects during application and their prevention
– craters 120–121
– dirt 119–120
– film thickness/throwing power 122–123
– hash marks 123
– pinholes 123–124
– surface roughness 121–122
– water spotting 123
– design of cathodic electrocoating process 261
– anolyte system 111
– baking oven 118–119
– general functions and equipment 107–108
– integration process 106–107
– pretreatment 107
– replenishment and anode cells 111–114
– tanks, filters, heat exchanger, and power supply for 108–111
– ultrafiltration and rinsing zones 114–117
– film performance of cathodic electrocoatings
– chip resistance 104–105
– corrosion protection 102–104
– physical film data 101–102
– surface smoothness and appearance 105–106
– history 89–90

– physico-chemical basics of the 90–95
– quality control
  – solid content, solvent content and pH 97–98
  – voltage, current density, bath temperature and bath conductivity 96
  – wet film conductivity 96–97
– resins and formulation principles
  – anodic electrodepostion coating 99
  – cathodic electrodepostion coating 99–101
  – nature of resins 98
electro-discharge texturing (EDT) method 36
ElectroSTatic Application (ESTA) 308
– powder 166
emulsion polymerization process 216, 222–224, 241
environmental legistation 7
– pretreatment 86
– primer surfaces 132
– topcoats 187
epoxy resins, see resins
esterizing process 139
etching tests 422–424, 439
ethylenediaminetetraacetic acid (EDTA) 69
ethylene–propylene–diene–monomer (EPDM) 310
Euro-New Car Assessment Programme (NCAP) 22
European Enhanced Vehicle Safety Committee (EEVC) 22
European Integrated Pollution and Prevention Control Bureau (EIPPCB) 132
extender, in primer surfacers
– barium sulfate 144
– carbon blacks 146
– feldspar (China clay) 146
– silicon dioxide 144–145
– talc 144
exterior plastic coating
– car body color 320–324
– ecological aspects 305–306
– economical aspects 307
– plastic coating materials 316–318
– pretreatment 315–318
– process definitions
  – offline, inline, and online painting 307
  – process-related issues, advantages, and disadvantages 307–310
– substrates and parts
  – overview 310–311

– physical characteristics 311–315
– technical and design aspects 306–307
– technical demands and testing 324–329
– trends, challenges and limitations
  – materials 330–332
  – processes 332–334
  – substrates and parts 329–330

f
falling abrasive sand test 455–456
falling-weight test 448–449
Federal Motor Vehicle Safety Standards (FMVSS-214) 22
Federal Motor Vehicle Safety Standards (FMVSS) 212/208 regulations 371
fiber composites, in body construction
– applications 46–47
– forming behavior 45
– molding processes of reinforced plastic parts 46
– polyester resins 45
– pretreatment process of plastic parts 77
– share of plastics in motor vehicles 43
– thermoplastics 45
film thickness/throwing power, of body for cars and trucks 122–123
'finger print' methods 7, 204
flake orientation process 183–184
flash off 189, 288
flexible-structural bonding 354
flop index 182, 402
Florida, outdoor weathering in 406–408, 410
Florida exposure test 134
fluorination 317
foil technology 333–335
fusion welding methods 42

g
galvannealed sheets 59
glewing technology 5
gloss retention 3
GM SCAB-test 364
gravelometer test 133

h
hazardous air polluting substances (HAPS) 9
Health risk factors, to the painters 3
Heat exposure and weathering impact 407
Hem-flange bonding 368
1,6-hexamethylene diisocyanate (HDI) 232
hexmethoxymethylamine (HMMM) 139, 247
high solid base coats 181
hot-forming process 50

humidity exposure and weathering   408
hybrid method, of construction   19
hydroforming process   49–50
hydrogen peroxide   70
hydroprimers   322
N-beta-hydroxy alkyl diamides   252
hydroxycarboxylic acids   235
hydroxy-functional polyacrylic   223
N-hydroxymethylol groups   238
hydroxylamine   70
hydroxypivalic acid neopentyl glycol ester   225

i

illuminant metamerism   397
integrated paint processes   263, 480
  – primerless coating process   209–210
  – wet-on-wet application   208–209
interior plastic coating
  – adhension   342
  – concept   334–335
  – laser coatings   337–339
  – performances   341–346
  – properties of different classes of paints   343
  – raw materials for   346–347
  – surfaces and effects   335–337
interpenetrating networks (IPN), of polymer classes   239
IR (infrared) radiators   289
isophorone diisocyanate (IPDI)   233, 249, 321

j

jacksonville etching   198
joining methods, in automobile body construction
  – bonding   53–54
  – clinching   55–56
  – evaluation   53
  – laser welding   54–55
  – riveting   56
  – roller hemming   56–57

k

ketazine and ketamine processes   237
Kroll process, of titanium   42
Kubelka–Munk theory   402
'Kyoto' protocol   8

l

laboratory weathering trial   419–421
Land Rover   39
laser coatings   5, 337–339
laser MIG hybrid welding   19

laser scanning confocal fluorescence microscopy   316
laser texturing (Lasertex) method   36
laser welding technique   54–55
Lateral Impact NewCar Assessment Program (LINCAP)   22
layout of coating lines   266, 298
legislations
  – in coating processes   7–9
  – for metal surface treatment chemicals   86
  – for primers   132
  – transportation and safety regulations for phosphate accelerators   74
  – VOC restrictions   100, 175
Lewis acids   153, 230, 244, 251
light fastness test (aging from hot light)   421
lightness flop   176
light-weight vehicles, design principles   22–23
linear polyesters   226
liquid clear coats, *see* clear coats
liquid primer surfacer
  – application   158–160
  – formula principles   152–156
  – manufacturing process   156–157
low solid base coats   185

m

magnesium and body construction   40–42
  – pretreatment process   76–77
maleinized polybutadiene resins   3
manufacturing methods, of automobile bodies, *see also* body construction, of automobile
  – hydroforming   49–50
  – metal foam   50–51
  – press hardening   50
  – roll forming to shape   52
  – sandwich structures   51
  – tailored products   47–49
manufacturing of resins (*see also* polyacrylic resins)   213–245
mass polymerization   224
mass production and coating process   1
medium solid base coats   185
melamine crosslinkers   3, 247
melamine resin   154
Melt process   228
Mercedes concept car   25
Metal foams   50–51
metallic and interference colors, visual evaluation of   397–398

metallic appearance 178, 180
methoxymethyl-functional melamines 246
N-methyl caprolactam 235
N-methylmorpholine-*N*-oxide (NMMO) 71
N-methyl pyrrolidone 235
mica effect pigments 176–177
modular method, of construction 19–20
monocoats 180–181
monocoque design, of construction 15–17
Morgan 18

n

nanoscratch test 460–461
neopentyl glycol 225
nitrilo-triacetic acid (NTA) 69
nitroguanidine (CN4) 71
nonionic surfactants 66
Norrish–Trommsdorf effect 224

o

observer and color perceptions, *see* visual perception
OEM coating, of passenger cars 5
– market 10
Ohm's law 94
online painting 307
Opel 194
'Orange peel', evaluation of 384–390
organic polyurethanes 154
organophosphates 180
outdoor weathering
  – in Florida and Arizona 406–410
  – procedures 411–413
oxidizing compounds 70–71

p

paint color-changer 275
paint finishes, specification and testing of, *see* specification and testing, for paint finishes
painting process
  – coating facilities
    – automation in paint application 269–271
    – process technology 268–269
  – coating process steps
    – cavity preservation 265–266
    – electrocoating (EC) 261–262
    – paint application 263–265
    – pretreatment 261
    – sealing and underbody protection 262–263
  – conveyor equipment 287–288
  – drying 288–290
  – economic aspects
    – full automation 296–297
    – layout 296
    – robot interior painting with high-speed rotation 299–300
  – in a fully automated paint shop 259
  – layout of paint shop 266–267
  – quality aspects
    – automated quality assurance 293–296
    – control technology 290–292
    – process monitoring and regulation 292–293
    – process optimization 296–297
  – supply system
    – circulation line system 282–283
    – container group 280–282
    – for industrial sector 280
    – paint-mix room 280
    – for smaller circulation system 283–284
    – for smaller consumption quantities 283
    – for special colors 284–285
    – voltage block systems 285–287
  – technology
    – atomizer 272–274
    – dosing technique 277–279
    – paint color changer 275–276
    – pigging technique 275–276
  – used by original equipment manufacturers (OEMs) 260
paint-mix room 275, 280
passenger safety standards 22
passivation mechanism 75, 85
patchwork blanks 48
pH of the solvent, defined 97
phosphating process 84–85
pigging technology 277–278
pigments 319
  – and color 175
  – aluminum 179
  – aluminum flake effects 178
  – attributes describing color 176
  – classification 176
  – global production data 176
  – inorganic 177–178
  – modern 180
  – organic classes 179
  – in primer surfacers
    – factors affecting paint 141
    – refractive index, impact of 141–143
    – titanium dioxide 143–144
  – technical properties 177

plasma treat process   360–361
plastic parts   43, 77, 310–315
polyacrylic carbonic acids   154
polyacrylic microparticles   223
polyacrylic resins, as coating materials
    – manufacturing   218–224
    – mass polymerization   224
    – property profile   214–218
polyamides   311
polyester, as coating materials
    – manufacturing   228–231
    – property profile   224–228
polyester modified epoxy chain   100
polyisocyanates   249–252
polymerization   222–224
    – aqueous   222–224
    – in solution   218
    – mass   224
polyurethane-based crosslinkers   3
polyurethane polyacrylic polymers, in coating materials
    – manufacturing   240–241
    – property profile   239
polyurethane–reinforced-reaction-inmold (PUR–RRIM)   308
polyurethanes, as coating material
    – manufacturing   234–238
    – property profile   232–234
powder clear coats   195–198, 264
powder coatings   4, see also clear coats
powder primer surfacer   130–131
    – application   167–170
    – formula primers   160–163
    – manufacturing process   163–166
powder slurries, see water-dispersed powder clear coat systems
prelubes   63
press hardening process   50
pretreatment process   261
    – activation   69–70
    – aluminum   75–76
    – construction materials   82
    – degreasing   65–69
    – environmental legislations for metal surface treatment chemicals   86
    – of exterior plastic   315–318
    – lines   77–80
      – continuous horizontal spray/dip   80
      – cycle box spray   82
      – cycle dip   82
      – cycle immersion   82
      – RoDip3   80–81
      – spray   80
      – vario shuttle   81
– magnesium   76–77
– passivation   75
– plastic parts   77
– process
    – activation   84
    – deionized water rinsing   85
    – dip degrease   83–84
    – electrocoat line   85
    – passivation   85
    – phosphating   84–85
    – precleaning stage   82–83
    – rinsing   84, 85
    – spray degrease   83
    – zinc phosphating   70
– requirements and specifications for zinc phosphate conversion layers   85–86
– sequence   65
– steel structures   75–76
– surface conditions and contaminations, in body assembly   63–64
– zinc phosphating   70–74
primerless coating process   130, 209–210, 263–264
primer surfacer   129, 261
    – application with powder material   264
    – cost and environmental aspects   172–173
    – film properties   133
    – liquid
      – application   158–160
      – formula principles   152–156
      – manufacturing process   156–157
    – market potential   129
    – polyester-epoxy   170
    – polyurethane-based resin   213
    – powder   130–131
      – application   166–170
      – formula primers   160–162
      – manufacturing process   162–166
    – process sequence   170–171
    – quality and process reliability   173
    – raw materials
      – additives   146–150
      – pigments and extenders   141–146
      – resin components   139–141
      – solvents   150–152
    – requirement profile
      – legislative   132
      – technological   133–138
    – solventborne   321
    – technical data for technologies   139

– volatile organic compound (VOC) standards in the United States 130
– waterborne 131–132
pull-off test 441
pyrogenic silicic acid 145

q

quality assurance 465
quality control 288–293
quantum leap 24
QUV tester 417–419

r

radiation and weathering impact 406–407
radical polymerization 214–215, 219–220, 241
repairs
– after pretreatment and electrocoat application 377
– after primer surfacer 378
– end-of-line 380
– top-coat 378–379
resins 318–319
– benzoguanamine 212, 246
– for the electrocoating process 98
– epoxy 3, 213
– manufacturing 243–244
– property profile 242–243
– hexamethoxymethylmelamine (HMMM) type melamine 186
– maleinized polybutadiene 3
– melamine 217
– in primer surfacer 139–141
reverse injection moldings (RIM) 318
reversible addition fragmentation chain transfer (RAFT) 219
rheology 149–150, 153–156
– base coat 184–185
– of solventborne base coat without rheology control 184
– of waterborne and high solids base coats 185, 187
rinsing process 84–85
riveting technique 56
robots 6, 270–271, 301–302
roller hemming technique 56–57
roll forming to shape process 52
roughness, of primer coating 135

s

sandwich designs 23
scanning electronic micrographs (SEMs), of the zinc phosphate coatings 71–72
scratch resistance

– car wash simulation test 458–459
– crockmeter test 456–457
– nanoscratch test 460–461
– of clear coats 201
– wet-scrub abrasion test 457–458
sealing process 262–263
secondary-ion mass spectrometry (SIMS) 316
selective catalytic reduction (SCR) technology 25
sheet-metal parts, designing of 21
sheet molding compounds (SMC) 310
side crash velocity 22
simulation of acid rain tests 422
single-stone impact test 133
sodium chlorate 71
sodium nitrite 70
sodium nitrobenzenesulfonate (SNIBS) 71
soft kill option (SKO) method 24
solid colors, visual evaluation of 393–397
solvent-borne paint formulations 8
solvents
– in plastic coatings 319
– in primer surfacer 150–152
space-frame concept 18–19
– interior passenger space and cargo space 20
specification and testing, for paint finishes
– abrasion
– falling abrasive sand test 455–456
– Taber Abraser procedure 454–455
– adhesion and flexibility examinations
– bending 445–446
– cross-cut test 441–444
– cupping test 446–447
– falling-weight test 448–449
– pull-off test 441
– steam jet tests 444–445
– color and appearance
– appearance measurement techniques 392–393
– applications of color control in the automotive industry 400–401
– color measurement outlook 403–405
– distinctness-of-image 385–386
– dullness measurement 385–386
– evaluation of 'orange peel' 386–390
– measurement of colors 398–405
– metallic and interference colors, visual evaluation of 397–398
– for multiangle color measurement of interference pigments 403–405

– solid colors, visual evaluation of 393–397
– specular gloss measurement 382–384
– visual evaluation of appearance 379–380
– visual perception of a surface 390–392
– corrosion protection
  – tests for edge, contact and inner part protection 432–433
  – tests for surface protection 429–431
  – total body testing 433–434
  – types 428–429
– mechanical examinations 435–440
– scratch resistance
  – car wash simulation test 458–459
  – crockmeter test 456–457
  – nanoscratch test 460–461
  – wet-scrub abrasion test 457–458
– standards of tests 462
– stone-chip resistance
  – single-impact test methods 451–454
  – standardized multi-impact test methods 450–451
– uniformity 381
– weathering resistance of automotive coatings
  – artificial weathering 414–421
  – environmental impacts 406–421
  – natural weathering 408–410
  – procedures of outdoor weathering 411–413
  – testing 421–426
specular gloss measurement 382–384
spinning nozzle inert flotation (SNIF) 35
sports cars 18
spot weld bonding 369
spray application techniques 3
spray degreasing 83
stannic acid 230
steam jet tests 444–445
steel
  – areas of use 27
  – characteristics and application of different types of cold-rolled high-strength steels 28
  – cold-rolled thin sheets 28
  – complex-phase (CP) 29
  – deepdrawing steels 29
  – electrogalvanized and hot-dip galvanized steel sheets 62

– higher-strength 29
  – bake-hardening 30
  – interstitial free (IF) 30–31
  – isotropic steel 31
  – microalloyed 30
  – phosphor-alloyed 30
– high-grade 32
– high-strength
  – complex-phase (CP) 32
  – dual-phase 31
  – martensite phase (MS) 32
  – rest austenite (RA) 31–32
– low-carbon deep drawing 29
– manganese–boron 32–33
– modern passenger cars 26
– passivation effect on 75
– twinning-induced plasticity (TWIP) 33–34
stone-chip resistance test methods 133, 156
  – single-impact 451–454
  – standardized multi-impact 450–451
stress-controlled growth principles 24
stripping process 101
sulfonic-acid-based compounds 233
Superleggera method, of construction 18
suppliers, of paint 10–11
supply concepts
  – of paint materials 279–286
  – quality assurance 467–468
  – supply chain 468–473
surface conditions and contaminations, in body assembly 63–64
  – roughness of electrocoat surfaces 121–122
surface protection
  – corrosion prevention in design phase 59–60
  – precoating of sheets 59
surface tension 313–315
surfactants 65–66

t

Taber Abraser procedure 454–455
tailored blanks 47
tailored strips 48
tailored tubes 48
testing, for paint finishes, *see* specification and testing, for paint finishes
2,2,6,6-tetramethylpiperidin-*N*-oxide (TEMPO) 219
tetramethyl xylylene diisocyanate (TMXDI) 233, 236
thermal joining processes 19

thermoplastic polyolefins (TPOs)   310
titanium and body construction   42–43
Titanium Firebird car   43
titanium phosphate activators   69–70
titration methods   67–68
tolerance compensation   356
topcoats   175
    – single staged   180
TPO bumpers, painting in   320–321
triglycidyl isocyanurate (TGIC)   252–253
trimethyl hexamethylene diisocyanate   232
Trim shop bonding   371–374
tris(alkoxycarbonylamino)-1,3,5-triazine (TACT)   248–249
tris(aminocarbonyl)triazine   227
two-component (2K-) formulation   3

u

ultrafiltration devices   68
ultralight steel auto body–advanced vehicle concepts (ULSAB–AVC)   57–58
ultralight steel auto closures (ULSAC) project   57
ultraviolet radiation   5
underbody protection   262–263
US Insurance Institute for Highway Safety (IIHS)   22
UV-curing clear coats   206–208
UV-durability   199

v

VDA humidity test   134
viscoelastic materials   311
viscosity   153, 154
visual perception
    – of color   394–395
        – inherited types of color vision defects   394–395
    – of metallic finishes   397–401
    – of a surface   390–392
volatile organic compounds (VOC), emission of   8
Volkswagen   41
voltage block systems   285–287
Volvo   207, 385, 450

w

washing oils   64
waterborne coats
    – base   3, 186–188
    – clear   194–195, 264
    – conversion, global   187
waterborne paints   2
waterborne polyester dispersions   230
waterborne primer surfacers   131–132, 154, 156
water-dispersed powder clear coat systems   264
water immersion test   134
waviness measurements   388–390
weathering impact, on automotive coatings
    – artificial   414–421
    – correlation between artificial and natural   426–427
    – environmental impacts   406–423
        – heat exposure   407
        – humidity exposure   408
        – outdoor weathering in Florida and Arizona   408, 410
        – radiation   406–407
    – with Fresnel mirror systems   410
    – natural   408–410
    – testing equipment and technology
        – cyclic corrosion tests (CCT)   421–422
        – etching tests   422–424, 439
        – laboratory weathering trial   419–421
        – light fastness test (aging from hot light)   421
        – QUV tester   417–419
        – simulation of acid rain tests   422
        – standards   423–426
        – xenon arc lamps   415–417
wet film conductivity, defined   96
wet-on-wet coating technology   3, 208
wet-scrub abrasion test   457–458

x

xenon arc lamps   415–417
x-ray photoelectron spectroscopy (XPS)   316

z

zinc phosphate conversion coatings   86
zinc phosphating process   70–74
Zr-based passivation   85